Lecture Notes in Computer Science   5557

Commenced Publication in 1973
Founding and Former Series Editors:
Gerhard Goos, Juris Hartmanis, and Jan van Leeuwen

Yeow Meng Chee   Chao Li   San Ling
Huaxiong Wang   Chaoping Xing (Eds.)

Coding
and Cryptology

Second International Workshop, IWCC 2009
Zhangjiajie, China, June 1-5, 2009
Proceedings

 Springer

Volume Editors

Yeow Meng Chee
San Ling
Huaxiong Wang
Chaoping Xing

Division of Mathematical Sciences
School of Physical and Mathematical Sciences
Nanyang Technological University
21 Nanyang Link, Singapore 637371

E-mail:{ymchee; lingsan; hxwang; xingcp}@ntu.edu.sg

Chao Li
National University of Defense Technology,
410073 Changsha Hunan, China

E-mail: lichao_nudt@sina.com

Library of Congress Control Number: Applied for

CR Subject Classification (1998): E.3-4, G.2.1, C.2, J.1

LNCS Sublibrary: SL 4 – Security and Cryptology

ISSN 0302-9743

ISBN 978-3-642-01813-8 Springer Berlin Heidelberg New York

Typesetting: Camera-ready by author, data conversion by Scientific Publishing Services, Chennai, India
Printed on acid-free paper   SPIN: 12674015   06/3180   5 4 3 2 1 0

Preface

The biennial International Workshop on Coding and Cryptology (IWCC) aims to bring together many of the world's greatest minds in coding and cryptology to share ideas and exchange knowledge related to advancements in coding and cryptology, amidst an informal setting conducive for interaction and collaboration.

It is well known that fascinating connections exist between coding and cryptology. Therefore this workshop series was organized to facilitate a fruitful interaction and stimulating discourse among experts from these two areas.

The inaugural IWCC was held at Wuyi Mountain, Fujian Province, China, during June 11-15, 2007 and attracted over 80 participants. Following this success, the second IWCC was held June 1-5, 2009 at Zhangjiajie, Hunan Province, China. Zhangjiajie is one of the most scenic areas in China.

The proceedings of this workshop consist of 21 technical papers, covering a wide range of topics in coding and cryptology, as well as related fields such as combinatorics. All papers, except one, are contributed by the invited speakers of the workshop and each paper has been carefully reviewed. We are grateful to the external reviewers for their help, which has greatly strengthened the quality of the proceedings.

IWCC 2009 was co-organized by the National University of Defense Technology (NUDT), China and Nanyang Technological University (NTU), Singapore. We acknowledge with gratitude the financial support from NUDT.

We would like to express our thanks to Springer for making it possible for the proceedings to be published in the *Lecture Notes in Computer Science series*. We also thank Zhe-xian Wan for his great encouragement and constant support for this workshop series. We are also grateful to the staff and students from both NUDT and NTU for the administrative and technical support they have rendered to the conference and the proceedings. Special thanks go to Longjiang Qu for taking care of the website of the workshop, and Yang Ding for assistance on matters related to LaTeX.

Yeow Meng Chee
Chao Li
San Ling
Huaxiong Wang
Chaoping Xing

Organization

Chair of Organizing Committee

Zhe-xian Wan   Academy of Mathematics and System Sciences, CAS, China

Organizing Committee

Yeow Meng Chee	Nanyang Technological University, Singapore
Chao Li	National University of Defense Technology, China
San Ling	Nanyang Technological University, Singapore
Huaxiong Wang	Nanyang Technological University, Singapore
Zhengming Wang	National University of Defense Technology, China
Chaoping Xing	Nanyang Technological University, Singapore
Jianming Zhu	National University of Defense Technology, China

Table of Contents

An Infinite Class of Balanced Vectorial Boolean Functions with Optimum Algebraic Immunity and Good Nonlinearity

Claude Carlet[1] and Keqin Feng[2]

[1] Universities of Paris 8 and Paris 13; CNRS, UMR 7539 LAGA
University of Paris 8, Department of Mathematics, 2 rue de la liberté
93526 Saint-Denis cedex 02, France
claude.carlet@inria.fr
[2] Department of Mathematical Sciences, Tsinghua University, Beijing China 100084
kfeng@math.tsinghua.edu.cn

Abstract. In this paper, we study the cryptographic properties of an infinite class of balanced vectorial Boolean functions recently introduced by Feng, Liao and Yang. These functions provably achieve an optimum algebraic immunity. We give a simpler proof of this fact and we prove that these functions have also an optimum algebraic degree and a non-weak nonlinearity.

1 Introduction

Let $n \geqslant 2$ and let $\mathbb{B}_n$ be the ring of Boolean functions in n variables.

$$f = f(X) = f(x_1, \cdots, x_n) : \mathbb{F}_2^n \to \mathbb{F}_2.$$

We have the ring isomorphism $\mathbb{B}_n = \mathbb{F}_2[x_1, \cdots, x_n]/(x_1^2 - x_1, \cdots, x_n^2 - x_n)$. Namely, each function $f \in \mathbb{B}_n$ has a unique representation as a multivariate polynomial over $\mathbb{F}_2$, called the algebraic normal form (ANF), of the special form:

$$f(x_1, \cdots, x_n) = \sum_{I \subseteq \{1, \cdots, n\}} a_I \prod_{i \in I} x_i \qquad (a_I \in \mathbb{F}_2).$$

The algebraic degree of $f \neq 0$ is defined as:

$$\deg(f) = \max\{|I| : I \subseteq \{1, \cdots, n\}, a_I = 1\}.$$

In this paper we shall need another representation of Boolean functions, by univariate polynomials over the field $\mathbb{F}_{2^n}$. Choosing a basis $(\beta_1, \cdots, \beta_n)$ of $\mathbb{F}_{2^n}$ over $\mathbb{F}_2$, we can identify the field $\mathbb{F}_{2^n}$ and the vector space $\mathbb{F}_2^n$ by

$$\mathbb{F}_2^n \xrightarrow{\sim} \mathbb{F}_{2^n}, \quad (x_1, \cdots, x_n) \mapsto x_1 \beta_1 + \cdots + x_n \beta_n.$$

Every function $f : \mathbb{F}_{2^n} \to \mathbb{F}_{2^n}$ (and in particular every Boolean function $f : \mathbb{F}_{2^n} \to \mathbb{F}_2$) can then be uniquely represented as a polynomial

$$f = f(x) = \sum_{j=0}^{2^n - 1} a_j x^j \qquad (a_j \in \mathbb{F}_{2^n}) \tag{1}$$

C. Xing et al. (Eds.): IWCC 2009, LNCS 5557, pp. 1–11, 2009.

In fact,

$$f = \sum_{a \in \mathbb{F}_{2^n}} f(a)(1 + (x+a)^{2^n-1}) \tag{2}$$

Then function f is Boolean if and only if $f(x)^2 = f(x)$ which means that $a_0, a_{2^n-1} \in \mathbb{F}_2$ and for $1 \leqslant i \leqslant 2^n - 2$, $a_{2j} = (a_j)^2$ where $2j$ is taken mod $2^n - 1$. It is well-known that the algebraic degree of the Boolean function $f \neq 0$ expressed by (1) is

$$\deg(f) = \max\{w_2(j) : a_j \neq 0, 0 \leqslant j \leqslant 2^n - 1\}$$

where, given the 2-adic expansion $j = j_0 + j_1 2 + \cdots + j_{n-1} 2^{n-1}$, with $j_0, \cdots, j_{n-1} \in \{0,1\}$, we have $w_2(j) = j_0 + j_1 + \cdots + j_{n-1}$; $w_2(j)$ is called the 2-weight of j.

Boolean functions used in cryptographic systems must have large algebraic degrees. Several other cryptographic properties for Boolean functions have been raised in passed years to allow the ciphers using them to resist continuously invented attack methods. We list them below.

A Boolean function $f \in \mathbb{B}_n$ is called balanced if $|f^{-1}(0)| = |f^{-1}(1)|(= 2^{n-1})$ where for $b \in \mathbb{F}_2$,

$$f^{-1}(b) = \{a = (a_1, \cdots, a_n) \in \mathbb{F}_2^n : f(a) = b\}.$$

The nonlinearity of $f \in \mathbb{B}_n$ is defined as:

$$nl(f) = \min\{d_H(f,g) : g \in A_n\}$$

where A_n is the set of affine functions $f(x_1, \cdots, x_n) = \lambda_1 x_1 + \cdots + \lambda_n x_n + \lambda$ $(\lambda_i, \lambda \in \mathbb{F}_2)$ and for $f, g \in \mathbb{B}_n$, $d_H(f,g)$ is the Hamming distance between f and g defined as

$$d_H(f,g) = \#\{a \in \mathbb{F}_2^n : f(a) \neq g(a)\}.$$

It is known that

$$nl(f) = 2^{n-1} - \frac{1}{2} \max\{|W_f(\lambda)| : \lambda = (\lambda_1, \cdots, \lambda_n) \in \mathbb{F}_2^n\} \tag{3}$$

where $W_f : \mathbb{F}_2^n \to \mathbb{Z}$ is the Walsh transform of $f(x)$ given by

$$W_f(\lambda) = \sum_{x \in \mathbb{F}_2^n} (-1)^{f(x)+\lambda \cdot x} \qquad (\lambda \in \mathbb{F}_2^n)$$

and $\lambda \cdot x$ is an inner product in $\mathbb{F}_2^n$. For a Boolean function expressed by (1), we can take $\lambda \cdot x = tr(\lambda x)$ for $\lambda, x \in \mathbb{F}_{2^n}$ where tr is the trace mapping:

$$tr : \mathbb{F}_{2^n} \to \mathbb{F}_2, \quad tr(\alpha) = \sum_{i=0}^{n-1} \alpha^{2^i} \qquad (\alpha \in \mathbb{F}_{2^n}).$$

Then the Walsh transform of a Boolean function $f : \mathbb{F}_{2^n} \to \mathbb{F}_2$ can be expressed as:

$$W_f(\lambda) = \sum_{x \in \mathbb{F}_{2^n}} (-1)^{f(x)+tr(\lambda x)} \qquad (\lambda \in \mathbb{F}_{2^n}) \tag{4}$$

For a Boolean function $0 \neq f = f(x_1, \cdots, x_n) \in \mathbb{B}_n$, the algebraic immunity of f is defined as

$$
\begin{aligned}
AI(f) &= \min\{\deg(g) : 0 \neq g \in \mathbb{B}_n, g\,f = 0 \text{ or } g(f+1) = 0\} \\
&= \min\{\deg(g) : 0 \neq g \in \mathbb{B}_n, g|_{f^{-1}(1)} = 0 \text{ or } g|_{f^{-1}(0)} = 0\}
\end{aligned}
$$

where, for a subset S of $\mathbb{F}_2^n$, $g|_S : S \to \mathbb{F}_2$ is the restriction of g to S.

A Boolean function used in cryptographic systems is required to be balanced, with large algebraic degree (close to n), nonlinearity (close to the maximum $2^{n-1} - 2^{n/2-1}$) and algebraic immunity (close to the maximum $\lceil n/2 \rceil$). There are tremendous developments to construct Boolean functions having ideal specific cryptographic property. But, it is a difficult challenge to find functions achieving all of the necessary criteria and the research of such functions has taken a significant delay with respect to cryptanalyses. It is proved that for $0 \neq f \in \mathbb{B}_n$ and $n \geqslant 2$,

(1) If f is balanced, then $\deg f \leqslant n - 1$;
(2) $AI(f) \leqslant \deg(f)$ and $AI(f) \leqslant \lceil \frac{n+1}{2} \rceil$;
(3) $nl(f) \leqslant 2^{n-1} - 2^{\frac{n}{2}-1}$.

In [4], the authors constructed an infinite class of balanced functions with optimum algebraic immunity, the largest algebraic degree for balanced function and potentially good nonlinearity.

Lemma 1 ([4]). *Let $n \geqslant 2$ and α be a primitive element of the field $\mathbb{F}_{2^n}$. For each integer s, $0 \leqslant s \leqslant 2^n - 2$, let $f = f(x) : \mathbb{F}_{2^n} \to \mathbb{F}_2$ defined by*

$$
f(x) = \begin{cases} 1, \text{ if } x \in \{0, \alpha^s, \alpha^{s+1}, \cdots, \alpha^{s+2^{n-1}-2}\} \\ 0, \text{ if } x \in \{\alpha^{s+2^{n-1}-1}, \alpha^{s+2^{n-1}}, \cdots, \alpha^{s+2^n-2}\} \end{cases}
$$

Then f is a balanced Boolean function in $\mathbb{B}_n$, $\deg(f) = n - 1$, $AI(f) = \lceil \frac{n+1}{2} \rceil$ and $nl(f) \geqslant 2^{n-1} - n \cdot \ln 2 \cdot 2^{\frac{n}{2}} - 1$. □

The exact values of the nonlinearity computed for small values of n are much better than what gives this lower bound. Such situation, where the observed nonlinearity is much better than what can be proved, happens also in the framework of sequences, see e.g. [5].

Remark: Lemma1 has been proved in [4] for $s = 0$. The proof obviously works for each s, $0 \leqslant s \leqslant 2^n - 2$, but the resulting function is then linearly equivalent to the original one.

This single-output Boolean function (usable in stream ciphers) was an item of a class of vectorial (multi-output) Boolean functions (usable as S-boxes in stream or block ciphers). In this paper we generalize this result to this general class of vectorial functions.

2 Generalization to Vectorial Boolean Functions

From now on, we assume that $n \geqslant 2$ and $1 \leqslant m \leqslant n$. We consider vectorial Boolean (n, m)-functions (that is, functions from $\mathbb{F}_2^n$ to $\mathbb{F}_2^m$):

$$F = (f_1, \cdots, f_m) : \mathbb{F}_2^n \to \mathbb{F}_2^m \qquad (f_i \in \mathbb{B}_n, 1 \leqslant i \leqslant m).$$

The following definitions of cryptographic properties of a vectorial Boolean functions are generalizations of the ordinary Boolean function case ($m = 1$). For these definitions and known results we refer to [3].

An (n, m)-function $F = (f_1, \cdots, f_m) : \mathbb{F}_2^n \to \mathbb{F}_2^m$ is called balanced if it takes every value $b \in \mathbb{F}_2^m$ the same number 2^{n-m} of times. Namely, $|F^{-1}(b)| = 2^{n-m}$ for each $b = (b_1, \cdots, b_m) \in \mathbb{F}_2^m$ where

$$\begin{aligned} F^{-1}(b) &= \{x \in \mathbb{F}_2^n \mid F(x) = b\} \\ &= \{x \in \mathbb{F}_2^n \mid f_i(x) = b_i \quad (1 \leqslant i \leqslant m)\}. \end{aligned}$$

It is known (see e.g. [3]) that F is balanced if and only if the so-called component function $v \cdot F$ (that is, $v_1 f_1 + \cdots + v_m f_m$ if the inner product in $\mathbb{F}_2^m$ is the usual one) is balanced for every $0 \neq v \in \mathbb{F}_2^m$.

The nonlinearity of an (n, m)-function F is defined as

$$nl(F) = \min\{nl(v \cdot F) : 0 \neq v \in \mathbb{F}_2^m\}$$

$$= 2^{n-1} - \frac{1}{2} \max\left\{ \left| \sum_{x \in \mathbb{F}_2^n} (-1)^{v \cdot F(x) + u \cdot x} \right| : 0 \neq v \in \mathbb{F}_2^m, u \in \mathbb{F}_2^n \right\}.$$

It is proved (see e.g. [3]) that $nl(F) \leqslant 2^{n-1} - 2^{\frac{n}{2}-1}$ for every n and m and that $nl(F) = 2^{n-1} - 2^{\frac{n}{2}-1}$ is possible only if n is even and $m \leqslant \frac{n}{2}$. If $m = n$, then $nl(F) \leqslant 2^{n-1} - 2^{\frac{n-1}{2}}$.

The algebraic degree of F is defined as

$$\mathrm{Deg}(F) = \max\{\deg(f_i) : 1 \leqslant i \leqslant m\} = \max\{\deg(v \cdot F) : 0 \neq v \in \mathbb{F}_2^m\}.$$

Another notion of degree, called the minimum degree in [3], is defined as

$$\deg(F) = \min\{\deg(v \cdot F) : 0 \neq v \in \mathbb{F}_2^m\}.$$

It is obvious that $\deg(F) \leqslant \mathrm{Deg}(F)$. If F is balanced, then the f_i's ($1 \leqslant i \leqslant m$) being balanced, we have $\mathrm{Deg}(F) \leqslant n - 1$.

There exist several possible generalizations of the notion of algebraic immunity to (n, m)-functions. The main one, defined by Armknecht and Krause in [1] is:

$$AI(F) = \min\{\deg(g) : 0 \neq g \in \mathbb{B}_n, \text{ there exists } b \in \mathbb{F}_2^m \text{ s.t. } g|_{F^{-1}(b)} = 0\}.$$

It satisfies the following upper bound: let $d = d(n, m)$ be the smallest integer such that

$$\sum_{i=0}^{d} \binom{n}{i} > 2^{n-m},$$

then [1], we have $AI(F) \leqslant d$ for every (n, m)-function F.

Another notion of algebraic immunity of an (n, m)-function F is given by Ars and J.-C. Faugère [2], but it is proved [7] that the two definitions of algebraic immunity in [1] and [2] are the same. Further notions are introduced in [1] and [2] but we do not study them in the present paper.

In [6] is introduced an infinite class of (n, m)-functions with optimum algebraic immunity. Let α be a primitive element of the field $\mathbb{F}_{2^n}$ so that $\mathbb{F}_{2^n} = \{0\} \cup \{\alpha^i \mid 0 \leqslant i \leqslant 2^n - 2\}$. For any fixed integer s, $0 \leqslant s \leqslant 2^n - 2$, $\mathbb{F}_{2^n}$ is a disjoint union of the following 2^m subsets:

$$S_0 = \{\alpha^l \mid s \leqslant l \leqslant s + 2^{n-m} - 2\} \cup \{0\}$$
$$S_b = \{\alpha^l \mid s + 2^{n-m}b - 1 \leqslant l \leqslant s + 2^{n-m}(b+1) - 2\}; \ 1 \leqslant b \leqslant 2^m - 1 \quad (5)$$

Each integer b, $0 \leqslant b \leqslant 2^m - 1$ has a 2-adic expansion

$$b = b_0 + b_1 2 + \cdots + b_{m-1} 2^{m-1} \qquad (b_0, \cdots, b_{m-1} \in \{0, 1\})$$

and corresponds to the vector $\bar{b} = (b_0, \cdots, b_{m-1}) \in \mathbb{F}_2^m$. For each integer i, $0 \leqslant i \leqslant m - 1$, we define the Boolean function $f_i : \mathbb{F}_{2^n} \to \mathbb{F}_2$ by

$$f_i(x) = \begin{cases} 1, \text{ if } x \in \bigcup_{\substack{0 \leqslant b \leqslant 2^m - 1 \\ b_i = 1}} S_b \\ 0, \text{ otherwise.} \end{cases} \quad (6)$$

Then for the (n, m)-function

$$F = (f_0, \cdots, f_{m-1}) : \mathbb{F}_{2^n} \to \mathbb{F}_2^m,$$

we have, for each $\bar{b} = (b_0, \cdots, b_{m-1}) \in \mathbb{F}_2^m$ and $b = \sum_{i=0}^{m-1} b_i 2^i$,

$$x \in F^{-1}(\bar{b}) \Leftrightarrow f_i(x) = b_i \qquad (0 \leqslant i \leqslant m - 1)$$
$$\Leftrightarrow x \in \bigcap \{S_c \mid 0 \leqslant c \leqslant 2^m - 1, c_i = b_i\} \qquad (0 \leqslant i \leqslant m - 1)$$
$$\Leftrightarrow x \in S_b.$$

Therefore the (n, m)-function F can be characterized by

$$F^{-1}(\bar{b}) = S_b \qquad (\text{for each } b, 0 \leqslant b \leqslant 2^m - 1) \quad (7)$$

Theorem 1. *(1). The function F is balanced.*

(2). $\text{Deg}(F) = n - 1$ and $\deg(F) = n - 1$ if and only if $(\frac{\alpha}{1+\alpha})^{2^i}$ $(i = 0, 1, \cdots, m - 1)$ are linearly independent over $\mathbb{F}_2$.

Proof. (1) is obvious since, for each $\bar{b} \in \mathbb{F}_2^m$, the size $F^{-1}(\bar{b}) = S_b$ is 2^{n-m}.

(2). Let $0 \leqslant b \leqslant 2^m - 1$ and $g_b : \mathbb{F}_{2^n} \to \mathbb{F}_2$ be the Boolean function defined as

$$g_b(x) = \begin{cases} 1, \text{ if } x \in S_b \\ 0, \text{ otherwise.} \end{cases}$$

Let $q = 2^n$, we have, for $1 \leqslant b \leqslant 2^m - 1$,

$$
\begin{aligned}
g_b(x) &= \sum_{a \in S_b} (1 + (x + a)^{q-1}) \qquad \text{(by (2))} \\
&= \sum_{a \in S_b} \frac{x^q + a^q}{x + a} \\
&= \sum_{a \in S_b} \sum_{j=0}^{q-1} x^j a^{q-1-j} \\
&= \sum_{a \in S_b} \sum_{j=1}^{q-2} x^j a^{q-1-j}
\end{aligned}
$$

Then by (6), for $0 \leqslant i \leqslant m - 1$,

$$
\begin{aligned}
f_i(x) &= \sum_{\substack{(b_0, \cdots, b_{m-1}) \in \mathbb{F}_2^m \\ b_i = 1}} g_b(x) \qquad \left(b = \sum_{\lambda=0}^{m-1} b_\lambda 2^\lambda \right) \\
&= \sum_{\substack{(b_0, \cdots, b_{m-1}) \in \mathbb{F}_2^m \\ b_i = 1}} \sum_{j=1}^{q-1} x^j a^{q-1-j} \\
&= \sum_{j=1}^{q-1} c_j x^j
\end{aligned}
$$

where, for $1 \leqslant j \leqslant q - 1$,

$$
c_j = \sum_{\substack{(b_0, \cdots, b_{m-1}) \in \mathbb{F}_2^m \\ b_i = 1}} \sum_{a \in S_b} a^{q-1-j}
$$

Therefore $c_{q-1} = 2^{n-1} = 0$ and

$$
\begin{aligned}
c_{q-2} &= \sum_{\substack{(b_0, \cdots, b_{m-1}) \in \mathbb{F}_2^m \\ b_i = 1}} \sum_{a \in S_b} a \\
&= \sum_{b_0, \cdots, b_{i-1}, b_{i+1}, \cdots, b_{m-1} \in \mathbb{F}_2} \sum_{l=0}^{2^{n-m}-1} \alpha^{s + 2^{n-m}(b_0 + \cdots + b_{i-1} 2^{i-1} + 2^i + b_{i+1} 2^{i+1} + \cdots + 2^{m-1}) - 1 + l} \text{(by (5))} \\
&= \alpha^{s-1} \left(\sum_{l=0}^{2^{n-m}-1} \alpha^l \right) \\
&\qquad \left[(1+\alpha)(1+\alpha^2) \cdots (1 + \alpha^{2^{i-1}}) \alpha^{2^i} (1 + \alpha^{2^{i+1}}) \cdots (1 + \alpha^{2^{m-1}}) \right]^{2^{n-m}} \\
&= \alpha^{s-1} (1+\alpha)^{2^{n-m}-1} \left(\frac{(1+\alpha)^{2^m} \alpha^{2^i}}{(1+\alpha)^{2^i+1}} \right)^{2^{n-m}} \\
&= \alpha^{s-1} (1+\alpha)^{2^n-1} \left(\frac{\alpha}{(1+\alpha)} \right)^{2^{n-m+i}} \neq 0
\end{aligned}
$$

$$\tag{8}$$

Therefore $\text{Deg}(F) = n-1$. Moreover, $\deg(F) \leqslant n-1$ and by (8) and the definition of $\deg(F)$:

$$\deg(F) = n - 1 \Leftrightarrow$$
$$\sum_{i=0}^{m-1} v_i \left(\frac{\alpha}{1+\alpha}\right)^{2^{n-m+i}} \neq 0 \text{ for every } 0 \neq (v_0, \cdots, v_{m-1}) \in \mathbb{F}_2^m \Leftrightarrow$$
$$\left(\frac{\alpha}{1+\alpha}\right)^{2^i} \quad (i = 0, 1, \cdots, m-1) \text{ are linear independent over } \mathbb{F}_2.$$

This completes the proof of Theorem 1.

Remark: From the conclusion (2) of Theorem 1 we may ask the following question: for an integer $n \geqslant 2$, is there a primitive element α in $\mathbb{F}_{2^n}$ such that $\left\{\left(\frac{\alpha}{1+\alpha}\right)^{2^i} : i = 0, 1, \cdots, n-1\right\}$ is a (so called normal) basis of $\mathbb{F}_{2^n}$ over $\mathbb{F}_2$? If the question has affirmative answer, then $\deg(F) = n - 1$ for each m, $1 \leqslant m \leqslant n$. It is easy to see that if $n = p$ is a prime number and $2^p - 1$ is a (Mersenne) prime number, then the answer of the question is affirmative since we have a normal basis $\{\beta^{2^i} \mid 0 \leqslant i \leqslant n-1\}$ in $\mathbb{F}_{2^n}$ and $\alpha = \frac{\beta}{1+\beta} \neq 1$ so that α is a primitive element of $\mathbb{F}_{2^n}$.

Next, we show that the (n, m)-function F has maximum algebraic immunity.

Theorem 2. $AI(F) = d$ where $d = d(n, m)$ is the smallest integer such that $\sum_{i=0}^{d} \binom{n}{i} > 2^{n-m}$.

Proof. Suppose that $g \in \mathbb{B}_n$, $\deg(g) \leqslant d - 1$ and $g|_{F^{-1}(b)} = 0$ for some b, $0 \leqslant b \leqslant 2^m - 1$. We need to show $g = 0$. Let $q = 2^n$ and

$$g(x) = \sum_{\lambda=0}^{q-1} c_\lambda x^\lambda \in \mathbb{F}_q[x].$$

By the definition of $d = d(n, m)$ we know that $\sum_{i=0}^{d-1} \binom{n}{i} \leqslant 2^{n-m}$. Let

$$N = \#\{\lambda \mid 0 \leqslant \lambda \leqslant q-1, c_\lambda \neq 0\}$$

be the number of nonzero coefficients c_λ of $g(x)$. From $\deg(g) \leqslant d - 1$ we know that

$$N \leqslant \#\{\lambda \mid 0 \leqslant \lambda = \sum_{j=0}^{m-1} \lambda_j \cdot 2^j \leqslant 2^m - 1, \lambda_0 + \cdots + \lambda_{m-1} \leqslant d - 1\}$$
$$= \sum_{i=0}^{d-1} \binom{n}{i} \leqslant 2^{n-m}. \tag{9}$$

Suppose that $g \neq 0$. For $1 \leqslant b \leqslant 2^m - 1$, the assumption $g|_{F^{-1}(b)} = g|_{S_b} = 0$ and (7) imply that $g(\alpha^l) = 0$ for sequential 2^{n-m} values of l, $s + 2^{n-m}b - 1 \leqslant l \leqslant s +$

$2^{n-m}b + 2^{n-m} - 2$. By the BCH bound [8] in coding theory, we get $N \geqslant 2^{n-m}+1$ which contradicts (9). For $b = 0$, from $0 \in S_0$ and $g|_{F^{-1}(0)} = g|_{S_0} = 0$ we know that $0 = g(0) = c_0$ so that $N \leqslant \sum_{i=1}^{d-1} \binom{n}{i} \leqslant 2^{n-m} - 1$. Moreover, for $s \leqslant l \leqslant s + 2^{n-m} - 2$ we have $\alpha^l \in S_0$ so that $g(\alpha^l) = 0$. Then the BCH bound implies that $N \geqslant 2^{n-m}$ which contradicts $N \leqslant 2^{n-m} - 1$. Therefore $g = 0$ and the proof of Theorem 2 is completed.

At last, we present a lower bound of the nonlinearity of the (n,m)-function F.

Theorem 3. $nl(F) \geqslant 2^{n-1} - \frac{2^{\frac{n}{2}+m}}{\pi} \ln(\frac{4(2^n-1)}{\pi}) - 1 \sim 2^{n-1} - \frac{\ln 2}{\pi} 2^{\frac{n}{2}+m} \cdot n$.

Proof. By definition,

$$nl(F) = 2^{n-1} - \frac{1}{2} \max\{|W_{v \cdot F}(\alpha^t)| : 0 \neq v \in \mathbb{F}_2^m, 0 \leqslant t \leqslant 2^n - 2\}$$

where

$$\begin{aligned} W_{v \cdot F}(\alpha^t) &= \sum_{x \in \mathbb{F}_{2^n}} (-1)^{v \cdot F(x) + Tr(\alpha^t x)} \\ &= \sum_{b=0}^{2^m-1} (-1)^{\bar{b} \cdot v} \sum_{x \in S_b} (-1)^{Tr(\alpha^t x)} \end{aligned} \tag{10}$$

Let $q = 2^n$, $\zeta_{q-1} = e^{\frac{2\pi i}{q-1}}$ and $\chi : \mathbb{F}_q \to \mathbb{C}$ be the multiplicative character of $\mathbb{F}_q$ defined as $\chi(\alpha) = \zeta_{q-1}$ and $\chi(0) = 0$. We have the Gauss Sums

$$G(\chi^\mu) = \sum_{x \in \mathbb{F}_q} \chi^\mu(x)(-1)^{Tr(x)} \qquad (0 \leqslant \mu \leqslant q - 2).$$

It is known that $|G(\chi^\mu)| = \sqrt{q}$ for $1 \leqslant \mu \leqslant q-2$ and $G(\chi^0) = -1$. By the orthogonal relations of the characters χ^μ $(0 \leqslant \mu \leqslant q - 2)$ we get

$$(-1)^{tr(x)} = \frac{1}{q-1} \sum_{\mu=0}^{q-2} G(\chi^\mu) \bar{\chi}^\mu(x)$$

Therefore,

$$\begin{aligned} \sum_{x \in S_b} (-1)^{tr(\alpha^t x)} &= \frac{1}{q-1} \sum_{x \in S_b} \sum_{\mu=0}^{q-2} G(\chi^\mu) \bar{\chi}^\mu(\alpha^t x) \\ &= \frac{1}{q-1} \sum_{\mu=0}^{q-2} G(\chi^\mu) \bar{\chi}(\alpha)^{\mu t} \sum_{x \in S_b} \bar{\chi}^\mu(x) \\ &= -\frac{M_b}{q-1} + \frac{1}{q-1} \sum_{\mu=1}^{q-2} G(\chi^\mu) \bar{\chi}(\alpha)^{\mu t} \sum_{x \in S_b} \bar{\chi}^\mu(x) \end{aligned} \tag{11}$$

where $M_b = \sum\limits_{x \in S_b} \bar{\chi}^0(x) = 2^{n-m}$ for $1 \leqslant b \leqslant 2^m - 1$ and $M_0 = 2^{n-m} - 1$. By (10) and (11) we get, for $0 \neq v = (v_0, v_1, \cdots, v_{m-1}) \in \mathbb{F}_2^m$,

$$
\begin{aligned}
W_{v \cdot F}(\alpha^t) &= \frac{1}{q-1} \sum_{\mu=1}^{q-2} G(\chi^\mu) \bar{\chi}(\alpha)^{\mu t} \sum_{b=0}^{2^m-1} (-1)^{\bar{b} \cdot v} \sum_{x \in S_b} \bar{\chi}^\mu(x) \\
&\quad - \frac{1}{q-1} \sum_{b=0}^{2^m-1} (-1)^{\bar{b} \cdot v} M_b \\
&= \frac{1}{q-1} \sum_{\mu=1}^{q-2} G(\chi^\mu) \bar{\chi}(\alpha)^{\mu t} \sum_{b=0}^{2^m-1} (-1)^{\bar{b} \cdot v} \sum_{x \in S_b} \bar{\chi}^\mu(x) + \frac{1}{q-1}
\end{aligned}
\tag{12}
$$

For $1 \leqslant \mu \leqslant q - 2$, we have

$$
\begin{aligned}
\sum_{x \in S_0} \bar{\chi}^\mu(x) &= \sum_{l=s}^{s+2^{n-m}-2} \bar{\chi}^\mu(\alpha^l) = (\bar{\chi}^\mu(\alpha))^s \sum_{t=0}^{2^{n-m}-2} (\bar{\chi}^\mu(\alpha))^t \\
&= (\bar{\chi}^\mu(\alpha))^s (1 - \bar{\chi}^\mu(\alpha)^{2^{n-m}-1}) \big/ (1 - \bar{\chi}^\mu(\alpha))
\end{aligned}
$$

and for $1 \leqslant b \leqslant 2^m - 1$,

$$
\begin{aligned}
\sum_{x \in S_b} \bar{\chi}^\mu(x) &= \bar{\chi}^\mu(\alpha)^{s+2^{n-m}b-1} \sum_{t=0}^{2^{n-m}-1} \bar{\chi}^\mu(\alpha^t) \\
&= \bar{\chi}^\mu(\alpha)^{s+2^{n-m}b-1} (1 - \bar{\chi}^\mu(\alpha)^{2^{n-m}}) \big/ (1 - \bar{\chi}^\mu(\alpha)).
\end{aligned}
$$

Therefore, let $\beta = \bar{\chi}^\mu(\alpha) = e^{\frac{-2\pi\mu\sqrt{-1}}{q-1}}$, we get

$$
\sum_{b=0}^{2^m-1} (-1)^{\bar{b} \cdot v} \sum_{x \in S_b} \bar{\chi}^\mu(x) =
$$

$$
\frac{\beta^s}{1 - \beta} \left[(1 - \beta^{2^{n-m}-1}) + (1 - \beta^{2^{n-m}}) \sum_{b=1}^{2^m-1} (-1)^{\bar{b} \cdot v} \beta^{(2^{n-m}b-1)} \right]
\tag{13}
$$

But

$$
\begin{aligned}
&\sum_{b=1}^{2^m-1} (-1)^{\bar{b} \cdot v} \beta^{(2^{n-m}b-1)} \\
&= \beta^{-1} \left[\sum_{b=0}^{2^m-1} (-1)^{\bar{b} \cdot v} \beta^{2^{n-m}b} - 1 \right] \\
&= \beta^{-1} \left[-1 + \sum_{(b_0,\cdots,b_{m-1}) \in \mathbb{F}_2^m} (-1)^{b_0 v_0 + \cdots + b_{m-1} v_{m-1}} \beta^{2^{n-m}(b_0 + b_1 2 + \cdots + b_{m-1} 2^{m-1})} \right] \\
&= \beta^{-1} \left[-1 + (1 + (-1)^{v_0} \beta^{2^{n-m}})(1 + (-1)^{v_1} \beta^{2^{n-m+1}}) \cdots (1 + (-1)^{v_{m-1}} \beta^{2^{n-1}}) \right]
\end{aligned}
$$

From (13) we get, for $1 \leqslant \mu \leqslant q - 2$ and $\beta = \bar{\chi}^\mu(\alpha)$,

$$
\sum_{b=0}^{2^m-1} (-1)^{\bar{b} \cdot v} \sum_{x \in S_b} \bar{\chi}^\mu(x) =
$$

$$\frac{\beta^s}{1-\beta}\left[(1-\beta^{2^{n-m}-1})+(1-\beta^{2^{n-m}})\beta^{-1}(-1+\prod_{i=0}^{m-1}(1+(-1)^{v_i}\beta^{2^{n-m+i}}))\right]$$

$$\left|\sum_{b=0}^{2^m-1}(-1)^{\bar{b}\cdot v}\sum_{x\in S_b}\bar{\chi}^\mu(x)\right|\leqslant\frac{1}{|1-\bar{\chi}^\mu(\alpha)|}[2+2(2^m-1)]=2^{m+1}\Big/|1-\bar{\chi}^\mu(\alpha)|$$

By (12) we get, for $0\neq v\in\mathbb{F}_2^m$ and $0\leqslant t\leqslant q-2$,

$$\left|W_{v\cdot F}(\alpha^t)\right|\leqslant\frac{1}{q-1}+\frac{\sqrt{q}\cdot 2^{m+1}}{q-1}\sum_{\mu=1}^{q-2}\frac{1}{|1-\bar{\chi}^\mu(\alpha)|}$$

$$=\frac{1}{q-1}\left(1+2^{\frac{n}{2}+m+1}\sum_{\mu=1}^{\frac{q}{2}-1}\frac{1}{\sin\frac{\pi\mu}{q-1}}\right)\tag{14}$$

It can be checked easily that for $0\leqslant\theta<t$ and $t+\theta\leqslant\pi$:

$$\frac{1}{\sin(t-\theta)}+\frac{1}{\sin(t+\theta)}\geqslant\frac{2}{\sin t}.$$

Then we deduce

$$\int_{t-\frac{\theta}{2}}^{t+\frac{\theta}{2}}\frac{d\alpha}{\sin\alpha}\geqslant\frac{\theta}{\sin t}$$

and taking $\theta=\frac{\pi}{q-1}$:

$$\sum_{\mu=1}^{\frac{q}{2}-1}(\sin\frac{\pi\mu}{q-1})^{-1}\leqslant\frac{q-1}{\pi}\sum_{\mu=1}^{\frac{q}{2}-1}\int_{\frac{\pi\mu}{q-1}-\frac{\pi}{2(q-1)}}^{\frac{\pi\mu}{q-1}+\frac{\pi}{2(q-1)}}\frac{d\alpha}{\sin\alpha}$$

$$=\frac{q-1}{\pi}\int_{\frac{\pi}{2(q-1)}}^{\frac{\pi}{2}}\frac{d\alpha}{\sin\alpha}$$

$$=(q-1)\int_{\frac{1}{2(q-1)}}^{\frac{1}{2}}\frac{dt}{\sin\pi t}$$

$$=\frac{q-1}{\pi}\int_{\frac{\pi}{2(q-1)}}^{\frac{\pi}{2}}\frac{dt}{\sin\pi t}$$

$$=\frac{q-1}{\pi}\cdot\frac{1}{2}\ln(\tan(\frac{t}{2}))\Big|_{t=\frac{\pi}{2(q-1)}}^{t=\frac{\pi}{2}}$$

$$=\frac{q-1}{2\pi}\ln\left(\tan(\frac{\pi}{4(q-1)})^{-1}\right)$$

$$\leqslant\frac{q-1}{2\pi}\ln(\frac{4(q-1)}{\pi}).$$

where we use $\tan(x)\geqslant x$ for $0\leqslant x\leqslant\frac{\pi}{2}$. Then, by (14) we get

$$\left|W_{v\cdot F}(\alpha^t)\right|\leqslant\frac{1}{q-1}(1+2^{\frac{n}{2}+m}\frac{(q-1)}{\pi}\ln(\frac{4(q-1)}{\pi}))$$

$$\leqslant\frac{2^{\frac{n}{2}+m}}{\pi}\ln(\frac{4(q-1)}{\pi})+1$$

for all $0 \neq v \in \mathbb{F}_2^m$ and $0 \leqslant t \leqslant q - 2$.

Therefore $nl(F) \geqslant 2^{n-1} - \frac{2^{\frac{n}{2}+m-1}}{\pi} \ln(\frac{4(2^n-1)}{\pi}) - 1$. This completes the proof of Theorem 3.

Remark. A notion of generalized nonlinearity of Boolean functions over $\mathbb{F}_{2^n}$ is introduced in [9]. It equals the minimum distance between the Boolean function f and all functions $tr(ax^d)$, where d is co-prime with $2^n - 1$ (hence, it also equals the minimum distance between f and all functions $tr((ax)^d)$, and the minimum nonlinearity of all functions $f(x^c)$, where c is co-prime with $2^n - 1$). The related notion of generalized nonlinearity of a vectorial function is the minimum generalized nonlinearity of its component functions. The lower bound of Theorem 3 is still valid for this generalized nonlinearity, by replacing μ by $c\mu$ in the proof (note that the mapping $\mu \to c\mu$ is bijective since c is co-prime with $2^n - 1$).

Acknowledgement

We thank K. Khoo for the observation that our lower bound on the nonlinearity is valid for the generalized nonlinearity. The second author is supported by the 973 project of China no.2004CB3180004 and the State Key Lab. on Information Security" of China (SKLOIS).

References

1. Armknecht, F., Krause, M.: Constructing single- and multi-output Boolean functions with maximal algebraic immunity. In: Bugliesi, M., Preneel, B., Sassone, V., Wegener, I. (eds.) ICALP 2006. LNCS, vol. 4052, pp. 180–191. Springer, Heidelberg (2006)
2. Ars, G., Faugère, J.-C.: Algebraic immunity of functions over finite fields, INRIA, No. 5532, 17 pages (2005)
3. Carlet, C.: Vectorial Boolean Functions for Cryptography. In: Crama, Y., Hammer, P. (eds.) Boolean Methods and Models. Cambridge Univ. Press, Cambridge (to appear)
4. Carlet, C., Feng, K.: An infinite class of balanced functions with optimum algebraic immunity, good immunity to fast algebraic attacks and good nonlinearity. In: Carlet, C., Feng, K. (eds.) ASIACRYPT 2008. LNCS. Springer, Heidelberg (2008)
5. Dmitriev, D., Jedwab, J.: Bounds on the growth rate of the peak sidelobe level of binary sequences. Advances in Mathematics of Communications 1, 461–475 (2007)
6. Feng, K., Liao, Q., Yang, J.: Maximal values of generalized algebraic immunity. Designs, Codes and Cryptography (to appear)
7. Liao, Q., Feng, K.: A note on algebraic immunity of vectorial Boolean functions (preprint, 2008)
8. MacWilliams, F.J., Sloane, N.J.: The theory of error-correcting codes. North-Holland, Amsterdam (1977)
9. Youssef, A.M., Gong, G.: Hyper-bent functions. In: Pfitzmann, B. (ed.) EURO-CRYPT 2001. LNCS, vol. 2045, pp. 406–419. Springer, Heidelberg (2001)

Separation and Witnesses

Gérard Cohen

ENST and CNRS, Paris, France
cohen@enst.fr

Abstract. We revisit two notions of difference between codewords, namely separation and the existence of small witnesses, and explore their links.

1 Introduction

Let Q be an alphabet of size q. A subset C of Q^n with $|C| = M$ is an $(n, M)_q$ or (n, M)-*code*. Elements $c = (c_1, \ldots, c_n)$ of C are *codewords*. Let $R = R(C) = \log_q M/n$ denote the rate of C.

Coding theory asks for codes (or sets) C such that every codeword $c \in C$ is as "different" as possible from all the others. The usual requirement is a large minimum Hamming distance between codewords; the associated question is to determine the maximum size of such a code.

We survey here two relaxations of this problem, namely, the notions of separation and witness, and their interplay.

In the first one, separation, we look for some minimum distance between disjoint *subsets* of codewords (instead of merely -singletons of- codewords).

In the second relaxation, dealing with the existence of a witness, we look for a small subset $W \subset [n]$ of coordinates such that c differs from every other codeword in W. In other words, c can be singled out from all the other codewords by observing only a small subset of coordinates.

We then establish links between some separating and witness codes and conclude with a few open problems.

2 Separation

As an introductory illustration before the general case, consider hashing, central in Computer Science and Coding, see, e.g., [12] and its references.

For a parameter $t \geq 2$ a code C is called t-*hashing* if for any t distinct codewords $c^1, \ldots, c^t \in C$ there is a coordinate $1 \leq i \leq n$ such that all values c_i^j, $1 \leq j \leq t$ are distinct.

An obvious necessary condition for the existence of a t-hashing family is $q \geq t$; it turns out to be sufficient too (see [11], [22], [23], [29] for bounds on the rate of t-hashing families of growing length).

An extension of hashing was introduced in [6].

C. Xing et al. (Eds.): IWCC 2009, LNCS 5557, pp. 12–21, 2009.
© Springer-Verlag Berlin Heidelberg 2009

Definition 1. *Let $2 \leq t < u$ be integers. A subset $C \subset Q^n$ is (t, u)-hashing if for any two subsets T, U of C such that $T \subset U$, $|T| = t$, $|U| = u$, there is some coordinate $i \in \{1, \dots, n\}$ such that for any $x \in T$ and any $y \in U, y \neq x$, we have $x_i \neq y_i$.*

The concept of (t, u)-hashing is easily seen to generalize the standard one and some variants of separation. Indeed, when $u = t + 1$, a (t, u)-hashing family is $(t + 1)$-hashing; when $t = 1, u = 3$, we get $(1, 2)$-separation (see later).

The main use of (t, u)-hashing codes is as a tool to show the existence of high rate parent-identifying codes ([6,2], that we now describe).

2.1 Parent Identifying Codes

Let C be an (n, M)-code. Suppose $X \subseteq C$. For any coordinate i define the *projection*

$$P_i(X) = \bigcup_{x \in X} \{x_i\}.$$

Define the *envelope* $e(X)$ of X by:

$$e(X) = \{x \in Q^n : \forall i, x_i \in P_i(X)\}.$$

Elements of $e(X)$ are *descendants* of X. Observe that $X \subseteq e(X)$ and $e(X) = X$ if $|X| = 1$.

Given a descendant $s \in Q^n$, we want to identify at least one member of X (a parent). From [6,3], we have the following definition, generalizing case $t = 2$ from [18].

Definition 2. *For any $s \in Q^n$ let $\mathcal{H}_t(s)$ be the set of subsets $X \subset C$ of size at most t such that $s \in e(X)$. We shall say that C has the* identifiable parent property of order t *(or is a t-identifying code, or has the t-IPP, for short) if for any $s \in Q^n$, either $\mathcal{H}_t(s) = \emptyset$ or*

$$\bigcap_{X \in \mathcal{H}_t(s)} X \neq \emptyset.$$

Parent identifying codes are motivated by their connection to digital fingerprinting and software piracy, see, e.g., [10], [9], [34].

Let $R_q(t) := \liminf_{n \to \infty} \max R(C_n)$, where the maximum is computed over all t-identifying codes C_n of length n. In [6], answering a question of [34], the following is proved:

Theorem 1. *$R_q(t) > 0$ if and only if $t \leq q - 1$.*

The proof is based on a connection between (t, u)-hashing and t-IPP:

Lemma 1. *Let $u = \lfloor (t/2 + 1)^2 \rfloor$. If C is (t, u)-hashing then it is t-identifying.*

By the probabilistic method [1], one can obtain a lower bound on the rate of $(t, u = \lfloor (t/2 + 1)^2 \rfloor)$-hashing families, and thus of t- identifying codes:

Theorem 2. *Let* $u = \lfloor (t/2 + 1)^2) \rfloor$. *We have*

$$R_q(t) \geq \frac{1}{u-1} \log_q \frac{(q-t)!q^u}{(q-t)!q^u - q!(q-t)^{u-t}}.$$

The rate guaranteed by Theorem 2 is further improved in [2], where explicit constructions of high rate (t, u)-hashing and t- separating families are given.

2.2 Generalized Separation

Interest in separating codes comes mainly from digital fingerprinting [9]. A vendor distributes digital copies of a copyrighted work, and wants to prevent the users from making illegal copies. Watermarking can be used to give every sold copy a unique ID, a digital fingerprint, identifying the buyer. If an illegal copy subsequently appears, the user guilty of copying may be identified.

An interesting combinatorial problem arises when facing coalitions of pirates. If several users collude, they may compare their copies. Every differing bit is assumed to be part of the fingerprint and these are the only ones prone to modification (by the so-called *Marking Assumption*).

The fingerprints the pirates are able to forge, based on the set X they hold, form the so-called *feasible* set or envelope previously defined.

If the set (code) of valid fingerprints still makes it possible to trace at least one guilty pirate out of a coalition of size t or less, we have the already discussed t-identifiable parent property.

If the pirates are able to forge the fingerprint of an innocent user, we say that this user is framed. Codes which prevent framing are called *frameproof codes*.

Definition 3. *A sequence* $(T_1, \ldots, T_z)$ *of pairwise disjoint sets of words is called a* $(t_1, \ldots, t_z)$-*configuration if* $\#T_j = t_j$ *for all j. Such a configuration is separated if there is a position i, such that for all $l \neq l'$ every word of T_l is different from every word of $T_{l'}$ on position i.*

A code is $(t_1, \ldots, t_z)$-*separating if every* $(t_1, \ldots, t_z)$-*configuration is separated. A t-separating code is also called a* **t-SS** *(separating system).*

A few properties are covered by our general definition of $(t_1, \ldots, t_z)$-separation: with $z = 2$ we have the (t, u)-separation; when $t_i = 1$ for each i, we recover z-hashing; when $z = t + 1$, $t_z = u - t$, and $t_i = 1$ for $i < z$, (t, u)-hashing.

In earlier works on watermarking, (t, t)-separating codes have been called t-SFP (secure frameproof) [33,34]. The current terminology is older though [32]. The t-frameproof codes from [33,34] are just $(t, 1)$-separating codes.

Non-binary $(2, 1)$- and $(2, 2)$-SS appear in [30].

Separating codes have also been studied in a set-theoretic framework, e.g. [24]. Reference [21] gives various problems equivalent to $(2, 1)$-separation.

2.3 A Sufficient Condition for Separation

For any word $c = (c_1, \ldots, c_n) \in Q^n$ we define the support to be

$$\chi(c) := \{i \mid c_i \neq 0\}.$$

For any subset $S \subset V$, the support is

$$\chi(S) := \bigcup_{c \in S} \chi(c).$$

We define the weight of subsets and codewords to be the size of their support, and denote it $w(c) := \#\chi(c)$ or $w(S) := \#\chi(S)$.

We write $\boldsymbol{t} = (t_1, \ldots, t_z)$. Given a $\boldsymbol{t}$-configuration $(T_1, \ldots, T_z)$, we define the separating set $\Theta(T_1, \ldots, T_z)$ to be the set of coordinate positions where $(T_1, \ldots, T_z)$ is separated. Let $\theta(T_1, \ldots, T_z) := \#\Theta(T_1, \ldots, T_z)$ be the separating weight. Clearly $\theta(T_1, \ldots, T_z) \geq 1$ is equivalent to $(T_1, \ldots, T_z)$ being separated. The minimum $\boldsymbol{t}$-separating weight $\theta_{\boldsymbol{t}(C)}$ is the least separating weight of any $\boldsymbol{t}$-configuration of C, previously studied in [32]. Clearly $\theta_{1,1}(C) = d(C)$.

Define

$$P(t_1, \ldots, t_z) := \sum_{i=1}^{z-1} \sum_{j=i+1}^{z} t_i t_j.$$

Note that if $t_j = 1$ for all j, then

$$P(t_1, \ldots, t_z) = \binom{z}{2},$$

and if $z = 2$, then $P(t_1, t_2) = t_1 t_2$. The following sufficient condition on minimum distance for separability from [14] generalizes various results on separating codes and perfect hashing families. Write $(n, M, d)_q$ for a q-ary code of minimum distance d.

Proposition 1. *An $(n, M, d)_q$ code Γ is $\boldsymbol{t}$-separating if*

$$\frac{d}{n} > 1 - \frac{1}{P(\boldsymbol{t})}.$$

Proof. Consider any $\boldsymbol{t}$-configuration $(T_1, \ldots, T_z)$ from Γ, and define the sum

$$\Sigma := \sum_{i=1}^{z-1} \sum_{j=i+1}^{z} \sum_{(x,y) \in T_i \times T_j} d(x, y).$$

This is the sum of $P(t_1, \ldots, t_z)$ distances in the code, so

$$\Sigma \geq P(t_1, \ldots, t_z)d. \tag{1}$$

Each coordinate can contribute at most $P(t_1, \ldots, t_z)$ to the sum Σ; if any coordinate does contribute that much, then it separates. Hence we get that

$$\Sigma \leq n(P(\boldsymbol{t}) - 1) + \theta_{\boldsymbol{t}}. \tag{2}$$

The proposition follows by combining the upper and lower bounds (1) and (2).

For sufficiently large alphabets, good separating codes are constructible from, e.g., algebraic geometry (AG).

Theorem 3 (The AG Codes). *[35] For any $\alpha > 0$ there are constructible, infinite families of codes $A(N)$ with parameters $[N, NR, N\delta]_q$ for $N \geq N_0(\alpha)$ and*

$$R + \delta \geq 1 - (\sqrt{q} - 1)^{-1} - \alpha.$$

2.4 Concatenation

For small alphabets, we can resort to concatenation to build infinite families of separating codes. Though this construction is well-known in various special cases from the literature, we give a general statement below for the sake of completeness. The outer codes used in concatenation will often be AG codes.

Definition 4 (Concatenation). *Let C_1 be an $(n_1, Q)_q$ (the inner code) and let C_2 be an $(n_2, M)_Q$ code (the outer code). Then the concatenated code $C_1 \circ C_2$ is the $(n_1 n_2, M)_q$ code obtained by taking the words of C_2 and mapping every symbol on a word from C_1.*

Proposition 2. *Let Γ_1 be a $(n_1, M)_{M'}$ code with minimum t-separating weight $\theta_{t(1)}$, and let Γ_2 be a $(n_2, M')_q$ code with separating weight $\theta_{t(1)}$. Then the concatenated code $\Gamma := \Gamma_2 \circ \Gamma_1$ has minimum separating weight $\theta_t = \theta_{t(1)} \cdot \theta_{t(2)}$.*

Proof. Consider a t-configuration $(T_1, \ldots, T_z)$ in Γ. Then there is a corresponding configuration in Γ_1, $(T_1'', \ldots, T_z'')$ which is separated on a set I of at least $\theta_{t(1)}$ positions by assumption. Considering only the positions of Γ corresponding to a particular position $i \in I$ in Γ_2, we get a t'-configuration $(T_1', \ldots, T_z')$ in Γ_1 where $1 \leq t_j' \leq t_j$ for all j. Clearly, $(T_1', \ldots, T_z')$ must be separated on at least $\theta_{t(2)}$ positions, and consequently $\theta(T_1, \ldots, T_z) \geq \theta_{t(1)} \theta_{t(2)}$.

2.5 A Variation on the Tetracode

The ternary construction of [14] makes use of three ingredient codes, and applies twice the concatenation method. It gives an asymptotic family of codes which are $(2, 2)$-, $(3, 1)$-, and $(1, 1, 1)$-separating.

The first seed is the $(4, 3^2, 3)_3$ tetracode T, a simplex (all codewords are at distance 3 apart). It follows that T is 3-hashing and $(3, 1)$-separating from Proposition 1. The tetracode was first proved 3-hashing in [23]; combined with the $(2, 2)$-separation property, this yields the 2-IPP property ([18]).

Let R_1 be the $(9, (3^2)^3, 7)_{3^2}$ Reed-Solomon code, which is both $(2, 2)$- and $(1, 3)$-separating, and 3-hashing, again by Proposition 1. The concatenated code $T \circ R_1$ has parameters $(36, 3^6)_3$, and by Proposition 2, it is $(2, 2)$- and $(1, 3)$-separating, and 3-hashing. Concatenating it with $A(N)$ over $GF(3^6)$ results in $T \circ R_1 \circ A(N)$, an infinite family of ternary $(3, 1)$- and $(2, 2)$-separating and 3-hashing codes with rate $R'/6 \approx 0.0352$.

3   Witness

We now move to the second extension on the notion of minimum distance. We consider only the binary case in this section and follow [13].

For $x \in \{0,1\}^n$, and $W \subset [n]$, define the projection π_W

$$\pi_W : \{0,1\}^{[n]} \to \{0,1\}^W$$
$$x \mapsto (x_i)_{i \in W}$$

and say that W is a *witness set* (or a witness for short) for $c \in C$ if $\pi_W(c) \neq \pi_W(c')$ for every $c' \in C$, $c \neq c'$. Codes for which every codeword has a small witness set arise in a variety of contexts, in particular in machine learning theory [4,7,17] where a witness set is also called a specifying set or a discriminant: see [19, Ch. 12] for a short survey of known results and also [5,25] and references therein for a more recent discussion.

A code has the *w*-witness property, or is a *w-witness code*, if every one of its codewords has a witness set of size w. Our concern in [13] was to study the maximum possible cardinality $f(n,w)$ of a *w*-witness code of length n.

3.1   Easy Facts

Let C be the set of all n vectors of length n and weight 1. Then every codeword of C has a witness of size 1, namely its support. Note the dramatic change for the slightly different code $C \cup \{0\}$. Now the all-zero vector 0 has no witness set of size less than n. Bondy [7] shows however that if $|C| \leq n$, then C is a *w*-witness code with $w \leq |C| - 1$ and furthermore C is a *uniform w-witness code*, meaning that there exists a single subset of $[n]$ of size w that is a witness set for *all* codewords.

A trivial lower bound on $f(n,w)$ is based on a construction.

Proposition 3. *We have:* $f(n,w) \geq \binom{n}{w}$.

Proof. Let $C = \binom{[n]}{w}$ be the set of all vectors of weight w. Notice that for all $c \in C$, $W(c) = support(c)$ is a witness set of c.

Note that the problem is essentially solved for $w \geq n/2$; since $f(n,w)$ is increasing with w, we then have:
$$2^n \geq f(n,w) \geq f(n,n/2) \geq \binom{n}{n/2} \geq 2^n/(2n)^{1/2}.$$
Thus, only the case $w \leq n/2$ is considered here.

3.2   Upper Bounds

In [13], an upper bound is obtained which comes close to the lower bound of Proposition 3; the key result there is the following.

Theorem 4. *Let* $g(n,w) = f(n,w)/\binom{n}{w}$. *Then, for fixed* w, $g(n,w)$ *is a decreasing function of* n. *That is:*

$$n \geq v \geq w \quad \Rightarrow \quad g(n,w) \leq g(v,w).$$

Theorem 4 has a number of consequences.

Corollary 1. *For fixed w, the limit*

$$\lim_{n\to\infty} g(n,w) = \frac{f(n,w)}{\binom{n}{w}}$$

exists.

Corollary 2. *For $w \le n/2$, we have the upper bound:*

$$f(n,w) \le 2w^{1/2}\binom{n}{w}.$$

Set $w = \omega n$ and denote by $h(x)$ the binary entropy function

$$h(x) = -x \log_2 x - (1-x)\log_2(1-x).$$

Corollary 2 together with Proposition 3 yield:

Corollary 3. *We have*

$$\lim_{n\to\infty} \tfrac{1}{n} \log_2 f(n, \omega n) = h(\omega) \qquad \text{for } 0 \le \omega \le 1/2.$$

4 Links between $(1,2)$-Separation and Witnesses

We first summarize the known facts here. Denote by $R = R_2$ in this section the largest possible rate of a $(1,2)$- separating code. A lower bound on R is easily provided by the non-constructive approach (see, e.g., [16]), yielding $R \ge 1 - (1/2)\log 3$.

For constructions, we use the fact pointed out in [26] that shortened Kerdock codes $K'(m)$ for $m \ge 4$ are $(2,1)$-SS and concatenate them with the following ones (with $t = 2$) from [36].

Theorem 5. *Suppose that $q = p^{2r}$ with p prime, and that s is an integer such that $2 \le t \le \sqrt{q}-1$. Then there is an asymptotic family of $(t,1)$-separating codes with rate*

$$R_t = \frac{1}{t} - \frac{1}{\sqrt{q}-1} + \frac{1 - 2\log_q t}{t(\sqrt{q}-1)}.$$

Corollary 4. *There is a constructive asymptotic family of $(2,1)$-SS with $R = 0.2033$.*

Proof. Take an arbitary subcode of size 11^2 in $K'(4)$ which is a $(15, 2^7)$ $(2,1)$-SS. Concatenate with a code of Theorem 5 code $(t = 2)$ over $GF(11^2)$, with $R \approx 0.4355$.

Let us now rephrase the classical proof of the upper bound, to emphasize the relationship with witnesses. Consider *any* partition of $[n]$ into two parts P_1, P_2. Then, for *any* $c \in C$, P_1 or P_2 is a witness for c. Indeed, otherwise, c would be matched by some c^i on $P_i, i = 1, 2$ and c^1 and c^2 together would frame c. Denoting by U_i the subcode of C with witness $P_i, i = 1, 2$ and making the two parts (almost) equal, we get $|C| \leq |U_1| + |U_2| \leq 2.2^{\lceil n/2 \rceil}$. Summarizing:

Proposition 4. *The rate of the largest $(1, 2)$-separating code satisfies:*
$1 - (1/2) \log 3 \approx 0.207518 \leq R \leq 1/2.$

The gap between the lower and upper bounds is annoyingly wide, and narrowing it seems a difficult problem. Let us nevertheless explore a possible approach to improve the upperbound.

Consider a $(1, 2)$- separating code $C(n, \lfloor 2^{Rn} \rfloor)$. Let c and c' be two codewords at distance w; thus $|\chi(c + c')| = w$. Then, by the preceeding discussion, $P_1 := \chi(c + c')$ is a witness for c and for c'. The idea is now to expurgate C from its pairs of close-by codewords to end up with a subcode of increased minimum distance, while preserving the rate. More precisely:

Definition 5. *Denote by $f(n, w, \geq d)$ the maximal size of a w-witness code with minimum distance d. Let's go asymptotics and set*

$\limsup_{n \to \infty} \frac{1}{n} \log_2 f(n, \omega n, \geq \delta n) := \phi(\omega, \delta).$
Finally, set $\phi(\delta, \delta) := \phi(\delta).$

From Corollary 3, we know that for $0 \leq \delta \leq 1/2$, $\phi(\delta) \leq h(\delta)$. As long as $\phi(\delta) < R$, we keep expurgating. Note that there are at most $f(n, i, \geq i)$ pairs of codewords in C at distance i apart. This process results in a code with distance $\lfloor \delta n \rfloor$ retaining the original rate.

Define $\epsilon_\delta \geq 0$ by setting $\phi(\delta) := h(\delta - \epsilon_\delta)$; or equivalently:

$\delta = h^{-1}(R) + \epsilon_\delta.$
We get the following "win-win" result:

Proposition 5. *If $\epsilon_\delta > 0$, then i) or ii) hold:*

i) $R = 1/2$ and the expurgated code satifies $\delta = h^{-1}(R) + \epsilon_\delta > h^{-1}(1 - R)$, i.e. lies **above** *the Varshamov-Gilbert bound!*
ii) $R < 1/2$.

5   Open Problems

The size of optimal w-witness codes is asymptotically known. A few issues remain open, among which:

- When is the sphere $S_w(0)$ the/an optimal w-witness code?
- Do we have $f(n, w) = \binom{n}{w}$ for $w \leq n/2$?
- In the asymptotic case, can Corollary 3 be improved for $\delta \leq \omega \leq 1/2$ to $\phi(\omega, \delta) < h(\omega)$?

– A conjecture of [31] says that R_2 obtained in Theorem 5 could be improved to $R_2 = (1/2) - (2(q^{1/2} - 1))^{-1}$. This would give, with $q = 11^2$ like in Corollary 4, $R_2 = 0.45$ and through concatenation a constructive rate of $R = (3/50) \log_2 11 \approx 0.207565$, thus beating the non-constructive bound (by an incredibly small quantity, though)!

References

1. Alon, N., Spencer, J.: The probabilistic method. Wiley Interscience, Hoboken (2000)
2. Alon, N., Cohen, G., Krivelevitch, M., Litsyn, S.: Generalized hashing and parent-identifying codes. J. Combin. Theory Ser. A 104, 207–215 (2003)
3. Alon, N., Fischer, E., Szegedy, M.: Parent-identifying codes. J. Combin. Theory Ser. A 95, 349–359 (2001)
4. Anthony, M., Brightwell, G., Cohen, D., Shawe-Taylor, J.: On exact specification by examples. In: 5th Workshop on Computational learning theory, pp. 311–318 (1992)
5. Anthony, M., Hammer, P.: A Boolean Measure of Similarity. Discrete Applied Mathematics 154(16), 2242–2246 (2006)
6. Barg, A., Cohen, G., Encheva, S., Kabatiansky, G., Zémor, G.: A hypergraph approach to the identifying parent property. SIAM J. Disc. Math. 14, 423–432 (2001)
7. Bondy, J.A.: Induced subsets. J. Combin. Theory (B) 12, 201–202 (1972)
8. Boneh, D., Franklin, M.: An efficient public-key traitor-tracing scheme. In: Wiener, M. (ed.) CRYPTO 1999. LNCS, vol. 1666, pp. 338–353. Springer, Heidelberg (1999)
9. Boneh, D., Shaw, J.: Collusion-secure fingerprinting for digital data. IEEE Trans. Inf. Theory 44, 480–491 (1998)
10. Chor, B., Fiat, A., Naor, M.: Tracing traitors. In: Desmedt, Y.G. (ed.) CRYPTO 1994. LNCS, vol. 839, pp. 257–270. Springer, Heidelberg (1994)
11. Fredman, M., Komlós, J.: On the size of separating systems and perfect hash functions. SIAM J. Algebraic and Disc. Meth 5, 61–68 (1983)
12. Cormen, T.H., Leiserson, C.E., Rivest, R.L.: Introduction to Algorithms, ch. 12. MIT Press, Cambridge (1990)
13. Cohen, G., Randriam, H., Zémor, G.: Witness sets. In: Barbero, A. (ed.) ICMCTA 2008. LNCS, vol. 5228, pp. 37–45. Springer, Heidelberg (2008)
14. Cohen, G., Schaathun, H.G.: Upper bounds on separating codes, http://personal.cs.surrey.ac.uk/personal/st/H.Schaathun/research/reports.html
15. Cohen, G., Schaathun, H.G.: Upper bounds on separating codes. IEEE Trans. Inf. Theory 50, 1291–1294 (2004)
16. Cohen, G., Zémor, G.: Intersecting codes and independent families. IEEE Trans. Inf. Theory 40, 1872–1881 (1994)
17. Goldman, S.A., Kearns, M.J.: On the complexity of teaching. In: 4th Workshop on Computational learning theory, 303–315 (1991)
18. Hollmann, H.D.L., van Lint, J.H., Linnartz, J.-P., Tolhuizen, L.M.G.M.: On codes with the identifiable parent property. J. Combin. Theory Ser. A 82, 121–133 (1998)
19. Jukna, S.: Extremal Combinatorics. Springer Texts in Theoretical Computer Science (2001)

20. Katsman, G.L., Tsfasman, M.A., Vlăduţ, S.G.: Modular curves and codes with a polynomial construction. IEEE Trans. Inform. Theory 30, 353–355 (1984)
21. Körner, J.: On the extremal combinatorics of the Hamming space. J. Combin. Theory Ser. A 71(1), 112–126 (1995)
22. Körner, J.: Fredman-Komlós bounds and information theory. SIAM J. Algebraic and Disc. Methods 7, 560–570 (1986)
23. Körner, J., Marton, K.: New bounds for perfect hashing via information theory. Europ. J. Combinatorics 9, 523–530 (1988)
24. Körner, J., Simonyi, G.: separating partition systems and locally different sequences. SIAM J. Discrete Math. 1, 355–359 (1998)
25. Kushilevitz, E., Linial, N., Rabinovitch, Y., Saks, M.: Witness sets for families of binary vectors. J. Combin. Theory (A) 73, 376–380 (1996)
26. Krasnopeev, A., Sagalovitch, Y.: The Kerdock codes and separating systems. In: Eight International Workshop on Algebraic and Combinatorial Theory, September 8-14, 2002, pp. 165–167 (2002)
27. Kumar, R., Rajagopalan, S., Sahai, A.: Coding constructions for blacklisting problems without computational assumptions. In: Wiener, M. (ed.) CRYPTO 1999. LNCS, vol. 1666, pp. 609–623. Springer, Heidelberg (1999)
28. MacWilliams, F.J., Sloane, N.J.A.: The Theory of Error-Correcting Codes. North-Holland, Amsterdam (1977)
29. Nilli, A.: Perfect hashing and probability. Combinatorics, Probability and Computing 3, 407–409 (1994)
30. Pinsker, M., Sagalovitch, Y.: A lower bound on the size of automata state codes. Problems Inform. Transmission 8(3), 59–66 (1972)
31. Randriam, H.: Personal communication
32. Sagalovich, Y.L.: Separating systems. Problems of Information Transmission 30(2), 105–123 (1994)
33. Stinson, D.R., Wei, R.: Combinatorial properties and constructions of traceability schemes and frameproof codes. SIAM J. Discrete Math. 11, 41–53 (1998)
34. Staddon, J.N., Stinson, D.R., Wei, R.: Combinatorial properties of frameproof and traceability codes. IEEE Trans. Information Theory 47, 1042–1049 (2001)
35. Tsfasman, M.A., Vlăduţ, S.G., Zink, T.: Modular curves, Shimura curves, and Goppa codes, better than Varshamov-Gilbert bound. Math. Nachr. 109, 21–28 (1982)
36. Xing, C.: Asymptotic bounds on frameproof codes. IEEE Trans. Inform. Th. 40, 2991–2995 (2002)

Binary Covering Arrays and Existentially Closed Graphs

Charles J. Colbourn[1] and Gerzson Kéri[2]

[1] Computer Science and Engineering, Arizona State University, P.O. Box 878809,
Tempe, AZ 85287, U.S.A.
`Charles.Colbourn@asu.edu`
[2] Computer and Automation Research Institute, Hungarian Academy of Sciences
H-1111 Budapest, Kende u. 13-17, Hungary
`keri@sztaki.hu`

Abstract. Binary covering arrays have been extensively studied in many different contexts, but the explicit construction of small binary covering arrays with strength larger than three remains poorly understood. Connections with existentially closed graphs and Hadamard matrices are examined, particularly those arising from the Paley graphs and tournaments. Computational results on arrays generated by column translation, such as the Paley graphs, lead to substantial improvements on known existence results for binary covering arrays of strengths four and five.

1 Introduction

Let $\mathcal{F} = \{F_1, \ldots, F_k\}$ be a set of k *factors*, and for each $F_f \in \mathcal{F}$ let V_f be the set of possible *levels* or *values* for factor F_f. A *t-tuple* or *t-way interaction* is a set F of t factors, and a value $\nu_f \in V_f$ for each factor $F_f \in F$. When $|V_f| = v_f$ (that is, v_f is the number of levels for factor F_f), a *mixed-level covering array*, $\mathsf{MCA}(N; t, (v_1, \ldots, v_k))$, is an $N \times k$ array in which, in every $N \times t$ subarray, every possible t-tuple occurs in at least one row. Then t is the *strength* of coverage and k is the number of factors. We suppose that $t \leq k$ to avoid degenerate situations. The exponential notation $g_1^{u_1} \cdots g_\ell^{u_\ell}$ (with $\sum_{i=1}^\ell u_i = k$) is used to specify that there are u_j factors having g_j levels. In the special case that $v_1 = \cdots = v_k = v$, the array is a *covering array* $\mathsf{CA}(N; t, k, v)$ (or $\mathsf{CA}(N; t, k, v^k)$). When 'at least' is replaced by 'exactly', this defines an *orthogonal array* [1], $\mathsf{OA}(N; t, k, v)$. We denote by $\mathsf{CAN}(t, k, v)$ the minimum N for which a $\mathsf{CA}(N; t, k, v)$ exists. The determination of $\mathsf{CAN}(t, k, v)$ has been the subject of much research; see [2–5] for survey material. However, only in the case of $\mathsf{CAN}(2, k, 2)$ is an exact determination known (see [3]). Applications in which experimental factors interact employ covering arrays [3, 4, 6].

While asymptotic results determine the growth rate of $\mathsf{CAN}(t, k, v)$ for fixed t and v as a function of k (see [7], for example), the explicit construction of covering arrays is required for testing applications. Among recursive techniques, Roux [8] developed a simple (but effective) doubling construction for binary covering arrays of strength three. Roux-type constructions operate by juxtaposing copies

C. Xing et al. (Eds.): IWCC 2009, LNCS 5557, pp. 22–33, 2009.

of smaller covering arrays, sometimes with smaller strength. Such constructions have been explored for strength three [2, 9, 10], strength four [4, 10], and arbitrary strength [4, 11, 12]. For strengths three and four, these constructions often yield the smallest known covering arrays for $v \leq 25$ and $k \leq 10000$ [13]. For strengths five and larger, they appear to be less effective at present. A further class of recursive constructions instead selects columns from a smaller covering array [4, 14–17], using the easy observation that any t columns from a covering array of strength t cover all t-way interactions.

However, the effective use of recursive techniques relies on the construction of 'small' covering arrays whose size is at or near the minimum. Orthogonal arrays provide a number of specific examples [1]. Computational methods produce many more. For example, simulated annealing [9, 18], tabu search [19], backtracking [20], integer programming [16], and constraint satisfaction [21] have proved successful for strengths three and four, but are less applicable to larger strengths at present. Compact representations of covering arrays as 'permutation vectors' can be used to extend heuristic search methods to these larger strengths [22, 23], but they restrict to a subset of the admissible parameter sets. Indeed for strength at least five, greedy methods [24–27] and random methods [28] are often the only ones for which competitive computational results can be obtained in a reasonable amount of time.

In order to reduce the size of the computational search, in Section 2 we examine covering arrays whose rows are produced by cyclic translation from a set of 'starter' rows. Easy computations then establish the existence of covering arrays with fewer rows than those currently tabulated. Section 3 specializes translation covering arrays to those obtained from the quadratic residues in the finite field, permitting many more computational constructions. The application of the quadratic residues establishes a connection with the Paley graphs, and hence with a well-studied study of existentially closed graphs and tournaments, discussed in Section 4. Generalizing from Paley graphs in a different direction, Section 5 examines the construction of binary covering arrays from Hadamard matrices. Finally in Section 6 we explore the consequences of the constructions given in terms of covering array numbers.

2   Translation Covering Arrays

In order to simplify the search for covering arrays computationally, we consider covering arrays in which the rows are obtained by cyclically translating a single base row. Variants on this basic strategy are also examined. In each case, V is a set of v elements equipped with addition, in which 0 is the additive identity.

Lemma 1. *Suppose there is a vector $R = (r_0, \ldots, r_{k-1})$ of length k with elements from $V = \{0, \ldots, v-1\}$ for which, for every set of distinct elements $\{i_1, \ldots, i_t\} \subseteq \{0, \ldots, k-1\}$ and every tuple $(c_1, \ldots, c_t)$ with $c_j \in V$ for $1 \leq j \leq t$, there is an integer $s \in \{0, \ldots, k-1\}$ and an element $\sigma \in V$ with for which $r_{s+i_j \bmod k} = c_j + \sigma \bmod v$ for each $1 \leq j \leq t$. Then $\mathsf{CAN}(t, k, v) \leq vk$.*

Proof. Let $R + i = (r_0 + i \bmod v, \ldots, r_{k-1} + i \bmod v)$ for each $i \in V$. Then for each of $R + 0, \ldots, R + (v - 1)$, form all k cyclic translates of the vector R. The resulting kv vectors form the rows of a $\mathsf{CA}(vk; t, k, v)$. ∎

When for every set of distinct elements $\{i_1, \ldots, i_{t-1}\} \subseteq \{0, \ldots, k-1\}$ and every tuple $(c_1, \ldots, c_{t-1})$ with $c_j \in V$ for $1 \le j \le t - 1$, there is an integer $s \in \{0, \ldots, k-1\}$ for which $r_{s+i_j \bmod k} = c_j$ for each $1 \le j \le t-1$, the vector R has *property ($\star$)*. When R has property ($\star$), so does $R + i$ for each $i \in V$.

Lemma 2. *Suppose there is a vector $R = (r_0, \ldots, r_{k-1})$ meeting the conditions of Lemma 1 and having property ($\star$). Then $\mathsf{CAN}(t, k + 1, v) \le vk$.*

Proof. Apply Lemma 1. Then append a new column and place i in each row of $R + i$ for $i \in V$. ∎

Corollary 1. $\mathsf{CAN}(4, k + 1, 2) \le 2k$ *for* $k = 23$ *and* $27 \le k \le 33$.

Proof. Apply Lemma 2 to the vectors:

k	Vector
23	00000111101011001100101
27	000110100011011100111010101000
28	0001101000100011110110011010
29	0001101000100010111001101011
30	0001101000100011101111001011101
31	0001101000100011100111111010010
32	0001101000100011011111010100101110
33	0001101000100011011010110011111101

∎

Corollary 2. $\mathsf{CAN}(4, k, 2) \le 2k$ *for* $k \in \{21, 24, 25, 26\}$.

Proof. Apply Lemma 1 to the vectors:

k	Vector
21	000011101000100100001
24	000110011110010010101000
25	0001101010110010000000111
26	00011010000110101011100100

∎

We weaken the requirements somewhat to obtain further examples. It may happen that the cyclic translates fail to cover only those orbits of t-way interactions in which all symbols are the same. When for every set of distinct elements $\{i_1, \ldots, i_{t-1}\} \subseteq \{0, \ldots, k-1\}$ and every tuple $(c_1, \ldots, c_{t-1})$ with $c_j \in V$ for $1 \le j \le t-1$ except possibly for $(0, 0, \ldots, 0)$, there is an integer $s \in \{0, \ldots, k-1\}$ for which $r_{s+i_j \bmod k} = c_j$ for each $1 \le j \le t-1$, the vector R has *property ($\star\star$)*.

Lemma 3. *Suppose there is a vector $R = (r_0, \ldots, r_{k-1})$ of length k with elements from $V = \{0, \ldots, v-1\}$ for which, for every set of distinct elements $\{i_1, \ldots, i_t\} \subseteq \{0, \ldots, k-1\}$ and every tuple $(c_1, \ldots, c_t)$ with not all symbols equal and having $c_j \in V$ for $1 \leq j \leq t$, there is an integer $s \in \{0, \ldots, k-1\}$ and an element $\sigma \in V$ with for which $r_{s+i_j \bmod k} = c_j + \sigma \bmod v$ for each $1 \leq j \leq t$. Then $\mathsf{CAN}(t, k, v) \leq vk + v$. If, in addition, R has property $(\star\star)$, then $\mathsf{CAN}(t, k+1, v) \leq vk + v$.*

Proof. The proof proceeds as for Lemmas 1 and 2, but to the final covering array v rows are added: For each $i \in V$, a row consisting of k (or $k+1$) copies of i is added. ∎

Corollary 3. $\mathsf{CAN}(4, k+1, 2) \leq 2k + 2$ for $k \in \{11, 19, 21, 22, 24, 25, 26\}$.

Proof. Apply Lemma 3 to the vectors:

11	00010110111
19	0000101011110010011
21	000100011111100101101
22	0001110111011010011100
24	000111011101100010010110
25	0001110011011011010101000
26	00011101101001101110001010

In each case, the vector has property $(\star\star)$. ∎

In Table 1, we give the current best bound on $\mathsf{CAN}(4, k, 2)$ for small k, along with the bound implied by the corollaries here. We consider just those cases with $k \leq 34$ in the table, although $\mathsf{CAN}(4, k, 2) \leq 2k - 2$ was previously known only for $k \geq 50$. As we generalize, the reason for this restriction on k will become

Table 1. Improvements for $\mathsf{CAN}(4, k, 2)$

k	Old Bound	New Bound	k	Old Bound	New Bound
5	16 [1]		6	21 [29]	
11	24 [20]	24	12	24 [20]	24
13	34 [18]		14	44 [26]	40
15	46 [26]	40	16	48 [26]	40
17	49 [26]	40	18	52 [26]	40
19	55 [10]	40	20	55 [10]	40
21	57 [10]	42	22	57 [10]	44
23	61 [10]	46	24	61 [10]	46
25	65 [26]	50	26	68 [27]	52
27	71 [27]	54	28	71 [27]	54
29	72 [27]	56	30	73 [26]	58
31	73 [26]	60	32	73 [26]	62
33	76 [27]	64	34	78 [27]	66

clear. The extent of the improvements is perhaps surprising; while it indicates in part the limitations of existing methods, it also suggests that the study of covering arrays obtained by translation is a useful one.

3   Quadratic Residues

In order to apply Lemmas 1, 2, and 3, one needs to find vectors that meet the conditions specified. However, for larger t and hence larger k, the number of candidate vectors to consider is too large for practical computation. Therefore we examine a specific construction. Let k be a prime, and let ω be a primitive element of $\mathbb{F}_k$. Then the *quadratic residue vector* $Q_k = (q_0, \ldots, q_{k-1})$ has $q_i = 0$ when $i = 0$; $q_i = 0$ when $i = \omega^j$ and j is even; and $q_i = 1$ when $i = \omega^j$ and j is odd.

Although one must still check whether the conditions of the lemmas are met or not, restricting to the quadratic residue vectors avoids the search through an exponential number of candidates. A simple computation now establishes the following results.

Lemma 4

1. $\mathsf{CAN}(5, p+1, 2) \le 2p + 2$ *for* $p \in \{67, 71, 79, 83\}$.
2. $\mathsf{CAN}(5, p, 2) \le 2p$ *for* $p = 103$.
3. $\mathsf{CAN}(5, p+1, 2) \le 2p$ *for* $p \in \{89, 97, 101\}$, *and for* $107 \le p \le 773$, p *prime*.

Proof. Apply Lemma 1, 3, or 2 to the quadratic residue vector Q_p.   ∎

Lemma 5

1. $\mathsf{CAN}(6, p+1, 2) \le 2p + 2$ *for* $p \in \{359, 503\}$.
2. $\mathsf{CAN}(6, p, 2) \le 2p$ *for* $p = 379$.
3. $\mathsf{CAN}(6, p+1, 2) \le 2p$ *for* $p \in \{431, 463, 467, 487, 491, 499, 509, 521\}$.

Proof. Apply Lemma 1, 3, or 2 to the quadratic residue vector Q_p.   ∎

Lemma 6

1. $\mathsf{CAN}(4, p, 2) \le p$ *for* $67 \le p \le 773$ *a prime*, $p \notin \{73, 103\}$.
2. $\mathsf{CAN}(5, p, 2) \le p$ *for* $p \in \{359, 431, 463, 467, 487, 491, 499, 503, 509, 521\}$.

Proof. Starting with a $\mathsf{CA}(N; t, k, v)$, select any column and choose a symbol that appears the fewest times, say N', in that column. Restricting attention to the rows that contain the chosen symbol in the chosen column, and deleting the chosen column, yields a $\mathsf{CA}(N'; t-1, k-1, v)$, a so-called *derived* CA. Evidently $N' \le \lfloor N/v \rfloor$. Apply to the results of Lemmas 4(3) and 5(3). Now using Lemma 4(1), $\mathsf{CAN}(4, p, 2) \le p+1$ for $p \in \{67, 71, 79, 83\}$. However, the p row translates of Q_p generate a $\mathsf{CA}(p; 4, p, 2)$ (in other words, in the derivation from the $\mathsf{CA}(2p+2; 4, p+1, 2)$ the constant row is redundant). Hence $\mathsf{CAN}(4, p, 2) \le p$ for $p \in \{67, 71, 79, 83\}$. Similarly, $\mathsf{CAN}(5, p, 2) \le p$ for $p \in \{359, 503\}$.   ∎

4   Existentially Closed Graphs and Tournaments

Let $G = (V, E)$ be a graph. When for every $A, B \subset V$ with $|A \cup B| = t$ and $A \cap B = \emptyset$, there is a vertex $v \in V \setminus (A \cup B)$ for which v is adjacent to every vertex of A and not adjacent to any vertex of B, the graph G is t-existentially closed (t-ec). An oriented graph is a directed graph $G = (V, E)$ for which, whenever $(x, y) \in E$, $(y, x) \notin E$. It is a tournament if for every $x, y \in V$ with $x \neq y$, exactly one of $\{(x, y), (y, x)\}$ is in E. An oriented graph $G = (V, E)$ is t-existentially closed if for every $A, B \subset V$ with $|A \cup B| = t$ and $A \cap B = \emptyset$, there is a vertex $v \in V \setminus (A \cup B)$ for which $(x, v) \in E$ for every $x \in A$ and $(v, y) \in E$ for every $y \in B$.

The study of t-existentially closed graphs and tournaments has a long history. Erdős [30] first studied such tournaments in solving a puzzle of Schütte, using an elementary probabilistic argument. Indeed Erdős and Rényi [31] showed that for every fixed t, random graphs with edges chosen with probability p $(0 < p < 1)$ are almost all t-ec. (See also [32, 33].) Since that time, the search for explicit constructions has been extensive. Our reason for interest in them is the following:

Theorem 1. *The adjacency matrix of a t-existentially closed graph or oriented graph on n vertices is a* CA$(n; t, n, 2)$.

Explicit and probabilistic constructions of existentially closed graphs and tournaments are numerous[34–40]. Constructions from Bush-type Hadamard matrices [41], strongly regular graphs [42], Steiner triple systems [43], and affine geometries [44] have been studied. However, Paley graphs and tournaments provide the best-known examples. Let q be an odd prime power. Form a $q \times q$ matrix $P = (p_{ij})$ with rows and columns indexed by the elements of $\mathbb{F}_q$. Set $p_{ij} = 0$ when $i = j$; $p_{ij} = 1$ when $i \neq j$ and $i - j$ is a square in $\mathbb{F}_q$, and $p_{ij} = 0$ when $i \neq j$ and $i - j$ is not a square. Then P is the adjacency matrix of a Paley graph P_q when $q \equiv 3 \pmod 4$, and of a Paley tournament P_q when $q \equiv 1 \pmod 4$. Using techniques from character theory, a strong existence result has been proved:

Theorem 2. *[45–48] When* $q > t^2 2^{2t-2}$, *P_q is t-existentially closed.*

A similar statement in different vernacular is proved in [49]. Compare Theorem 2 and Lemma 6. By Theorem 2, CA$N(4, q, 2) \leq q$ when $q > 4^2 \cdot 2^6 = 1024$ and q is a prime power and that CA$N(5, q, 2) \leq q$ when $q > 5^2 \cdot 2^8 = 6400$ and q is a prime power. In both cases, Lemma 6 establishes the desired result for many smaller values of q. Theorem 2 provides an explicit construction of binary covering arrays for arbitrarily large strength, however. Indeed by derivation it establishes the following theorem, where $d(x, \tau) = x$ when $\tau = 0$, and $d(x, \tau) = \left\lceil \frac{d(x, \tau-1)}{2} \right\rceil$ when $\tau \geq 1$ is an integer.

Theorem 3. CA$N(t, q - \tau, 2) \leq d(q, \tau) \leq \lfloor \frac{q}{2^\tau} \rfloor$ *when* $q > (t + \tau)^2 2^{2(t+\tau-1)}$ *and q is an odd prime power.*

The constraints imposed by the construction of Paley matrices leads to a requirement that the numbers of rows and columns coincide. However for the construction of covering arrays, this is artificial. To overcome this, let $G = (V_1 \cup V_2, E)$

be a bipartite graph and let (V_1, V_2) be the bipartition of the vertices so that all edges have one endpoint in V_1 and the other in V_2. It is *t-covering* (of V_2 by V_1, implicitly) if for every $T \subseteq V_2$ with $|T| = t$, and every partition of T into two classes T_1 and T_2, there is a vertex $v \in V_1$ which is adjacent to every vertex in T_1 but none in T_2. The following equivalence is immediate.

Lemma 7. *A* $\mathsf{CA}(N; t, k, 2)$ *is equivalent to a t-covering bipartite graph* $(V_1 \cup V_2, E)$ *with* $|V_1| = N$ *and* $|V_2| = k$.

Paley-type constructions for such *t*-covering bipartite graphs have been employed in [50–54], while a somewhat more restricted variant is considered in [55]. In each case, however, constructions are given when $|V_1| = |V_2|$, and hence the same covering array numbers are examined as above. Explicit constructions of *t*-covering bipartite graphs in which the two classes have quite different sizes appear not to have been studied.

5   Hadamard Matrices

We have seen that Paley graphs provide examples of binary covering arrays of high strength. The adjacency matrix of the Paley graph of order $q \equiv 3 \pmod 4$, q a prime power, underpins a classical construction of Hadamard matrices.

A *Hadamard matrix* is a square matrix whose entries are either $+1$ or -1 and whose rows are mutually orthogonal, that is, have inner product 0. Hadamard matrices of order n exist only if $n = 1$, $n = 2$, or $n \equiv 0 \pmod 4$ (see [1], for example); they are unique (up to equivalence) for $n \leq 12$, while the number of inequivalent Hadamard matrices of order n for $n = 16, 20, 24, 28$ is 5, 3, 60, and 487, respectively (see, for example, [56]). A *Paley Hadamard matrix* is obtained from the adjacency matrix of the Paley graph by changing all 0 entries to -1, and adding a headline and sideline of all -1s.

A Hadamard matrix H of order n is *t-strong* when $\left(\begin{smallmatrix} H \\ -H \end{smallmatrix} \right)$ is a $\mathsf{CA}(2n; t, n, 2)$; when $t = 4$, the term *strong* is employed. Evidently for every t, and every $q > t^2 2^{2t-2}$ the Paley Hadamard matrix of order $q+1$ is *t*-strong. This suggests that Hadamard matrices form a natural place to look for binary covering arrays.

Every Hadamard matrix is 3-strong. For H a Hadamard matrix of order $n \geq 4$, $\left(\begin{smallmatrix} H & H \\ H & -H \end{smallmatrix} \right)$ is a Hadamard matrix of order $2n$ that is 3-strong but not strong (consider columns $i, j, n+i, n+j$ for $1 \leq i < j \leq n$, for example). The existence spectrum of *t*-strong Hadamard matrices for $t \geq 4$ is a natural question in terms of the construction of binary covering arrays.

Theorem 4

1. *Every Hadamard matrix of order n is strong when $n \equiv 4 \pmod 8$ and $n > 4$.*
2. *The Hadamard matrices of orders four and eight are not strong.*
3. *None of the five Hadamard matrices of order $n = 16$ is strong.*
4. *Exactly two of the 60 Hadamard matrices of order $n = 24$ are strong.*
5. *There is a strong Hadamard matrix of order 32.*

Proof. Statement (1) is essentially given in [57, Theorem 8.9]. We include a proof here. Let H be a Hadamard matrix of order $n > 4$ that is not strong. Consider four columns (c_1, c_2, c_3, c_4) of $\begin{pmatrix} H \\ -H \end{pmatrix}$ in which some 4-tuple is not covered. Without loss of generality, in H, column c_1 contains only the symbol 1 (simply negate rows of H to ensure this). We use the fact that H^T is also a Hadamard matrix. Each of $(1, 1, 1)$, $(1, 1, -1)$, $(1, -1, 1)$, or $(1, -1, -1)$ appears in columns (c_1, c_2, c_3) the same number $\frac{n}{4}$ of times. Then let a, b, c, d be the numbers of rows of H in which we find entry 1 in column c_4 and $(1, 1, 1)$, $(1, 1, -1)$, $(1, -1, 1)$, or $(1, -1, -1)$, respectively, in columns (c_1, c_2, c_3). Now $a + b + c + d = \frac{n}{2}$, $a + b - c - d = 0$, and $a - b + c - d = 0$; hence $a = d$ and $b = c = \frac{n}{4} - a$. Because some 4-tuple fails to appear in columns (c_1, c_2, c_3, c_4) of $\begin{pmatrix} H \\ -H \end{pmatrix}$, we must have $a \in \{0, \frac{n}{4}\}$. Negating column c_4 if necessary, without loss of generality $a = 0$. Then consider a fifth column c_5 and define a', b', c', d' to be the numbers of rows of H with 1 in column c_5 and $(1, 1, 1)$, $(1, 1, -1)$, $(1, -1, 1)$, or $(1, -1, -1)$, respectively, in columns (c_1, c_2, c_3). Then $a' = \frac{n}{4} - b' = \frac{n}{4} - c' = d'$ as above. But $-a' + b + c - d' = 0$ and hence $a' = b' = c' = d' = \frac{n}{8}$, so $n \equiv 0 \pmod 8$ is required.

For (2)-(4), the arrays tested are from [57, DVD Supplement]; for (4), the arrays numbered 16 and 60 are the only strong ones of order 24 (among the transposes of the matrices given, only those of numbers 59 and 60 are strong). For (5), matrices 32P02, P13, P15, P17, and P19 from [58] are the only strong ones among the 19 given there (32P02 is the Paley Hadamard matrix).   ∎

6   Consequences

In Figure 1(a), bounds on $\mathsf{CAN}(4, k, 2)$ for $0 \le k \le 10000$ are plotted, graphing the size of the array vertically and $\log_{10}(k)$ horizontally. The best previous bound [13] is shown as diamonds, while the bounds computed here, along with their consequences using constructions in [10, 15], are shown as squares. Figure 1(b) shows the same plot restricted to the range $0 \le k \le 100$.

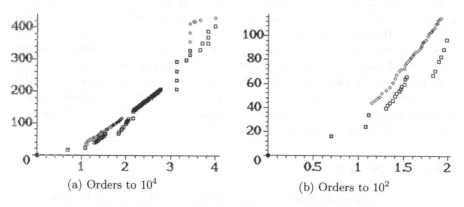

(a) Orders to 10^4 (b) Orders to 10^2

Fig. 1. Known and Improved Bounds on $\mathsf{CAN}(4, k, 2)$

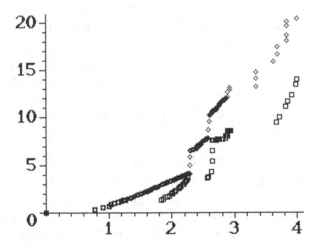

Fig. 2. Known and Improved Bounds on $\mathsf{CAN}(5, k, 2)$

Figure 2 shows an even more pronounced improvement for strength 5; here the vertical axis is marked in hundreds of rows. The discontinuities in the graph result from limitations of the constructions known. While Theorem 2 affords powerful constructions for large strengths, within the range of interest here, results from this general construction are not competitive. (For example, $\mathsf{CAN}(4,1031,2)$ ≤ 1031 and $\mathsf{CAN}(4,6421,2) \leq 3210$ from Theorem 3 are far from the best known.) Similar results arise for strength 5 and 6. Nevertheless, the explicit computations of Sections 2 and 3 demonstrate that Paley graphs and tournaments often lead to useful improvements on our knowledge about $\mathsf{CAN}(t, k, 2)$.

7   Conclusions

The connections of binary covering arrays to existentially closed graphs and Hadamard matrices developed here open many avenues of research. Moreover, as shown by the Paley graphs, these constructions can lead not only to effective explicit constructions for large strength, but also to competitive upper bounds on covering array numbers. In this paper we have focussed on binary covering arrays, but those based on larger alphabets are also amenable to the techniques developed here. While most of the literature at present concentrates on the binary case (with the notable exception of [60]), the use of cyclotomic classes more generally affords constructions for covering arrays with larger alphabets. This is a current topic of investigation.

Acknowledgements

Thanks to Hadi Kharaghani and Patric Östergård for helpful pointers to the literature. Research of the first author is supported by DOD grant N00014-08-1-1070.

Research of the second author is supported by the Hungarian National Research Fund, OTKA, Grant No. K 60480.

References

1. Hedayat, A.S., Sloane, N.J.A., Stufken, J.: Orthogonal Arrays. Springer, New York (1999)
2. Chateauneuf, M.A., Kreher, D.L.: On the state of strength-three covering arrays. J. Combin. Des. 10(4), 217–238 (2002)
3. Colbourn, C.J.: Combinatorial aspects of covering arrays. Le Matematiche (Catania) 58, 121–167 (2004)
4. Hartman, A.: Software and hardware testing using combinatorial covering suites. In: Golumbic, M.C., Hartman, I.B.A. (eds.) Interdisciplinary Applications of Graph Theory, Combinatorics, and Algorithms, pp. 237–266. Springer, Norwell (2005)
5. Hartman, A., Raskin, L.: Problems and algorithms for covering arrays. Discrete Math 284(1-3), 149–156 (2004)
6. Cohen, D.M., Dalal, S.R., Fredman, M.L., Patton, G.C.: The AETG system: an approach to testing based on combinatorial design. IEEE Transactions on Software Engineering 23(7), 437–444 (1997)
7. Godbole, A.P., Skipper, D.E., Sunley, R.A.: t-covering arrays: upper bounds and Poisson approximations. Combinatorics, Probability and Computing 5, 105–118 (1996)
8. Roux, G.: k-Propriétés dans les tableaux de n colonnes: cas particulier de la k-surjectivité et de la k-permutivité. PhD thesis, Université de Paris (1987)
9. Cohen, M.B., Colbourn, C.J., Ling, A.C.H.: Constructing strength three covering arrays with augmented annealing. Discrete Math. 308, 2709–2722 (2008)
10. Colbourn, C.J., Martirosyan, S.S., van Trung, T., Walker II, R.A.: Roux-type constructions for covering arrays of strengths three and four. Designs, Codes and Cryptography 41, 33–57 (2006)
11. Martirosyan, S.S., Colbourn, C.J.: Recursive constructions for covering arrays. Bayreuther Math. Schriften 74, 266–275 (2005)
12. Martirosyan, S.S., van Trung, T.: On t-covering arrays. Des. Codes Cryptogr. 32(1-3), 323–339 (2004)
13. Colbourn, C.J.: Covering arrays 2005–present, http://www.public.asu.edu/~ccolbou/src/tabby
14. Bierbrauer, J., Schellwat, H.: Almost independent and weakly biased arrays: efficient constructions and cryptologic applications. In: Bellare, M. (ed.) CRYPTO 2000. LNCS, vol. 1880, pp. 533–543. Springer, Heidelberg (2000)
15. Colbourn, C.J.: Distributing hash families and covering arrays. J. Combin. Inf. Syst. Sci. (to appear)
16. Sloane, N.J.A.: Covering arrays and intersecting codes. Journal of Combinatorial Designs 1, 51–63 (1993)
17. Tang, D.T., Chen, C.L.: Iterative exhaustive pattern generation for logic testing. IBM Journal Research and Development 28(2), 212–219 (1984)
18. Soriano, P.P.: Private communication by e-mail (March 2008)
19. Nurmela, K.: Upper bounds for covering arrays by tabu search. Discrete Applied Mathematics 138(9), 143–152 (2004)
20. Yan, J.: A backtracking search tool for constructing combinatorial test suites. Technical Report ISCAS-LCS-07-04, Institute of Software, Chinese Academy of Sciences, Beijing, China (2007)

21. Hnich, B., Prestwich, S., Selensky, E., Smith, B.M.: Constraint models for the covering test problem. Constraints 11, 199–219 (2006)
22. Sherwood, G.B., Martirosyan, S.S., Colbourn, C.J.: Covering arrays of higher strength from permutation vectors. J. Combin. Des. 14(3), 202–213 (2006)
23. Walker II, R.A., Colbourn, C.J.: Tabu search for covering arrays using permutation vectors. J. Stat. Plann. Infer. 139, 69–80 (2009)
24. Bryce, R.C., Colbourn, C.J.: A density-based greedy algorithm for higher strength covering arrays. Software Testing, Verification, and Reliability (to appear)
25. Lei, Y., Kacker, R., Kuhn, D.R., Okun, V., Lawrence, J.: IPOG: A general strategy for t-way software testing. In: Fourteenth Int. Conf. Engineering Computer-Based Systems, pp. 549–556 (2007)
26. Kuliamin, V.V.: Private communication by e-mail (February 2007)
27. Linnemann, D., Frewer, M.: Computations with the density algorithm (private communication by e-mail) (October 2008)
28. Kuhn, D.R., Lei, Y., Kacker, R., Okun, V., Lawrence, J.: Paintball: A fast algorithm for covering arrays of high strength. Internal Tech. Report, NISTIR 7308 (2007)
29. Johnson, K.A., Entringer, R.: Largest induced subgraphs of the n-cube that contain no 4-cycles. J. Combin. Theory Ser. B 46(3), 346–355 (1989)
30. Erdős, P.: On a problem in graph theory. Math. Gaz. 47, 220–223 (1963)
31. Erdős, P., Rényi, A.: Asymmetric graphs. Acta Math. Acad. Sci. Hungar. 14, 295–315 (1963)
32. Blass, A., Harary, F.: Properties of almost all graphs and complexes. J. Graph Theory 3(3), 225–240 (1979)
33. Caccetta, L., Erdős, P., Vijayan, K.: A property of random graphs. Ars Combin. 19(A), 287–294 (1985)
34. Szekeres, E., Szekeres, G.: On a problem of Schütte and Erdős. Math. Gaz. 49, 290–293 (1965)
35. Ananchuen, W., Caccetta, L.: On tournaments with a prescribed property. Ars Combin. 36, 89–96 (1993)
36. Exoo, G.: On an adjacency property of graphs. J. Graph Theory 5(4), 371–378 (1981)
37. Exoo, G., Harary, F.: The smallest graphs with certain adjacency properties. Discrete Math. 29(1), 25–32 (1980)
38. Ananchuen, W., Caccetta, L.: On constructing graphs with a prescribed adjacency property. Australas. J. Combin. 10, 73–83 (1994)
39. Ananchuen, W., Caccetta, L.: A note on graphs with a prescribed adjacency property. Bull. Austral. Math. Soc. 51(1), 5–15 (1995)
40. Bonato, A., Cameron, K.: On an adjacency property of almost all graphs. Discrete Math. 231(1-3), 103–119 (2001)
41. Bonato, A., Holzmann, W.H., Kharaghani, H.: Hadamard matrices and strongly regular graphs with the 3-e.c. adjacency property. Electron. J. Combin. 8(1) (2001); Research paper 1, 9 pp. (electronic)
42. Cameron, P.J., Stark, D.: A prolific construction of strongly regular graphs with the n-e.c. property. Electron. J. Combin. 9(1) (2002); Research Paper 31, 12 pp. (electronic)
43. Forbes, A.D., Grannell, M.J., Griggs, T.S.: Steiner triple systems and existentially closed graphs. Electron. J. Combin. 12 (2005); Research Paper 42, 11 pp. (elecronic)
44. Baker, C.A., Bonato, A., Brown, J.M.N., Szőnyi, T.: Graphs with the n-e.c. adjacency property constructed from affine planes. Discrete Math. 308(5-6), 901–912 (2008)

45. Graham, R.L., Spencer, J.H.: A constructive solution to a tournament problem. Canad. Math. Bull. 14, 45–48 (1971)
46. Bollobás, B.: Random graphs. Academic Press Inc., London (1985)
47. Blass, A., Exoo, G., Harary, F.: Paley graphs satisfy all first-order adjacency axioms. J. Graph Theory 5(4), 435–439 (1981)
48. Ananchuen, W., Caccetta, L.: On the adjacency properties of Paley graphs. Networks 23(4), 227–236 (1993)
49. Peralta, R.: On the distribution of quadratic residues and nonresidues modulo a prime number. Math. Comp. 58(197), 433–440 (1992)
50. Alon, N., Goldreich, O., Håstad, J., Peralta, R.: Simple constructions of almost k-wise independent random variables. Random Structures Algorithms 3(3), 289–304 (1992)
51. Razborov, A.A.: Applications of matrix methods to the theory of lower bounds in computational complexity. Combinatorica 10(1), 81–93 (1990)
52. Babai, L., Gál, A., Kollár, J., Rónyai, L., Szabó, T., Wigderson, A.: Extremal bipartite graphs and superpolynomial lower bounds for monotone span programs. In: Proceedings of the Twenty-eighth Annual ACM Symposium on the Theory of Computing (Philadelphia, PA, 1996), pp. 603–611. ACM, New York (1996)
53. Babai, L., Gál, A., Wigderson, A.: Superpolynomial lower bounds for monotone span programs. Combinatorica 19(3), 301–319 (1999)
54. Gál, A.: A characterization of span program size and improved lower bounds for monotone span programs. Comput. Complexity 10(4), 277–296 (2001)
55. Borowiecki, M., Grytczuk, J., Hałuszczak, M., Tuza, Z.: Schütte's tournament problem and intersecting families of sets. Combin. Probab. Comput. 12(4), 359–364 (2003)
56. Kimura, H.: Classification of Hadamard matrices of order 28. Discrete Math. 133(1-3), 171–180 (1994)
57. Kaski, P., Östergård, P.R.J.: Classification algorithms for codes and designs. Springer, Berlin (2006)
58. Seberry, J.R.: Library of Hadamard matrices, http://www.uow.edu.au/~jennie/hadamard.html
59. Tonchev, V.D.: Hadamard matrices of order 36 with automorphisms of order 17. Nagoya Math. J. 104, 163–174 (1986)
60. Azar, Y., Motwani, R., Naor, J.: Approximating probability distributions using small sample spaces. Combinatorica 18(2), 151–171 (1998)

A Class of Three-Weight and Four-Weight Codes

Cunsheng Ding

Department of Computer Science and Engineering,
The Hong Kong University of Science and Technology,
Clear Water Bay, Kowloon, Hong Kong, China
cding@ust.hk
http://www.cse.ust.hk/faculty/cding/

Abstract. In this paper, a class of three-weight linear codes and a class of four-weight linear codes over GF(q) are presented and their weight distributions are determined. These codes are punctured from the irreducible cyclic codes, and contain optimal codes. Their duals contain also optimal codes.

1 Introduction

Throughout this paper, let $q = p^s$ and $r = q^m$, where s and m are positive integers and p is a prime, let $N > 1$ be a positive integer dividing $r - 1$ and define $\ell = (r-1)/N$. A linear $[\ell, k, d]$ code over GF(q) is a k-dimensional subspace of GF(q)$^\ell$ with minimum (Hamming) distance d.

Let α be a primitive element of GF(r) and let $\theta = \alpha^N$. The set

$$\mathcal{C} = \{(\text{Tr}(\beta), \text{Tr}(\beta\theta), ..., \text{Tr}(\beta\theta^{\ell-1})) : \beta \in \text{GF}(r)\} \tag{1}$$

is called an *irreducible cyclic* $[\ell, m_0]$ *code* over GF(q), here and hereafter Tr is the trace function from GF(r) onto GF(q) and m_0 divides m.

The weight distribution of irreducible cyclic codes is quite complicated in general [13]. However, in certain special cases the weight distribution is known. We summarize these cases below.

1. When $N|(q^j + 1)$ for some j being a divisor of $m/2$, which is called the *semi-primitive case*, the codes have two weights. These codes were studied by Delsarte and Goethals [5], McEliece [15], and Baumert and McEliece [1].
2. When N is a prime with $N \equiv 3 \pmod 4$ and $\text{ord}_q(N) = (N - 1)/2$, the weight distribution was determined by Baumert and Mykkeltveit [2].
3. When $q \equiv 1 \pmod N$ and $\gcd(m, N) = 1$, the weight distribution is known [6].
4. When N is even, $\gcd(n, N) = 1$, $q - 1 \equiv N/2 \pmod N$ and $\gcd((r - 1)/(q - 1) \bmod N, N) = 2$, the weight distribution is known [6].
5. When $N = 2$, the weight distribution was found by Baumert and McEliece [1].
6. When $N = 3$ and $N = 4$, the weight distribution is settled [6].

C. Xing et al. (Eds.): IWCC 2009, LNCS 5557, pp. 34–42, 2009.

More information on the irreducible cyclic codes may be found in [5,9,11,12,16,18,20,21,23].

Let $D = \{d_1, d_2, \ldots, d_n\} \subseteq \mathrm{GF}(r)^*$. We define a linear code of length n over $\mathrm{GF}(q)$ by

$$\mathcal{C}_D = \{(\mathrm{Tr}(xd_1), \mathrm{Tr}(xd_2), \ldots, \mathrm{Tr}(xd_n)) : x \in \mathrm{GF}(r)\}. \tag{2}$$

This is indeed a generic construction of linear codes, as every linear code over $\mathrm{GF}(q)$ can be constructed in this way (i.e., the trace representation of linear codes). It may produce a good or bad code depending on the selection of the defining set D.

When D is a subset of the subgroup generated by α^N, where α is a generator of $\mathrm{GF}(r)^*$, $\mathcal{C}_D$ is punctured from the irreducible cyclic code of (1) up to coordinator permutations. In this paper, we restrict ourselves to this case. In the semiprimitive case, the code $\mathcal{C}_D$ punctured from the irreducible cyclic codes was dealt with in [8] and [7]. The objectives of this paper are to study the punctured codes $\mathcal{C}_D$ in two other cases which are not semiprimitive, and present a class of three-weight codes and a class of four-weight codes. The two classes of codes presented in this paper contain optimal codes. Their duals contain also optimal codes.

2   Auxiliary Results

To introduce the three-weight and four-weight codes, we need cyclotomic classes and Gaussian periods. We also present some auxiliary results which will be required in the sequel.

Let α be a fixed primitive element of $\mathrm{GF}(r)$. Define $C_i^{(N,r)} = \alpha^i \langle \alpha^N \rangle$ for $i = 0, 1, \ldots, N-1$, where $\langle \alpha^N \rangle$ denotes the subgroup of $\mathrm{GF}(r)^*$ generated by α^N. The cosets $C_i^{(N,r)}$ are called the *cyclotomic classes* of order N in $\mathrm{GF}(r)$. The *cyclotomic numbers* of order N are defined by

$$(i,j)^{(N,r)} = \left| (C_i^{N,r} + 1) \cap C_j^{(N,r)} \right|$$

for all $0 \leq i \leq N-1$ and $0 \leq j \leq N-1$ [22].

The *Gaussian periods* are defined by

$$\eta_i^{(N,r)} = \sum_{x \in C_i^{(N,r)}} \chi(x), \quad i = 0, 1, \ldots, N-1,$$

where χ is the canonical additive character of $\mathrm{GF}(r)$.

For our applications later, we need to determine the values of the Gaussian periods. In general, it is very hard to do so. But it can be done in certain special cases. The *period polynomials* $\psi_{(N,r)}(X)$ are defined by

$$\psi_{(N,r)}(X) = \prod_{i=0}^{N-1} \left(X - \eta_i^{(N,r)} \right).$$

It is known that $\psi_{(N,r)}(X)$ is a polynomial with integer coefficients [14]. We will need the following four lemmas whose proofs can be found in [14].

Lemma 1. *Let $N = 3$. Let c and d be defined by $4r = c^2 + 27d^2$, $c \equiv 1 \pmod 3$, and, if $p \equiv 1 \pmod 3$, then $\gcd(c, p) = 1$. These restrictions determine c uniquely, and d up to sign. Then we have*

$$\psi_{(3,r)}(X) = X^3 + X^2 - \frac{r-1}{3} X - \frac{(c+3)r - 1}{27}.$$

Lemma 2. *Let $N = 3$. We have the following results on the factorization of $\psi_{(3,r)}(X)$.*

(a) *If $p \equiv 2 \pmod 3$, then ms is even, and*

$$\psi_{(3,r)}(X) = \begin{cases} 3^{-3}(3X + 1 + 2\sqrt{r})(3X + 1 - \sqrt{r})^2 & \text{if } sm/2 \text{ even,} \\ 3^{-3}(3X + 1 - 2\sqrt{r})(3X + 1 + \sqrt{r})^2 & \text{if } sm/2 \text{ odd,} \end{cases}$$

(b) *If $p \equiv 1 \pmod 3$, and $sm \not\equiv 0 \pmod 3$, then $\psi_{(3,r)}(X)$ is irreducible over the rationals.*

(c) *If $p \equiv 1 \pmod 3$, and $sm \equiv 0 \pmod 3$, then*

$$\psi_{(3,r)}(X) = 3^{-3}(3X + 1 - c_1 r^{1/3})\left(3X + 1 + \frac{1}{2}(c_1 + 9d_1)r^{1/3}\right) \times$$

$$\left(3X + 1 + \frac{1}{2}(c_1 - 9d_1)r^{1/3}\right),$$

where c_1 and d_1 are given by $4p^{sm/3} = c_1^2 + 27d_1^2$, $c_1 \equiv 1 \pmod 3$ and $\gcd(c_1, p) = 1$.

Lemma 3. *Let $N = 4$. Let u and v be defined by $r = u^2 + 4v^2$, $u \equiv 1 \pmod 4$, and, if $p \equiv 1 \pmod 4$, then $\gcd(u, p) = 1$. These restrictions determine u uniquely, and v up to sign.*
   If n is even, then

$$\psi_{(4,r)}(X) = X^4 + X^3 - \frac{1}{8}(3r - 3)X^2 + \frac{1}{16}((2u - 3)r + 1)X +$$

$$\frac{1}{256}(r^2 - (4u^2 - 8u + 6)r + 1).$$

If n is odd, then

$$\psi_{(4,r)}(X) = X^4 + X^3 + \frac{1}{8}(r + 3)X^2 + \frac{1}{16}((2u + 1)r + 1)X +$$

$$\frac{1}{256}(9r^2 - (4u^2 - 8u - 2)r + 1).$$

Lemma 4. *Let $N = 4$. We have the following results on the factorization of $\psi_{(4,r)}(X)$.*

(a) *If $p \equiv 3 \pmod 4$, then ms is even, and*

$$\psi_{(4,r)}(X) = \begin{cases} 4^{-4}(4X + 1 + 3\sqrt{r})(4X + 1 - \sqrt{r})^3 & \text{if } sm/2 \text{ even,} \\ 4^{-4}(4X + 1 - 3\sqrt{r})(4X + 1 + \sqrt{r})^3 & \text{if } sm/2 \text{ odd,} \end{cases}$$

(b) If $p \equiv 1 \pmod 4$, and sm is odd, then $\psi_{(4,r)}(X)$ is irreducible over the rationals.

(c) If $p \equiv 1 \pmod 4$, and $sm \equiv 2 \pmod 4$, then

$$\psi_{(4,r)}(X) = 4^{-4} \left((4X+1)^2 + 2\sqrt{r}(4X+1) - r - 2\sqrt{ru} \right) \times \\ \left((4X+1)^2 - 2\sqrt{r}(4X+1) - r + 2\sqrt{ru} \right),$$

the quadratics being irreducible, the u is defined in Lemma 3.

(d) If $p \equiv 1 \pmod 4$, and $sm \equiv 0 \pmod 4$, then

$$\psi_{(4,r)}(X) = 4^{-4} \left((4X+1) + \sqrt{r} + 2r^{1/4}u_1 \right) \left((4X+1) + \sqrt{r} - 2r^{1/4}u_1 \right) \\ \times \left((4X+1) - \sqrt{r} + 4r^{1/4}v_1 \right) \left((4X+1) - \sqrt{r} - 4r^{1/4}v_1 \right)$$

where u_1 and v_1 are given by $p^{sm/2} = u_1^2 + 4v_1^2$, $u_1 \equiv 1 \pmod 4$ and $\gcd(u_1, p) = 1$.

3 The Weight Formula for the Code $\mathcal{C}_D$

To study the weights in the code $\mathcal{C}_D$ of (2), it is convenient to define for each $x \in \mathrm{GF}(r)$,

$$\mathbf{c}_x = (\mathrm{Tr}(xd_1), \mathrm{Tr}(xd_2), \ldots, \mathrm{Tr}(xd_n)).$$

The Hamming weight $w(\mathbf{c}_x)$ of $\mathbf{c}_x$ is $n - N_x(0)$, where

$$N_x(0) = |\{1 \le i \le n : \mathrm{Tr}(xd_i) = 0\}|$$

for each $x \in \mathrm{GF}(r)$.

It is easily seen that for any $D = \{d_1, d_2, \ldots, d_n\} \subseteq \mathrm{GF}(r)$ we have

$$qN_x(0) = \sum_{i=1}^n \sum_{y \in \mathrm{GF}(q)} \lambda(y\mathrm{Tr}(xd_i))$$

$$= n + \sum_{i=1}^n \sum_{y \in \mathrm{GF}(q)^*} \chi(yxd_i), \tag{3}$$

where λ is the canonical additive character of $\mathrm{GF}(q)$ and χ is the canonical additive character of $\mathrm{GF}(r)$.

For arbitrary $D \subseteq \mathrm{GF}(r)$, the dimension of $\mathcal{C}_D$ depends on the choice of D and can be determined as follows [8].

Theorem 1. [8] Let S be the $\mathrm{GF}(q)$-linear subspace of $\mathrm{GF}(r)$ spanned by D. Then the dimension of the linear code $\mathcal{C}_D$ is equal to the dimension of S.

4 The Class of Three-Weight Codes

Let $p \equiv 1$ (mod 3) and let $m \equiv 0$ (mod 3). Put $N = 3$. Then we have

$$\frac{r-1}{q-1} \bmod N = m \bmod 3 = 0.$$

Notice that every element of $\mathrm{GF}(q)^*$ can be expressed as $\alpha^{i(r-1)/(q-1)}$, where α is a fixed primitive element of $\mathrm{GF}(r)$. Hence we have $\mathrm{GF}(q)^* \subseteq C_0^{(N,r)}$.

Clearly, $C_0^{(N,r)}$ and $\mathrm{GF}(q)^*$ are generated by α^N and $\alpha^{(r-1)/(q-1)}$, respectively. Define $d_i = \alpha^{3(i-1)}$ for $i = 1, \ldots, n$, where $n = (r-1)/(3(q-1))$. Then $d_1, d_2, \ldots, d_n$ form a complete set of coset representatives of the factor group $C_0^{(N,r)}/\mathrm{GF}(q)^*$. Put

$$D = \{d_1, d_2, \ldots, d_n\}.$$

Theorem 2. *Let $p \equiv 1$ (mod 3) and $m \equiv 0$ (mod 3). Let c_1 and d_1 be given by $4q^{m/3} = c_1^2 + 27d_1^2$, $c_1 \equiv 1$ (mod 3) and $\gcd(c_1, p) = 1$. Then the set C_D is a $[(q^m - 1)/(3(q-1)), m]$ code with the following weight distribution:*

weight	frequency
0	1
$\frac{r - c_1 r^{1/3}}{3q}$	$\frac{r-1}{3}$
$\frac{r + \frac{1}{2}(c_1 + 9d_1)r^{1/3}}{3q}$	$\frac{r-1}{3}$
$\frac{r + \frac{1}{2}(c_1 - 9d_1)r^{1/3}}{3q}$	$\frac{r-1}{3}$

The dual code $C_D^\perp$ of the code C_D has parameters

$$\left[\frac{r-1}{3(q-1)}, \frac{r-1}{3(q-1)} - m, d^\perp \right],$$

where $d^\perp \geq 3$.

Proof. Since $d_1, d_2, \ldots, d_n$ form a complete set of coset representatives of the factor group $C_0^{(N,r)}/\mathrm{GF}(q)^*$, we have

$$\cup_{y \in \mathrm{GF}(q)^*} \{yd_1, yd_2, \cdots, yd_n\} = C_0^{(N,r)}.$$

By (3), we have

$$qN_x(0) - n = \sum_{i=1}^{n} \sum_{y \in \mathrm{GF}(q)^*} \chi(yxd_i) = \sum_{z \in C_0^{(N,r)}} \chi(xz) = \sum_{z \in C_j^{(N,r)}} \chi(z) = \eta_j^{(N,r)}$$

whenever $x \in C_j^{(N,r)}$. Hence

$$w(c_x) = n - N_x(0) = \frac{(q-1)n - \eta_j^{(N,r)}}{q}.$$

The weight distribution then follows from Lemma 2.

We now prove that the dimension of the code $\mathcal{C}_D$ is m. We first have $r - c_1 r^{1/3} \neq 0$, as $\gcd(c_1, p) = 1$. Note that $4r^{1/3} = c_1^2 + 27d_1^2$, $c_1 \equiv 1 \pmod 3$ and $\gcd(c_1, p) = 1$. One can verify that $r + \frac{1}{2}(c_1 \pm 9d_1)r^{1/3} \neq 0$. Hence, $w(\mathbf{c}_x) > 0$ for each nonzero $x \in \mathrm{GF}(r)$, and the dimension of $\mathcal{C}_D$ is m.

It suffices to prove that $d^\perp \geq 3$. This follows from the fact that any pair of distinct d_i and d_j are linearly independent over $\mathrm{GF}(q)$, as they are representatives of distinct cosets of the factor group $C_0^{(N,r)}/\mathrm{GF}(q)^*$.

Example 1. Let $q = 7$ and $m = 3$. Then $\mathcal{C}_D$ is a $[19, 3, 15]$ code over $\mathrm{GF}(7)$ with the following weight distribution:

$$1 + 114x^{15} + 114x^{16} + 114x^{18}.$$

It is optimal according to Brauer's tables. Its dual is a $[19, 16, 3]$ code over $\mathrm{GF}(7)$, and is also optimal.

5   The Class of Four-Weight Codes

Let $p \equiv 1 \pmod 4$ and let $m \equiv 0 \pmod 4$. Put $N = 4$. Then we have

$$\frac{r-1}{q-1} \bmod N = m \bmod 4 = 0.$$

Notice that every element of $\mathrm{GF}(q)^*$ can be expressed as $\alpha^{i(r-1)/(q-1)}$, where α is a fixed primitive element of $\mathrm{GF}(r)$. Hence we have $\mathrm{GF}(q)^* \subseteq C_0^{(N,r)}$.

Clearly, $C_0^{(N,r)}$ and $\mathrm{GF}(q)^*$ are generated by α^4 and $\alpha^{(r-1)/(q-1)}$, respectively. Define $d_i = \alpha^{4(i-1)}$ for $i = 1, \ldots, (r-1)/(4(q-1))$. Let $n = (r-1)/(4(q-1))$. Then $d_1, d_2, \ldots, d_n$ form a complete set of coset representatives of the factor group $C_0^{(N,r)}/\mathrm{GF}(q)^*$. Put

$$D = \{d_1, d_2, \ldots, d_n\}.$$

Theorem 3. *Let $p \equiv 1 \pmod 4$ and $m \equiv 0 \pmod 4$. Let u_1 and v_1 be given by $q^{m/2} = u_1^2 + 4v_1^2$, $u_1 \equiv 1 \pmod 4$ and $\gcd(u_1, p) = 1$. Then the set $\mathcal{C}_D$ is a $[(q^m - 1)/(4(q-1)), m]$ code with the following weight distribution:*

weight	frequency
0	1
$\dfrac{r+r^{1/2}+2u_1 r^{1/4}}{4q}$	$\dfrac{r-1}{4}$
$\dfrac{r+r^{1/2}-2u_1 r^{1/4}}{4q}$	$\dfrac{r-1}{4}$
$\dfrac{r-r^{1/2}+4v_1 r^{1/4}}{4q}$	$\dfrac{r-1}{4}$
$\dfrac{r-r^{1/2}-4v_1 r^{1/4}}{4q}$	$\dfrac{r-1}{4}$

The dual code $C_D^\perp$ of the code C_D has parameters

$$\left[\frac{r-1}{4(q-1)}, \frac{r-1}{4(q-1)} - m, d^\perp\right],$$

where $d^\perp \geq 3$.

Proof. By (3), we have

$$qN_x(0) - n = \sum_{i=1}^{n} \sum_{y \in GF(q)^*} \chi(yxd_i) = \eta_j^{(N,r)}$$

whenever $x \in C_j^{(N,r)}$. Hence

$$w(\mathbf{c}_x) = n - N_x(0) = \frac{(q-1)n - \eta_j^{(N,r)}}{q}.$$

The weight distribution then follows from Lemma 3.

We now prove that the dimension of the code is m. Since $\gcd(u_1, p) = 1$, $v_1 \neq 0$. Hence, $u_1^2 < \sqrt{r}$. It then follows that

$$r + r^{1/2} \pm 2u_1 r^{1/4} > r - r^{1/2} > 0.$$

Note that q is at least 5 and $m > 4$. We have that $r > 9$. Again since $\gcd(u_1, p) = 1$, $u_1 \neq 0$. We have then $2|v_1| < r^{1/4}$. It then follows that

$$r - r^{1/2} \pm 4v_1 r^{1/4} > r - 3r^{1/2} > 0.$$

Hence, $w(\mathbf{c}_x) > 0$ for each nonzero $x \in GF(r)$, and the dimension of C_D is m.

It suffices to prove that $d^\perp \geq 3$. This follows from the fact that any pair of distinct d_i and d_j are linearly independent over $GF(q)$, as they are representatives of distinct cosets of the factor group $C_0^{(N,r)}/GF(q)^*$.

Example 2. Let $q = 5$ and $m = 4$. Then C_D is a $[39, 4, 28]$ code over $GF(5)$ with the following weight distribution:

$$1 + 156x^{28} + 156x^{31} + 156x^{32} + 156x^{34}.$$

It is not optimal. But its dual is a $[39, 35, 3]$ code over $GF(5)$, and is optimal.

6   Concluding Remarks

Three-weight and four-weight codes are an interesting topic of study [3,4,10,19]. It would be interesting to find some applications of these codes in cryptography.

Acknowledgments

The author would like to thank the reviewer for his comments that improved the quality of this paper.

The research of the author is supported by the Research Grants Council of the Hong Kong Special Administrative Region, China, under Proj. No. HKUST 612405.

References

1. Baumert, L.D., McEliece, R.J.: Weights of irreducible cyclic codes. Information and Control 20, 158–175 (1972)
2. Baumert, L.D., Mykkeltveit, J.: Weight distributions of some irreducible cyclic codes. DSN Progress Report 16, 128–131 (1973)
3. Calderbank, A.R., Goethals, J.-M.: Three-weight codes and association schemes. Philips J. Res. 39, 143–152 (1984)
4. Calkin, N.J., Key, J.D., de Resmini, M.J.: Minimum weight and dimension formulas for some geometric codes. Designs, Codes and Cryptography 17, 105–120 (1999)
5. Delsarte, P., Goethals, J.M.: Irreducible binary cyclic codes of even dimension. In: Proc. Second Chapel Hill Conf. on Combinatorial Mathematics and Its Applications, Univ. North Carolina, Chapel Hill, NC, pp. 100–113 (1970)
6. Ding, C.: The weight distribution of some irreducible cyclic codes. IEEE Trans. Inform. Theory (to appear)
7. Ding, C., Luo, J., Niederreiter, H.: Two-weight codes punctured from irreducible cyclic codes. In: Li, Y., Ling, S., Niederreiter, H., Wang, H., Xing, C., Zhang, S. (eds.) Proc. of the First International Workshop on Coding Theory and Cryptography, pp. 119–124. World Scientific, Singapore (2008)
8. Ding, C., Niederreiter, H.: Cyclotomic linear codes of order 3. IEEE Trans. Inform. Theory 53, 2274–2277 (2007)
9. Goethals, J.M.: Factorization of cyclic codes. IEEE Trans. Inform. Theory 13, 242–246 (1967)
10. Griera, M.: On s-sums-sets (s odd) and three-weight projective codes. In: Proc. AAECC, pp. 68–76 (1986)
11. Helleseth, T., Kløve, T., Mykkeltveit, J.: The weight distribution of irreducible cyclic codes with block length $n_1((q^l - 1)/N)$. Discrete Math. 18, 179–211 (1977)
12. Langevin, P.: A new class of two weight codes. In: Cohen, S., Niederreiter, H. (eds.) Finite Fields and Applications, pp. 181–187. Cambridge University Press, Cambridge (1996)
13. MacWilliams, F., Seery, J.: The weight distributions of some minimal cyclic codes. IEEE Trans. Inform. Theory 27, 796–806 (1981)
14. Myerson, G.: Period polynomials and Gauss sums for finite fields. Acta Arith. 39, 251–264 (1981)
15. McEliece, R.J.: A class of two-weight codes. Jet Propulsion Laboratory Space Program Summary 37–41 IV, 264–266
16. McEliece, R.J.: Irreducible cyclic codes and Gauss sums. In: Combinatorics, Part 1: Theory of Designs, Finite Geometry and Coding Theory. Math. Centre Tracts, vol. 55, pp. 179–196. Math. Centrum, Amsterdam (1974)
17. McEliece, R.J., Rumsey Jr., H.: Euler products, cyclotomy, and coding. J. Number Theory 4, 302–311 (1972)

18. Moisio, M.J., Väänen, K.O.: Two recursive algorithms for computing the weight distribution of certain irreducible cyclic codes. IEEE Trans. Inform. Theory 45, 1244–1249 (1999)
19. Ray-Chaudhuri, D.K., Xiang, Q.: New necessary conditions for abelian Hadamard difference sets. J. of Statistical Planning and Inference 62, 69–79 (1997)
20. Schmidt, B., White, C.: All two-weight irreducible cyclic codes? Finite Fields Appl. 8, 1–17 (2002)
21. Segal, R., Ward, R.L.: Weight distributions of some irreducible cyclic codes. Mathematics of Computation 46, 341–354 (1986)
22. Storer, T.: Cyclotomy and Difference Sets. Markham, Chicago (1967)
23. Van der Vlugt, M.: On the weight hierarchy of irreducible cyclic codes. J. Comb. Theory Ser. A 71, 159–167 (1995)

Equal-Weight Fingerprinting Codes

Ilya Dumer

University of California, Riverside, USA
dumer@ee.ucr.edu

Abstract. We consider binary fingerprinting codes that trace at least one of t pirates using the marking assumption. Ensembles of binary equal-weight codes are considered along with a new efficient decoding algorithm. The design substantially increases the code rates of the former fingerprinting constructions. In particular, for large t, the new t-fingerprinting codes have code rate of $t^{-2} \ln 2$ and identify a pirate with an error probability that declines exponentially in code length n.

1 Introduction

To protect the copyrights, a distributor of digital content marks each copy of data with a unique *digital fingerprint*. This fingerprint is a set of redundant digits which are inserted across the information digits of the original content. The inserted positions remain the same for all users but are unknown to them. However, a group of t users can compare their copies of the data, detect the positions in which these copies disagree, and arbitrarily alter these (and only these) positions. In doing so, the users collectively produce a new, unregistered copy, which is marked with a new digital fingerprint.

We define binary *t-fingerprinting* codes of length n and size M as follows. Consider the set $\mathbb{F}^n = \{0,1\}^n$ and some subset $S \subseteq \mathbb{F}^n$. Let $\mathcal{K} = \{1, \ldots, K\}$ be a set of keys and $\mathcal{M} = \{1, \ldots, M\}$ be a set of M users. A *coalition* U of t users is an arbitrary t-subset of $\mathcal{M}$. The distributor first chooses a key $k \in \mathcal{K}$ according to some probability distribution $\pi(k)$ on $\mathcal{K}$. For every key k, the distributor employs a pair $(\mathcal{A}_k, \mathcal{D}_k)$ of encoding and decoding mappings:

$$\mathcal{A}_k : \mathcal{M} \to S, \quad \mathcal{D}_k : S \to \mathcal{M} \cup \{0\}. \tag{1}$$

The set S, the distribution $\pi(k)$, and possible mappings $(\mathcal{A}_k, \mathcal{D}_k)$ are known to all M users, whereas the specific key k and the corresponding code $\mathcal{A}_k$ are not. Given the key k and the code $\mathcal{A}_k$, the distributor assigns some fingerprint $a_k(i) \in \mathcal{A}_k$ to a user i. A *pirate* coalition U does not know the set $\mathcal{A}_k(U) = \{\mathbf{x}_1, \ldots, \mathbf{x}_t\}$ of the fingerprints assigned to the members of the coalition nor the specific code $\mathcal{A}_k$. However, the pirates can create a new fingerprint $\mathbf{y} \in S$ that is restricted by the *marking assumption*. Namely, the pirates must keep the symbol x_{1i} in any *undetectable* position i such that $x_{1i} = x_{2i} = \cdots = x_{ti}$ but can arbitrarily define $\mathbf{y}$ in the remaining positions. Thus, a coalition U creates a fingerprint $\mathbf{y}$ from the *envelope*

$$\mathcal{E}(U) = \{\mathbf{y} \in S \,|\, y_i = x_{1i}, \forall i \text{ undetectable}\}. \tag{2}$$

C. Xing et al. (Eds.): IWCC 2009, LNCS 5557, pp. 43–51, 2009.
© Springer-Verlag Berlin Heidelberg 2009

We allow any t-coalition U to randomly create any $\mathbf{y} \in \mathcal{E}(U)$ with some probability $V(\mathbf{y}|\mathcal{A}_k(U))$. More generally, we consider any *mixed* coalition U that includes a subset $U_* \subseteq U$ of $m \leq t$ pirates and a complementary subset of innocent users. In this case, the colluding pirates do not know the innocent users but still wish to create some generic fingerprint $\mathbf{y} \in S$ that may satisfy the marking assumption (2) with some probability $V(\mathbf{y}|\mathcal{A}_k(U_*))$.

Given an unregistered fingerprint $\mathbf{y} \in S \setminus \mathcal{A}_k$, the distributor performs decoding $\mathcal{D}_k(\mathbf{y})$, which attempts to trace at least one of the pirates or outputs 0 if no pirates are detected. By definition, a *randomized fingerprinting code* is a r.v. $(\mathcal{A}, \mathcal{D})$ that takes the values $\{(\mathcal{A}_k, \mathcal{D}_k), k \in \mathcal{K}\}$. The *rate* of this code is $R_t = (\log_2 M)/n$. For a given (mixed) coalition U and a strategy V, the probability of error averaged over the set $\mathcal{K}$ is equal to

$$e(\mathcal{A}, \mathcal{D}, U, V) = \sum_{k \in \mathcal{K}} \pi(k) \sum_{\mathbf{y} \in \mathcal{E}(U):\ \mathcal{D}_k(\mathbf{y}) \notin U_*} V(\mathbf{y}|\mathcal{A}_k(U_*)). \tag{3}$$

We then consider the error probability of a randomized code $(\mathcal{A}, \mathcal{D})$, where we take the maximum over all coalitions U and their strategies V :

$$e(\mathcal{A}, \mathcal{D}, t) = \max_{U:|U|=\tau} \max_{V} e(\mathcal{A}, \mathcal{D}, U, V). \tag{4}$$

Definition 1. *A randomized code $(\mathcal{A}, \mathcal{D})$ is a t-fingerprinting code with an ε-error if $e(\mathcal{A}, \mathcal{D}, t) \leq \varepsilon$. A sequence of randomized codes is t-fingerprinting if the probability of error vanishes as n increases.*

The above setting has one important deviation from the previous work. Namely, our encoding-decoding $(\mathcal{A}, \mathcal{D})$ employs any subset S instead of the full space $\mathbb{F}^n$ used before. Thus, the conventional setting arises only if $S = \mathbb{F}^n$. In this way, we restrict the set of encoded words, but wish to take advantage of this restriction by taking the forged fingerprints from the same set S. For the practical purposes, we assume here that some local test can disable any forged fingerprint $\mathbf{y} \notin S$, and make data copy unreadable. In particular, in this paper we use fingerprints of the same Hamming weight pn. In this case, we can take any set of positions that include n inserted fingerprint positions, calculate the Hamming weight on these positions, and then discard any data copy that fails the weight check. This local check requires no knowledge of a specific key k nor the actual fingerprint positions.

Randomized fingerprinting codes designed under marking assumption were pioneered by Boneh and Shaw [1]. The zero-error case $\varepsilon = 0$ was considered in [2] - [5]. This setting gives the codes with the *identifiable parent property*. Long fingerprinting codes with a non-vanishing code rate were designed in [6]. An important breakthrough was obtained by Tardos [7] who designed t-fingerprinting codes of rate $R_t \geq c/t^2$, where $c = 1/100$. In addition, the decoding of [7] has a low complexity of order M. Using [7], constant c was later refined in [8] to about $1/(4\pi^2)$.

Another approach proposed in [10] gives for $t = 2, 3$ the code rates $R_2 = 1/4$ and $R_3 = 1/12$ that exceed the best former rates (see [6] and [9]). This is done by tracing only typical pirate coalitions and disregarding (a vanishing fraction) of all other, atypical coalitions. However, the minimum distance decoding used in [10] results in a significant deterioration of code rates for $t \geq 4$.

In this paper, our main goal is to develop a new decoding technique that can be applied to the coalitions of any size t. We employ equal-weight fingerprinting codes and perform a new, more efficient decoding algorithm. One particular advantage of our technique is that similar proofs hold for both pure and mixed coalitions. Then we show in Theorems 1 and 2 that the new design gives codes of rate up to $\theta_t \sim t^{-2} \ln 2 \approx 0.69 t^{-2}$ for any t. For small t, the rate θ_t is improved further. Note that the rate θ_t increases almost 30 times the former code rate of [8].

Finally, an important recent result of [11] introduces new multi-weight fingerprinting codes. These codes are constructed in the entire space $\mathbb{F}^n$ and further increase attainable code rates to $C_t = t^{-2}/(2 \ln 2) \approx 0.72 t^{-2}$. This multi-weight design of [11] also requires a more complex decoding that increases $2^t/t$ times the complexity of our single-weight design. In addition to a new high-rate design, another important result of [11] is the proof that the fingerprinting capacity has the order of t^{-2}, with a conjecture that it tends to the above code rate C_t for large t.

2 Equal-Weight Fingerprinting Codes

Code ensemble. Let $n \to \infty$, $p \in (0, 1/2)$ and let pn be an integer. Consider a sphere $S \subset \mathbb{F}^n$ of radius pn, which consists of $L = \binom{n}{pn}$ binary vectors of Hamming weight pn. Our random encoding $\mathcal{A}$ chooses uniformly and independently $M = 2^{nR}$ vectors in S. Thus, we consider L^M different codes $\mathcal{A}_k$, chosen equally likely with probability $\pi(k) = L^{-M}$. Parameters p and M will be defined later; however, we will obtain codes of low rate $R < H(p)/2$, where $H(p)$ is a binary entropy function. Note that $M = o(L^{1/2})$ for the rates $R < H(p)/2$, and with high probability[1] codes $\mathcal{A}_k$ have M distinct vectors.

We also use the following definitions. Consider a ternary alphabet $\mathbb{F}_* = \{0, 1, *\}$. Given two vectors $\mathbf{x}, \mathbf{y} \in \mathbb{F}_*^n$, we define their *collusion product* $\mathbf{z} = \mathbf{x} \otimes \mathbf{y}$, where for each $j = 1, ..., n$ and each $s \in \mathbb{F}_*$

$$z_j = \begin{cases} s \text{ if } y_j = x_j = s \\ * \text{ if } y_j \neq x_j \end{cases}$$

Also, define the composition of any vector $\mathbf{z} \in \mathbb{F}_*^n$ using the two *weights*

$$g(\mathbf{z}) = |\{j : z_j = 0\}|, \quad h(\mathbf{z}) = |\{j : z_j = 1\}|$$

Finally, given any positive ε and any function $r(n)$, we write $f(n) \overset{\varepsilon}{\sim} r(n)$ for some function $f(n)$ if

$$r(n)(1 - \varepsilon) \leq f(n) \leq r(n)(1 + \varepsilon).$$

---

[1] Given a sequence of codes, we say that an event occurs with high probability if the probability of its complement vanishes exponentially in n as $n \to \infty$.

Given any coalition $U = \{u_1, ..., u_t\}$ of size t, we now consider t embedded subsets $\{u_1, ..., u_m\}$ and the corresponding subsets of m fingerprints

$$X_m = \{\mathbf{x}_1, ..., \mathbf{x}_m\}, \quad m = 1, ..., t.$$

For each set X_m, define the vector $\mathbf{z}_m = \otimes_{i=1}^{i=m} \mathbf{x}_i$. Then we have the following lemma.

Lemma 1. *Let $n \to \infty$. Then for any $\varepsilon > 0$ and for any set U of t users, all $2t$ weights $g(\mathbf{z}_m)$ and $h(\mathbf{z}_m)$ of vectors $\mathbf{z}_m$ satisfy asymptotic equalities*

$$g(\mathbf{z}_m) \overset{\epsilon}{\sim} n(1-p)^m, \quad h(\mathbf{z}_m) \overset{\epsilon}{\sim} np^m, \quad m = 1, ..., t \tag{5}$$

with a vanishing probability of failure that declines exponentially in n.

Proof. For any coalition U, a random set of m fingerprints X_m is chosen uniformly on S as a key k runs through $\mathcal{K}$. Then for $s = 0, 1$ the number of positions j such that $\mathbf{x}_{1j} = ... = \mathbf{x}_{mj} = s$ has a mean value of $n(1-p)^m$ and np^m, respectively. Moreover, conditions (5) can fail only for the large deviations $\varepsilon n(1-p)^m$ and εnp^m from the corresponding mean values. These deviations occur with an exponentially declining probability as $n \to \infty$. Thus, conditions (5) hold simultaneously for all m with high probability as $n \to \infty$. $\qquad \square$

Decoding procedure $\mathcal{D}$. Our decoder $\mathcal{D}$ sets up a small parameter $\varepsilon > 0$ and traces only ε-*typical* coalitions $U = \{u_1, ..., u_t\}$, which satisfy conditions (5). Given any key $k \in \mathcal{K}$ and the corresponding code $\mathcal{A}_k$, decoder $\mathcal{D}$ first verifies if there exists an ineligible fingerprint $\mathbf{y} \in S \setminus \mathcal{A}_k$. Given such $\mathbf{y}$, $\mathcal{D}$ performs screening of all $\binom{M}{t}$ coalitions U and discards all atypical coalitions. Let $\mathbf{y}_0 = \mathbf{y}$. For any ε-typical coalition $U = \{u_1, ..., u_t\}$, decoder recursively defines vectors

$$\mathbf{y}_m \overset{\text{def}}{=} \mathbf{y} \otimes \mathbf{z}_m = \mathbf{y}_{m-1} \otimes \mathbf{x}_m, \quad m = 1, ..., t, \tag{6}$$

calculates their weights

$$g_m = g(\mathbf{y}_m), \quad h_m = h(\mathbf{y}_m)$$

and finds the quantities

$$p_m = \binom{g_{m-1}}{g_m} \binom{h_{m-1}}{h_m} \binom{n-h_{m-1}-g_{m-1}}{pn-h_m-(g_{m-1}-g_m)} \Big/ \binom{n}{pn} \tag{7}$$

If $\mathbf{y}_t = \mathbf{z}_t$, then coalition U satisfies the marking assumption. Then decoding $\mathcal{D}$ is completed by accusing a user u_s whose fingerprint minimizes the quantity p_m :

$$\mathcal{D}(\mathbf{y}, U) = s \quad \text{if} \quad p_s = \min_{1 \le m \le t} p_m. \tag{8}$$

Remark. Recall that a vector $\mathbf{y}_{m-1}$ has g_{m-1} zeros and h_{m-1} ones. Now assume that a vector $\mathbf{x}_m \in S$ is chosen randomly and independently of $\mathbf{y}_{m-1}$. Then p_m is the probability to choose $\mathbf{x}_m$ that agrees with $\mathbf{y}_{m-1}$ in g_m zeros and h_m

ones. Thus, p_m is the probability of choosing a random $\mathbf{x}_m$ that converts a pair (g_{m-1}, h_{m-1}) into (g_m, h_m) :

$$p_m = \Pr\{\mathbf{x}_m : g(\mathbf{y}_m) = g_m,\ h(\mathbf{y}_m) = h_m \mid \mathbf{y}_{m-1}\} \tag{9}$$

We will use this observation to prove the following theorem.

Theorem 1. *Consider an (n,t)-fingerprinting code $(\mathcal{A}, \mathcal{D})$ with 2^{nR} fingerprints of Hamming weight pn. Let*

$$q = 1 - p^t - (1-p)^t, \quad q' = 1 - (1-p)^t,$$

$$\theta = \theta(p,t) = \frac{1}{t}\left[H(p) - qH\left(\frac{p - p^t}{q}\right)\right]. \tag{10}$$

Then decoding (8) has error probability that declines exponentially in n for any code rate $R < \theta$ as $n \to \infty$.

Proof. The decoding error can occur in the following two cases:

A. $\mathcal{D}$ misses an existing pirate coalition;

B. $\mathcal{D}$ makes false decoding and accuses an innocent user.

Case **A** is covered in Lemma 1. Indeed, all typical coalitions are traced by decoder $\mathcal{D}$, whereas atypical coalitions form an exponentially small fraction. A typical pirate coalition (if it exists) satisfies the marking assumption and is retrieved by the decoder. Then the decoder accuses one of the pirates in this coalition.

We proceed with case **B** in two steps. In the first step **B1**, we consider any fingerprint $\mathbf{y} \in S$. We then prove that any typical pirate-free coalition satisfies the marking assumption with an exponentially small probability. In the second step **B2**, we consider a coalition U that includes a smaller subset U_* of m pirates. Then we again prove that an innocent user can be falsely accused with an exponentially small probability. Case **B2** will also cover all coalitions of $m \le t$ pirates.

Case **B1.** Consider any typical collection of t random fingerprints. Then for any $\varepsilon > 0$ and $n \to \infty$, vector $\mathbf{z}_t$ has the weights

$$g \overset{\text{def}}{=} g(\mathbf{z}_t) \overset{\varepsilon}{\sim} n(1-p)^t, \quad h \overset{\text{def}}{=} h(\mathbf{z}_t) \overset{\varepsilon}{\sim} np^t. \tag{11}$$

Random vector $\mathbf{z}_t$ satisfies the marking assumption, if its g zeros and h ones belong to the corresponding subsets of $(1-p)n$ zeros and pn ones in the forged fingerprint $\mathbf{y}$. The probability to obtain such a vector $\mathbf{z}_t$ is

$$\Pr(g, h) = \frac{\binom{np}{h}\binom{n-np}{g}}{\binom{n}{h}\binom{n-h}{g}} = \frac{\binom{n-h-g}{np-h}}{\binom{n}{np}} \tag{12}$$

Therefore, given $M = 2^{nR}$ fingerprints, the total probability P of false decoding is upper-bounded as

$$P \le \binom{M}{t} \max_{g,h} \Pr(g, h), \quad \text{where} \quad g \overset{\varepsilon}{\sim} n(1-p)^t, \quad h \overset{\varepsilon}{\sim} np^t.$$

For brevity, we write $f(n) \asymp r(n)$ if $\frac{1}{n} \ln \frac{f(n)}{r(n)} \to 1$ as $n \to \infty$. Note that for $\varepsilon \to 0$ and $q = 1 - p^t - (1-p)^t$,

$$\max \Pr(g,h) \asymp \frac{\binom{np}{np^t}\binom{n-np}{n(1-p)^t}}{\binom{n}{np^t}\binom{n-np^t}{n(1-p)^t}} = \frac{\binom{nq}{np-np^t}}{\binom{n}{np}} \asymp 2^{-nt\theta}. \tag{13}$$

Therefore, for any $\theta' < \theta$, there exists sufficiently small $\varepsilon > 0$ such that for $n \to \infty$,

$$\max \Pr(g,h) \leq 2^{-nt\theta'}. \tag{14}$$

Given any code rate $R < \theta$, we then take $\theta' = (R+\theta)/2$ and see that the total probability P declines exponentially in $nt(\theta - R)/2$ as $n \to \infty$.

Case **B2.** We begin with the following numerical lemma.

Lemma 2. *Consider any vectors $G = (g_1, ..., g_t)$ and $H = (h_1, ..., h_t)$ with positive integer symbols. Let the numbers $g = g_t$ and $h = h_t$ satisfy condition (11). Then the minimum number p_s defined in (8) satisfies inequality*

$$\lim_{n\to\infty} \frac{1}{n} \log_2 p_s \leq -\theta. \tag{15}$$

Proof. We proceed with an indirect proof and define the quantities p_m as probabilities (9). Recall from (6) that any vector $\mathbf{y}_t = \mathbf{y} \otimes \mathbf{z}_t$ with g zeros and h ones is recursively obtained through the previous vectors $\mathbf{y}_m$. This recursion implies that the probability $\Pr(g,h)$ can be derived by the equality

$$\Pr(g,h) = \sum_{G,H} \prod_{m=1}^{t} \Pr(g_m, h_m \,|\mathbf{y}_{m-1}). \tag{16}$$

where the sum is taken over all pairs (G, H). Now let us consider generic equality (16) in the case when random vectors $\mathbf{x}_1, ..., \mathbf{x}_t$ are all chosen independently of vector $\mathbf{y}$ (and each other). Then conditional probabilities of (16) can be written as quantities p_m of (9). This gives the estimate

$$\Pr(g,h) = \sum_{G,H} p_1 \cdot ... \cdot p_t \geq \sum_{G,H} p_s^t.$$

Finally, we use estimate (14) for $\Pr(g,h)$, which shows that for any $\theta' < \theta$ and $n \to \infty$,

$$p_s \leq \max\left\{\Pr(g_t, h_t)^{1/t}\right\} \leq 2^{-n\theta'}.$$

This completes the proof of Lemma. $\qquad \square$

Case **B2** (continued). Assume that the decoder retrieves a mixed coalition U that satisfies the marking assumption but includes a subset U_* of pirates. Then Lemma 2 holds due to restrictions (11). Let us also assume that the accused user s with parameter p_s is innocent. Recall that in this case a fingerprint $\mathbf{x}_s$

is independent of $\mathbf{y}$, and the quantity p_s is the probability of choosing such $\mathbf{x}_s$. Thus, we see that for all $i = 1, ..., M$, any innocent user can be accused with a probability at most Mp_s. Given a code of rate $R < \theta$, we again take $\theta' = (R+\theta)/2$. Then (15) shows that the probability Mp_s declines exponentially in $n(\theta - R)/2$. □

Remark. Note that general case **B2** follows from **B1** almost entirely. Indeed, case **B1** is based on the fact that any innocent coalition satisfies the marking assumption with a probability $\Pr(g, h)$ that declines faster than M^{-t}. Then case **B2** holds due to the fact that the smallest conditional probability $\Pr(g_m, h_m | \mathbf{y}_{m-1})$ declines faster than $\Pr^{1/t}(g, h)$. In this way, our technique can be used for any algorithm that decomposes the entire probability $\Pr(g, h)$ into the sum of the products of the individual conditional probabilities.

Our next goal is to optimize parameter p in expression (10) for any given t and achieve the maximum $\theta_t = \max_p \theta(p, t)$.

Theorem 2. *Equal-weight t-fingerprinting codes designed in Theorem 1 achieve any code rate $R < \theta_t$, where*

$$\theta_3 \approx 0.087, \quad \theta_4 \approx 0.045,$$
$$\theta_t = t^{-2} \ln 2 - \mathcal{O}(t^{-3}) \geq 0.69t^{-2} - \mathcal{O}(t^{-3}), \quad t \to \infty. \tag{17}$$

Proof. The first two estimates can be obtained numerically from (10) and are achieved at $p_3 \approx 0.25$ and $p_4 \approx 0.16$, respectively. The estimate (17) is obtained using parameters q and q' of (10) as follows

$$\theta = \frac{1}{t} \left[H(p) - qH\left(\frac{p - p^t}{q}\right) \right] \geq \frac{1}{t} \left[H(p) - qH(p/q) \right]$$
$$\geq \frac{1}{t} \left[H(p) - q'H(p/q') \right] = \frac{1}{t} \left[H((1 - p)^t) - (1 - p)H((1 - p)^{t-1}) \right]. \tag{18}$$

Here the second inequality follows from the fact that for any given $p < 1/2$, the function $qH(p/q)$ monotonically increases in the interval $q \in [p, 1]$. Now we maximize the last expression in (18) as a function of p. It can be readily verified that for large t, this maximum is achieved at $p = (t - 1)^{-1} \ln 2$, in which case $(1 - p)^{t-1} = 1/2 - \mathcal{O}(t^{-1})$ and

$$\left[H((1 - p)^t) - (1 - p)H((1 - p)^{t-1}) \right] = t^{-1} \ln 2 - \mathcal{O}(t^{-2}).$$

This gives estimate (17). □

3   Concluding Remarks

The following remarks show that the equal-weight design can be extended to some other cases. First, let us assume that the forged fingerprint $\mathbf{y}$ has Hamming weight λn that differs from the given weight pn. In this case, our main estimate (13) is replaced with an estimate

$$\max \Pr(g, h) \asymp \frac{\binom{n\lambda}{np^t} \binom{n-n\lambda}{n(1-p)^t}}{\binom{n}{np^t} \binom{n-np^t}{n(1-p)^t}} \asymp 2^{-nt\theta(\lambda, t)}$$

where

$$\theta(\lambda, t) = \tfrac{1}{t} \left[H(\lambda) - qH\left(\frac{\lambda - p^t}{q}\right) \right]$$

It can be shown that for any $\lambda \in (p, q')$, the quantity $\theta(\lambda, t)$ exceeds the former code rate $\theta(p, t)$ obtained in (10). Therefore, Theorems 1 and 2 still hold for our equal-weight codes of weight pn if the forged fingerprints $\mathbf{y}$ have varying Hamming weights in the interval $[pn, q'n]$.

We also note the remark of G. Tardos who pointed to a different setting that can be related to the cover-free sets [12]. In this setting, a forged fingerprint $\mathbf{y}$ equals 0 in positions where all t fingerprints equal 0, but has no restrictions for the all-one positions. Thus, this one-side marking assumption yields the null *envelope*

$$\mathcal{E}(U) = \{\mathbf{y} : y_i = 0, \text{ if } x_{1i} = x_{2i} = \cdots = x_{ti} = 0\}.$$

In this case, our main estimate (13) is replaced with a simpler estimate

$$\max \Pr(g, h) \asymp \frac{\binom{n-np}{n(1-p)^t}}{\binom{n}{n(1-p)^t}} = \frac{\binom{n-nq'}{np}}{\binom{n}{np}} \asymp 2^{-n[H(p) - q'H(p/q')]}$$

which is already considered in inequalities (18). Thus, the maximum code rate $\theta_t = t^{-2} \ln 2 - \mathcal{O}(t^{-3})$ obtained for the two-side envelope will remain for a one-side envelope. In other words, the former restrictions on the all-one positions do not change our code rate θ_t for large t.

Finally, note that the accusation rule (8) can be replaced with threshold decoding. For example, for a code of rate $R < \theta$, we can take the threshold $\theta' = (R + \theta)/2$ and perform decoding

$$\mathcal{D}(\mathbf{y}, U) = s \quad \text{if} \quad p_s \leq 2^{-n\theta'}. \tag{19}$$

Indeed, the proof of Theorem 1 shows that condition (19) holds for some user s given the marking assumption but all innocent users satisfy this condition with a vanishing probability. Thus, the decoding stops once the threshold is met by one of the pirates. This threshold decoding can slightly reduce the average complexity of our algorithm (8); however, in the worst case the decoding is set to screen all typical coalitions until some coalition satisfies the marking assumption. As a result, our decoding complexity has a high exponential order of M^t. The multi-weight code design of [11] also has very high complexity that is set to screen all coalitions and all subsets of a given coalition U. Therefore, accusation rules (8) and (19) reduce decoding complexity of [11] by the order of $2^t/t$ times. However, our current complexity is still much higher than the former decoding complexity of [7] that is linear in the number of users M. Therefore, one important open problem is to design high-rate fingerprinting codes whose complexity grows linearly with the number of users.

Acknowledgment. This research was supported in part by NSF grants CCF-0622242 and CCF-0635339. The author thanks G. Tardos for helpful remarks.

References

1. Boneh, D., Shaw, J.: Collusion-secure fingerprinting for digital data. IEEE Trans. Inform. Theory 44(5), 1897–1905 (1998)
2. Chor, B., Fiat, A., Naor, M.: Tracing traitors. In: Desmedt, Y.G. (ed.) CRYPTO 1994. LNCS, vol. 839, pp. 257–270. Springer, Heidelberg (1994)
3. Hollmann, H.D.L., van Lint, J.H., Linnartz, J.-P., Tolhuizen, L.M.G.M.: On codes with the identifiable parent property. J. Combinat. Theory A 82, 121–133 (1998)
4. Barg, A., Cohen, G., Encheva, S., Kabatiansky, G., Zémor, G.: A hypergraph approach to the identifying parent property: The case of multiple parents. SIAM Journal Discrete Math. 14(3), 423–431 (2001)
5. Blackburn, S.R.: An upper bound on the size of a code with the k-identifiable property. J. Combinat. Theory A 102, 179–185 (2003)
6. Barg, A., Blakley, G.R., Kabatiansky, G.: Digital fingerprinting codes: Problem statements, constructions, identification of traitors. IEEE Trans. Inform. Theory 49(4), 852–865 (2003)
7. Tardos, G.: Optimal probabilistic fingerprint codes. In: Proc. 35th Annual ACM Symposium on Theory of Computing, pp. 116–125 (2003)
8. Škorić, B., Vladimirova, T.U., Celik, M., Talstra, J.C.: Tardos fingerprinting is better than we thought. IEEE Trans. Inform. Theory 54(8), 3663–3676 (2008)
9. Blakley, G.R., Kabatiansky, G.: Random coding technique for digital fingerprinting codes: fighting two pirates revisited. In: Proc. IEEE Intern. Symp. Inform. Theory, p. 203 (2004)
10. Anthapadmanabhan, N.P., Barg, A., Dumer, I.: Fingerprinting capacity under the marking assumption. IEEE Trans. on Inform. Theory 54(7), 2678–2689 (2008)
11. Amiri, E., Tardos, G.: High rate fingerprinting codes and the fingerprinting capacity. In: Proceedings of the 20th Annual ACM-SIAM Symposium on Discrete Algorithms (SODA 2009), January 2009, pp. 336–345 (2009)
12. Ruszinkó, M.: On the upper bound of the size of the r-cover-free families. J. Combin. Theory A 66(2), 302–310 (1994)

Problems on Two-Dimensional Synchronization Patterns*

Tuvi Etzion

Department of Computer Science, Technion, Haifa 32000, Israel
etzion@cs.technion.ac.il

Abstract. A synchronization pattern is a sequence of dots in which the out-of-phase autocorrelation function takes the values *zero* or *one*. These patterns have numerous applications in information theory. Recently, two-dimensional synchronization patterns have found application in key pre-distribution for wireless sensor networks. This application has raised some new questions. We will discuss some of the old and new questions in this area. We will describe several solution techniques and present some open problems.

1 Introduction

Many problems arising from different applications such as radar, sonar, physical alignment, and time-position synchronization, are formulated in terms of two-dimensional patterns of *ones* and *zeroes* for which the two-dimensional aperiodic autocorrelation function has minimum out-of-phase values. In other words, one desire to construct a two-dimensional array of *blanks* and *dots* (corresponding to the zeroes and ones, respectively), such that any two lines connecting two different pairs of dots are different in their length or slope. The problem has been considered extensively in the last forty years both from its practical and combinatorial point of view [5, 8, 9, 10, 11, 12, 14, 15, 16, 17, 18].

Recently, these arrays have found a new application in key predistribution for wireless sensor networks [1]. The application we are interested in is to networks consisting of a large number of sensor nodes arranged in a two-dimensional grid. Although the number of sensors is evidently finite in practice, it is convenient to model the physical location of the nodes by the set of points of an infinite two-dimensional grid $\mathcal{G}$. A *distinct difference configuration* is a two-dimensional pattern of blank and dots such that any two lines connecting two different pairs of dots are different in their length or slope. Distinct difference configuration is used in a key predistribution scheme in the following way.

Let $D = \{\mathbf{v}_1, \mathbf{v}_2, \ldots, \mathbf{v}_m\}$ be a distinct difference configuration. Allocate keys to nodes as follows:

- Label each node with its position in $\mathcal{G}$.
- For every 'shift' $\mathbf{u} \in \mathcal{G}$, generate a key $k_{\mathbf{u}}$ and assign $k_{\mathbf{u}}$ to the nodes labelled by $\mathbf{u} + \mathbf{v}_i$, for $i = 1, 2, \ldots, m$.

---

* This research was supported in part by the United States-Israel Binational Science Foundation (BSF), Jerusalem, Israel, under Grant 2006097.

More informally, we can think of the scheme as covering $\mathcal{G}$ with all possible translations of the dots in D. We generate one key per translation, and assign that key to all dots in the corresponding translation of D. Distributing keys in this manner ensures that each node stores m keys and each key is shared by m nodes. In addition, the distinct difference property of the configuration implies that any pair of nodes shares at most one key, since the vector representing the difference in two nodes' positions can occur at most once as a difference vector of D. This leads to an efficient distribution of keys, since for a fixed number of stored keys, the number of distinct pairs of nodes that share a key is maximized (see [1] for further discussion).

Keys are used to establish secure communication links between users. However, the sensors in a wireless sensor network have strictly limited battery power, which in turn limits the range over which they can feasibly communicate. This leads us to consider distinct difference configurations with bounds on the distance between any two dots in the configuration [2].

Definition 1. *A Euclidean square distinct difference configuration* $DD(m, r)$ *is a set of m dots placed in a square grid such that the following two properties are satisfied:*

1. *Any two of the dots in the configuration are at Euclidean distance at most r apart.*
2. *All the $\binom{m}{2}$ differences between pairs of dots are distinct either in length or in slope.*

Definition 2. *A square distinct difference configuration* $\overline{DD}(m, r)$ *is a set of m dots placed in a square grid such that the following two properties are satisfied:*

1. *Any two of the dots in the configuration are at Manhattan distance at most r apart.*
2. *All the $\binom{m}{2}$ differences between pairs of dots are distinct either in length or in slope.*

We define similar notation in the hexagonal model, as follows.

Definition 3. *A Euclidean hexagonal distinct difference configuration* $DD^*(m, r)$ *is a set of m dots placed in a hexagonal grid with the following two properties:*

1. *Any two of the dots in the configuration are at Euclidean distance at most r apart.*
2. *All the $\binom{m}{2}$ differences between pairs of dots are distinct either in length or in slope.*

Definition 4. *A hexagonal distinct difference configuration* $\overline{DD}^*(m, r)$ *is a set of m dots placed in a hexagonal grid with the following two properties:*

1. *Any two of the dots in the configuration are at hexagonal distance at most r apart.*

2. All the $\binom{m}{2}$ differences between pairs of dots are distinct either in length or in slope.

The goal of this paper is to present some of the most interesting and important open problems in this area. In Section 2 we discuss arrays which are permutation matrices and called Costas arrays. In Section 3 we consider upper and lower bounds on the number of dots in a distinct difference configuration of a given shape. In Section 4 we consider two-dimensional periodic arrays with the distinct differences property. In Section 5 we consider the k-hop properties of distinct difference configurations when they are used for distribution of keys, and form the wireless sensor netywork.

2   Costas Arrays

A *Costas array* of order n is an $n \times n$ distinct difference configuration with n dots, exactly one dot in each row and exactly one dot in each column. There are essentially two constructions for Costas arrays, the Welch Construction and the Golomb Construction [8, 9, 10].

The Welch Construction
Let α be a primitive root modulo a prime p and let $\mathcal{A}$ be a $(p-1) \times (p-1)$ array. For any two integers i and j, $1 \le i, j \le p-1$ there is a dot in $\mathcal{A}(i,j)$ if and only if $\alpha^j \equiv i \ (mod\ p)$.

The Golomb Construction
Let α and β be primitive elements in $GF(q)$ and let $\mathcal{A}$ be a $(q-2) \times (q-2)$ array. For any two integers i and j, $1 \le i, j \le q-1$ there is a dot in $\mathcal{A}(i,j)$ if and only if $\alpha^i + \beta^j = 1$.

If $\alpha = \beta$ in the Golomb Construction then the construction is known as the Lempel Construction. There are several variants of these two construction in which we remove or add several dots to obtain a Costas array of a smaller or larger order, respectively. The main interesting question concerning Costas arrays is their existence problem.

Problem 1: Is there a Costas array of order n for each positive integer n? Our conjecture is NO.

All Costas arrays up to order 27 were found by computer search [4]. For small orders there can be found arrays which were not constructed by the two main constructions and their variants. For larger orders no such arrays were found. This leads to the next existence question.

Problem 2: Are there Costas arrays of orders greater than 27 which are not constructed from or modified from the arrays obtained by the Welch Construction or the Golomb Construction? Our conjecture is No.

3   Bounds on the Number of Dots

Give an shape S we ask the question, what is the maximum number of dots that can be placed in S such that S will be a distinct difference configuration. There are several shapes which seems to be more interesting than other, e.g., a circle and a regular polygon. For each of these shapes it is not difficult to prove [2] that if A is the area of the shape S then we can place at most $\sqrt{A} + o(\sqrt{A})$ dots in S in a way that S will be a distinct difference configuration. The following upper bounds were proved in [2] are based on a technique given in [6].

Theorem 1. *If a $\overline{DD}(m,r)$ exists, then*

$$m \le \tfrac{1}{\sqrt{2}}r + (3/2^{4/3})r^{2/3} + O(r^{1/3}).$$

Theorem 2. *If a $\overline{DD}^*(m,r)$ exists, then*

$$m \le \tfrac{\sqrt{3}}{2}r + (3^{4/3}2^{-5/3})r^{2/3} + O(r^{1/3}).$$

Theorem 3. *If a $DD(m,r)$ exists, then*

$$m \le \tfrac{\sqrt{\pi}}{2}r + \tfrac{3\,\pi^{1/3}}{2^{5/3}}r^{2/3} + O(r^{1/3}).$$

Theorem 4. *If a $DD^*(m,r)$ exists, then*

$$m \le \tfrac{\sqrt{\pi}}{\sqrt{2}\,3^{1/4}}r + \tfrac{3^{5/6}\pi^{1/3}}{2^{4/3}}r^{2/3} + O(r^{1/3}).$$

It was also proved in [2] that one of these bounds is relatively tight.

Theorem 5. *For r large enough there exists a $\overline{DD}(m,r)$ such that*

$$m = \tfrac{1}{\sqrt{2}}r - o(r).$$

Using folding and tiling some interesting lower bounds were found in [7].

Theorem 6. *For r large enough there exists a $\overline{DD}^*(m,r)$ such that*

$$m \approx \tfrac{\sqrt{3}}{2}r - o(r).$$

Theorem 7. *For r large enough there exists a $DD(m,r)$ such that*

$$m = 0.84442r - o(r).$$

Theorem 8. *For r large enough there exists a $DD^*(m,r)$ such that*

$$m = 0.90739r - o(r).$$

Table 1. Upper and lower bounds on the number of dots in a distinct difference configuration

	lower bound	upper bound
$\overline{\mathrm{DD}}(m,r)$	$(1/\sqrt{2})r - o(r)$	$(1/\sqrt{2})r + O(r^{2/3})$
$\mathrm{DD}(m,r)$	$0.85433r - o(r)$	$0.88623r + O(r^{2/3})$
$\overline{\mathrm{DD}}^*(m,r)$	$\approx \frac{\sqrt{3}}{2}r - o(r)$	$\frac{\sqrt{3}}{2}r + O(r^{2/3})$
$\mathrm{DD}^*(m,r)$	$0.91803r - o(r)$	$0.95231r + O(r^{2/3})$

Summary of the lower and upper bounds is given in Table 1.

Many related questions can be asked. We will state the two problems which we consider to be the most important.

Problem 3: Let $\mathcal{S}$ be a regular polygon with area t^2. Is there a doubly periodic distinct difference configuration with the shape $\mathcal{S}$ and $t + o(t)$ dots? Our conjecture is that the answer depends on the polygon.

Problem 4: For a given r, is there a distinct difference configuration whose shape is circle with radius r, on the square grid, with $\sqrt{\pi}r + o(r)$ dots? Our conjecture is Yes.

4   Doubly Periodic Configurations

Periodic configurations are interesting from both theoretical and practical point of view. All the known main constructions for distinct difference configurations have some periodicity properties due to the fact that they are constructed from some cyclic group. Costas arrays constructed by one of the two main constructions have some periodicity properties. An $n \times \infty$ array $\mathcal{A}$ is called *singly periodic* Costas array if any n consecutive columns of $\mathcal{A}$ form a Costas array. The Welch Construction can be extended to form a singly periodic Costas array.

Problem 5: Are there singly periodic Costas arrays which cannot be obtained from the Welch Construction? Our conjecture is No.

Singly periodic Costas arrays are quite rare, but it appears that it is not too difficult to construct doubly periodic distinct difference configurations. Let $\mathcal{A}$ be a (generally infinite) array of dots in the square grid, and let η and κ be positive integers. We say that v is *doubly periodic* with period (η, κ) if $v(i,j) = v(i+\eta, j)$ and $v(i,j) = v(i, j+\kappa)$ for all integers i and j. We define the *density* of $\mathcal{A}$ to be $d/(\eta\kappa)$, where d is the number of dots in any $\kappa \times \eta$ sub-array of $\mathcal{A}$. Note that the period (η, κ) will not be unique, but that the density of $\mathcal{A}$ does not depend on the period we choose. Doubly periodic distinct difference configurations have several nice applications [2, 3], e.g., in obtaining lower bounds on the number of dots in a distinct difference configuration. These applications lead to a more general definition. We say that an infinite array $\mathcal{A}$ of dots in the grid $\mathcal{G}$ is *a doubly periodic distinct difference configuration with shape $\mathcal{S}$* if every set of points of $\mathcal{G}$

(with its dots) with the shape of S form a distinct difference configuration. The density of $\mathcal{A}$ is defined similarly as the average number of dots per a point in the grid. Some known and new patterns were given in [1].

The Periodic Welch Construction

Let α be a primitive root modulo a prime p and let $\mathcal{A}$ be the square grid. For any integers i and j, there is a dot in $\mathcal{A}(i,j)$ if and only if $\alpha^i \equiv j \pmod{p}$.

Theorem 9. *Let $\mathcal{A}$ be the array of dots from the Periodic Welch Construction. Then $\mathcal{A}$ is a doubly periodic $p \times (p-1)$ distinct difference configuration with period $(p-1, p)$ and density $1/p$.*

The Periodic Golomb Construction

Let α and β be two primitive elements in $\mathrm{GF}(q)$, where q is a prime power. For any integers i and j, there is a dot in $\mathcal{A}(i,j)$ if and only if $\alpha^i + \beta^j = 1$.

Theorem 10. *Let $\mathcal{A}$ be the array of dots from the Periodic Golomb Construction. Then $\mathcal{A}$ is a doubly periodic $(q-1) \times (q-1)$ distinct difference configuration with period $(q-1, q-1)$ and density $(q-2)/(q-1)^2$.*

Doubly periodic constructions involve another combinatorial structure called a B_2-sequence.

Definition 5. *Let G be an abelian group, and let $D = \{a_1, a_2, \ldots, a_m\} \subseteq G$ be a sequence of m distinct elements of G. We say that D is a B_2-sequence over G if all the sums $a_{i_1} + a_{i_2}$ with $1 \leq i_1 \leq i_2 \leq m$ are distinct.*

For a survey on B_2-sequences and their generalizations the reader is referred to [13].

The Doubly Periodic Folding Construction

Let n be a positive integer and $D = \{a_1, a_2, \ldots, a_m\}$ be a B_2-sequence in $\mathbb{Z}_n$. Let ℓ and k be integers such that $\ell \cdot k \leq n$. Let $\mathcal{A}$ be the square grid. For any integers i and j, there is a dot in $\mathcal{A}(i,j)$ if and only if $a_t \equiv i \cdot \ell + j \pmod{n}$ for some t.

Theorem 11. *Let $\mathcal{A}$ be the array of the Doubly Periodic Folding Construction. Then $\mathcal{A}$ is a doubly periodic $\ell \times k$ distinct difference configuration of period $(\frac{n}{g.c.d.(n,\ell)}, n)$ and density m/n.*

The Doubly Periodic LeeDD Construction

Let r be an integer, $R = \lfloor \frac{r}{2} \rfloor$, and let $D = \{a_1, a_2, \ldots, a_\mu\}$ be a B_2-sequence over $\mathbb{Z}_n$, where $n \geq 2R^2 + 2R + 1$. Let $f(i,j) \equiv iR + j(R+1) \bmod n$. Let $\mathcal{A}$ be the square grid. For each two integers i and j, there is a dot in $\mathcal{A}(i,j)$ if and only if $f(i,j) \in D$.

Theorem 12. *The array $\mathcal{A}$ constructed in the LeeDD Construction is doubly periodic with period (n,n) and density μ/n. The dots contained in any Lee sphere of radius R form a distinct difference configuration.*

These constructions were generalized and found more applications to other problems in coding theory and combinatorial geometry in [7]. But, generally we still have the following fundamental open problem.

Problem 6: Let $\mathcal{S}$ be a regular polygon with area t^2. Is there a doubly periodic distinct difference configuration with the shape $\mathcal{S}$ and density $\frac{1}{t} + o(\frac{1}{t})$? Our conjecture is that the answer depends on the polygon.

5   k-Hop Coverage

If a $DD(m, r)$ is used in the key predistribution scheme, each node is able to communicate directly with $m(m-1)$ other nodes, since there are $\binom{m}{2}$ ways of choosing a pair of distinct dots from the configuration, and two ways to orient the corresponding difference vector.

Nodes A and B that cannot communicate directly may still be able to communicate if there is a node C that can communicate with each of them, and thus forward messages between them. This is referred to as a *two-hop path* between A and B. We define the *two-hop coverage* of a $DD(m, r)$ to be the number of squares that can be reached from a given square by means of a two-hop path (where the hops correspond to vector differences between dots of the $DD(m, r)$). For the purposes of distributing keys, we would like to use distinct difference configurations in which the two-hop coverage is as large as possible. In a similar way we define the *k-hop path* and the *k-hop coverage* (and all the definitions and questions in this section can be given similarly on $\overline{DD}(m, r)$, $DD^*(m, r)$, and $\overline{DD}^*(m, r)$). The following question seem to be very natural in this context.

Problem 7: For a given m and k, what is the maximum k-hop coverage of a $DD(m, r)$? What is the smallest r for which this maximum is attained?

The following related bounds were given in [3]. Their proofs involve B_2-sequences and their generalizations.

Theorem 13. *Let k be a fixed integer such that $k \geq 2$. Define $c = (\pi/16)2^{1/k}$. Then there exists a $DD(m, r)$ with maximum k-hop coverage such that $m \sim cr^{1/k}$.*

Theorem 14. *Let k be a fixed integer such that $k \geq 2$. Define $c' = (\pi/16)2^{1/k} \left(\frac{2}{3}\right)^{1/2k}$. Then there exists a $DD^*(m, r)$ with maximum k-hop coverage such that $m \sim c'r^{1/k}$.*

For any fixed values of m and k, we define $r(m, k)$ to be the smallest value of r such that there exists a $DD(m, r)$ with maximum k-hop coverage. It is an important problem to determine $r(m, k)$. The construction in Theorem 13 provides an upper bound on $r(m, k)$, showing that when k is fixed and $m \to \infty$ we have $r(m, k) = O(m^k)$.

For the hexagonal grid, we denote the smallest r for which there exists a $DD^*(m, r)$ with maximum k-hop coverage by $r^*(m, k)$.

Theorem 15. *Let k be an integer such that $k \geq 2$. Then $\frac{m^k}{\sqrt{\pi k! \cdot k}} + o(m^k) \leq r(k, m) \leq \frac{1}{2} \left(\frac{16}{\pi} \right)^k m^k + o(m^k)$.*

Theorem 16. *If $k \geq 2$ then $\sqrt{\frac{3}{2}} \frac{m^k}{\sqrt{\pi k! \cdot k}} + o(m^k) \leq r^*(k, m) \leq \sqrt{\frac{3}{2}} \frac{1}{2} \left(\frac{16}{\pi} \right)^k m^k + o(m^k)$.*

These results lead to the next problem.

Problem 8: Improve the lower and upper bounds on $r(k, m)$ and $r^*(k, m)$.

For a given k and r we say that that a DD(m, r) has a complete k-hop coverage in shape S, with a center, if the sensor in the center of S can communicate through a k-hop path with all the sensors in S. The fundamental question in this context is given in the next problem.

Problem 9: For a given k and r, what is the largest radius R for which a node A can communicate with all nodes within radius R via a k-hop path by using a DD(m, r)?

Let p be a prime such that $p \geq 5$. A construction of a $(p + 1) \times (p + 2)$ distinct difference configuration with $p + 2$ dot, which achieves complete 2-hop coverage within a $(2p-3) \times (2p-1)$ rectangle, for a sensor at its center, was given in [3]. We will describe the construction to which we would like to find other constructions which achieve similar goals. The construction can be thought of as being based on the periodicity properties of a B_2 sequence in $\mathbb{Z}_{(p^2-p)}$ proposed by Ruzsa in [19], or as a consequence of a periodic generalization of the Welch construction of a Costas array [10]. Before describing the DD(m, r), we discuss some properties of a related doubly periodic array. We repeat the definition of the periodic Welch Construction for completeness.

Definition 6 (Welch Periodic Array). *Let α be a primitive root modulo a prime p. We define the* Welch periodic array *to be the set*

$$\mathcal{R}_p = \{(i, j) \in \mathbb{Z}^2 | \alpha^j \equiv i \bmod p\}.$$

This array is doubly periodic in the sense that if $\mathcal{R}_p$ contains a dot at position (i, j) then it also contains dots at all positions of the form $(i + \lambda p, j + \mu(p - 1))$ where $\lambda, \mu \in \mathbb{Z}$. It has a distinct difference property "up to periodicity": see the lemma below. We say that dots A and A' at positions (i, j) and (i', j') are *equivalent*, and we write $A \equiv A'$, if $i' = i + \lambda p$ and $j' = j + \mu(p - 1)$ for some $\lambda, \mu \in \mathbb{Z}$.

Lemma 1. *Let d and e be integers such that $d \not\equiv 0 \bmod p$ and $e \not\equiv 0 \bmod (p - 1)$. Suppose that $\mathcal{R}_p$ contains dots A and B at positions (i_1, j_1) and $(i_1 + d, j_1 + e)$ respectively, and dots A' and B' at positions (i_2, j_2) and $(i_2 + d, j_2 + e)$ respectively. Then $A \equiv A'$ and $B \equiv B'$.*

Proof: By the definition of $\mathcal{R}_p$ we have

$$i_1 \equiv \alpha^{j_1} \bmod p$$
$$i_2 \equiv \alpha^{j_2} \bmod p$$
$$i_1 + d \equiv \alpha^{j_1+e} \bmod p$$
$$i_2 + d \equiv \alpha^{j_2+e} \bmod p.$$

Eliminating i_1, i_2 and d from these equations we get

$$(\alpha^e - 1)(\alpha^{j_1} - \alpha^{j_2}) \equiv 0 \pmod{p}.$$

Since $e \not\equiv 0 \bmod (p-1)$, this implies that $j_1 \equiv j_2 \bmod (p-1)$. The first two equations above then imply that $i_1 \equiv i_2 \bmod p$.

We note that in addition, if $\mathcal{R}_p$ contains dots at (i,j) and $(i+d,j)$ then $d \equiv 0$ $(\bmod\ p)$ and if it contains dots at (i,j) and $(i,j+e)$ then $e \equiv 0$ $(\bmod\ p-1)$. Thus we see that a vector (d,e) can occur at most once as a difference between two of the dots of $\mathcal{R}_p$ that lie within any particular $(p-1) \times p$ rectangle.

We now define a DD(m) by choosing a finite subset of the dots in $\mathcal{R}_p$, as follows.

Construction 1. *Let p be an odd prime. Let $(i,j) \in \mathbb{Z}^2$ be such that $\mathcal{R}_p$ has dots at (i,j) and $(i+1,j+1)$. Note that such a position (i,j) exists. To see this, let i and j be integers such that*

$$\alpha^j \equiv i \equiv \frac{1}{\alpha - 1} \bmod p.$$

The right-hand side of this equality is well-defined and non-zero modulo p, and so there is a suitable choice for i and j. Clearly $\mathcal{R}_p$ has a dot at the position (i,j). But there is also a dot at $(i+1,j+1)$ since

$$\alpha^{j+1} \equiv \frac{\alpha}{\alpha - 1} \equiv \frac{1}{\alpha - 1} + 1 \equiv i + 1 \bmod p.$$

Consider the $(p-1) \times p$ rectangle S bounded by the positions (i,j), $(i+p-1,j)$, $(i,j+p-2)$ and $(i+p-1,j+p-2)$. By construction, $\mathcal{R}_p$ has $p-1$ dots in S. Due to its periodic nature, $\mathcal{R}_p$ also has dots at positions $(i,j+(p-1))$, $(i+p,j)$ and $(i+p+1,j+p)$. We construct a configuration $\mathcal{B}$ by adding these three dots to the set of dots in $\mathcal{R}_p \cap S$.

Our configuration $\mathcal{B}$ is shown in Fig. 1. The configuration is contained in a $(p+1) \times (p+2)$ rectangle. The *border region* of width 2 contains exactly 5 dots: A, A', A'', B and B'. The *central region* is a $(p-3) \times (p-2)$ rectangle. This region contains $p-3$ dots: one column is empty, but every other column and every row contains exactly one dot. Note that $A \equiv A' \equiv A''$ and $B \equiv B'$, but there are no other equivalent pairs of dots in $\mathcal{B}$.

Developing the properties we described here with some other interesting features of the Welch Construction the following theorem was proved in [3].

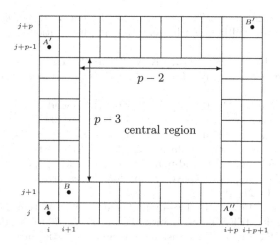

Fig. 1. The configuration $\mathcal{B}$. The five dots shown are the dots that lie the border of width 2 of the $(p+1) \times (p+2)$ rectangle containing the configuration.

Theorem 17. *Let p be a prime, $p \geq 5$. The distinct difference configuration $\mathcal{B}$ achieves complete two-hop coverage on a $(2p-3) \times (2p-1)$ rectangle relative to the central point of the rectangle.*

Let $\rho(r, k)$ be the largest radius of a circle for which our scheme achieves complete k-hop connectivity by using a $DD(m, r)$. Let $\overline{\rho}(r, k)$ be the largest radius of a Lee sphere for which our scheme achieves complete k-hop coverage by using a $\overline{DD}(m, r)$. Let $\sigma(r, k)$ be the largest width of a square for which our scheme achieves complete k-hop coverage by using a $\overline{DD}(m, r)$. Let $\rho^*(m, r, k)$ be the largest radius of an hexagonal sphere for which our scheme achieves complete k-hop coverage by using a $DD^*(m, r)$. Let $\overline{\rho}^*(m, r, k)$ be the largest radius of an hexagonal sphere for which our scheme achieves complete k-hop coverage by using a $\overline{DD}^*(m, r)$. Similarly, we can define more related function. From Theorem 17 we can infer some bounds on these functions. Other basic bounds can be attained from the discussion in [2, 3]. But, the problem has lot of potential for further research.

Problem 10: Find lower and upper bounds on $\rho(r, k)$, $\overline{\rho}(r, k)$, $\sigma(r, k)$, $\rho^*(m, r, k)$, $\overline{\rho}^*(m, r, k)$, and other related functions.

References

[1] Blackburn, S.R., Etzion, T., Martin, K.M., Paterson, M.B.: Efficient key predistribution for grid-based wireless sensor networks. In: Safavi-Naini, R. (ed.) ICITS 2008. LNCS, vol. 5155, pp. 54–69. Springer, Heidelberg (2008)
[2] Blackburn, S.R., Etzion, T., Martin, K.M., Paterson, M.B.: Two-Dimensional Patterns with Distinct Differences – Constructions, Bounds, and Maximal Anticodes (preprint)

[3] Blackburn, S.R., Etzion, T., Martin, K.M., Paterson, M.B.: Distinct Difference Configurations: Multihop Paths and Key Predistribution in Sensor Networks (preprint)

[4] Drakakis, K., Rickard, S., Beard, J.K., Caballero, R., Iorio, F., O'Brien, G., Walsh, J.: Results on Costas arrays of order 27. IEEE Trans. Inform. Theory IT-44, 730–743 (1998)

[5] Erdős, P., Graham, R., Ruzsa, I.Z., Taylor, H.: Bounds for arrays of dots with distinct slopes or lengths. Combinatorica 12, 39–44 (1992)

[6] Erdős, P., Turán, P.: On a problem of Sidon in additive number theory and some related problems. J. London Math. Soc. 16, 212–215 (1941)

[7] Etzion, T.: Folding, tiling, and multidimensional coding (preprint)

[8] Golomb, S.W.: Algebraic constructions for Costas arrays. J. Combin. Theory, Series A 37, 13–21 (1984)

[9] Golomb, S.W., Taylor, H.: Two-dimensional synchronization patterns for minimum ambiguity. IEEE Trans. Inform. Theory IT-28, 600–604 (1982)

[10] Golomb, S.W., Taylor, H.: Constructions and properties of Costas arrays. Proceedings of the IEEE 72, 1143–1163 (1984)

[11] Hamkins, J., Zeger, K.: Improved bounds on maximum size binary radar arrays. IEEE Trans. Inform. Theory IT-43, 997–1000 (1997)

[12] Lefmann, H., Thiele, T.: Point sets with distinct distances. Combinatorica 15, 379–408 (1995)

[13] O'Bryant, K.: A complete annotated bibliography of work related to Sidon sequences. The Electronic Journal of Combinatorics DS11, 1–39 (2004)

[14] Peile, R.E., Taylor, H.: Sets of points with pairwise distinct slopes. Computers and Mathematics 39, 109–115 (2000)

[15] Robinson, J.P.: Golomb rectangles. IEEE Trans. Inform. Theory IT-31, 781–787 (1985)

[16] Robinson, J.P.: Golomb rectangles as folded ruler. IEEE Trans. Inform. Theory IT-43, 290–293 (1997)

[17] Robinson, J.P.: Genetic search for Golomb arrays. IEEE Trans. Inform. Theory IT-46, 1170–1173 (2000)

[18] Robinson, J.P., Bernstein, A.J.: A class of binary recurrent codes with limited error propagation. IEEE Trans. Inform. Theory IT-13, 106–113 (1967)

[19] Ruzsa, I.Z.: Solving a linear equation in a set of integers. Acta Arith. 65, 259–282 (1993)

A New Client-to-Client Password-Authenticated Key Agreement Protocol

Deng-Guo Feng and Jing Xu

State Key Laboratory of Information Security,
Institute of Software, Chinese Academy of Sciences, Beijing, P.R.China
feng@is.iscas.ac.cn

Abstract. Client-to-client password-authenticated key agreement (C2C-PAKA) protocol deals with the authenticated key agreement process between two clients of different realms, who only share their passwords with their own servers. Recently, Byun et al. [13] proposed an efficient C2C-PAKA protocol and carried a claimed proof of security in a formal model of communication and adversarial capabilities. In this paper, we show that the protocol is insecure against *password-compromise impersonation* attack and the claim of provable security is seriously incorrect. To draw lessons from these results, we revealed fatal flaws in Byun et al.'s security model and their proof of security. Then, we modify formal security model and corresponding security definitions. In addition, a new *cross-realm* C2C-PAKA protocol is presented with security proof.

Keywords: Password-authenticated key agreement, cross-realm, cryptanalysis, provable security.

1 Introduction

Password-authenticated key agreement (PAKA) protocols are favorable to securely identifying remote users and communicating with each other in a network because of its easy-to-memorize property. Most password-authenticated key agreement protocols [1,2,3,4] in the literature are based on the client-server model which assumes a client has a secret password and a server has a corresponding password verifier in its database and provides password-authenticated key agreement between a client and a server.

However, with a variety of communication environments such as mobile network, it is considered as one of main concerns to establish a secure channel between clients registered in different servers. In fact, each organization would have his own trusted server to manage members' passwords and provide service for them. If a client shares his password with a server, we say that the client is in the realm of the server. So each server stores the passwords of all clients belonging to his realm. Such protocols are more popularly known as *cross-realm* C2C-PAKA protocols.

C. Xing et al. (Eds.): IWCC 2009, LNCS 5557, pp. 63–76, 2009.

1.1 Security Attributes

It is desirable for normal PAKA protocols to possess the following security attributes:

- **Forward secrecy:** If passwords of the entities are compromised, the secrecy of previously established session keys should not be affected.
- **Password-compromise impersonation resilience:** Compromising passwords of any client *Alice* should not enable outside adversary to share session key with *Alice* by masquerading as any other clients.
- **Unknown key share resilience:** Client *Alice* should not be coerced into sharing a key with any client *Carol* when in fact she thinks that she is sharing the key with client *Bob*.
- **Key control:** Any entities should not be able to force the session key to a preselected value.
- **Dictionary attack resilience:** All passwords in the protocol must be strongly protected against *off-line* dictionary attack and *undetectable on-line* dictionary attack, and even if an attacker is given one password, other passwords must be prevented from such attacks.

In addition to above basic properties, more properties [5,6,7,8,9,10,11,12] should be considered under the setting of *cross-realm* C2C-PAKA. More precisely, the descriptions of some properties should be modified according to the new framework in C2C-PAKA:

- **Forward secrecy:** If long-term private keys of the entities are compromised, the secrecy of previously established session keys should not be affected.
- **Password-compromise (Key-compromise) impersonation resilience:** Compromising passwords of any client *Alice* (or long-term private keys of any server S_A) in a realm should not enable outside adversary to share session key with *Alice* by masquerading as any other clients belongs to another realm.
- **Dictionary attack resilience:** Any entity (clients or/and servers) in one realm should not be able to mount a dictionary attack to other entities belongs to another realm.

1.2 Related Works and Our Contribution

The first *cross-realm* C2C-PAKA protocol is proposed by Byun et al.[7]. Subsequently, Chen [8] first pointed out that one malicious server can mount a *dictionary* attack to obtain the password of client who belongs to the other realm. Wang et al.[9] also showed three *dictionary* attacks on the same protocol, and Kim et al.[10] pointed out that the protocol was susceptible to *Dening-Sacco* attack. Again, Kim et al. proposed an improved C2C-PAKA protocol. However, Phan et al.[11] suggested *unknown key share* attacks on the improved protocol [10]. Yoon et al.[12] also pointed out that Kim et al.'s protocol [10] was susceptible to *one-way man-in-the-middle* attack and *password-compromise impersonation* attack, and presented an enhancement. Recently, Byun et al.[13] introduced a formal security model and corresponding security definitions, and proposed the first provably secure C2C-PAKA protocol.

Our contribution in this paper is twofold. Firstly, we demonstrate that Byun et al.'s protocol [13] is still insecure against the *password-compromise imper- sonation* attack, where the compromise of a client *Alice*'s password pwa will enable outside adversary $\mathcal{A}$ to share session key with *Alice* by masquerading as any other clients belonging to another realm. We also discuss why the provable security proofs failed to capture flaws in the protocol that allowed our attack to work. Secondly, we incorporate a reasonable notion of *password-compromise impersonation* resilience into security model, and then propose a new *cross-realm* C2C-PAKA protocol with security proof.

1.3   Organization

The remainder of this paper is organized as follows. The next section analyzes the vulnerabilities of Byun et al.'s C2C-PAKA protocol. In section 3, we introduce the formal model and security assumptions for C2C-PAKA protocol. Our new *cross-realm* C2C-PAKA protocol is proposed in section 4, along with its security and efficiency analysis. Finally, we conclude in section 5.

2   Security Flaws of Byun et al.'s Protocol

This section shows that Byun et al.'s protocol [13] is totally insecure and also discusses why their provable security proofs failed to capture flaws in the protocol that allowed our attack to work.

For convenience, the notations and definitions used in this paper are shown in Table 1.

Table 1. Notations and definitions

Notation	Definition
Alice, *Bob*	Clients
ID_A, ID_B	The identity of client Alice and client Bob respectively
S_A, S_B	Trusted authentication server of Alice and Bob respectively
pwa, pwb	Passwords of Alice and Bob
Pri_{S_A}, Pub_{S_A}	Private and public key of the trusted authentication server S_A
Pri_{S_B}, Pub_{S_B}	Private and public key of the trusted authentication server S_B
q	Sufficiently large prime
g	A generator of Z_q^*
K	Symmetric key shared between S_A and S_B
$MAC_k(\cdot)$	A message authentication code using key k
$[\cdot]_K$	Symmetric-key encryption with key K
$\{\cdot\}_{Pub}$	Asymmetric-key encryption with public key Pub
$\langle\cdot\rangle_{Pri}$	Digital signature with private key Pri
$\mathcal{H}(\cdot)$	a one-way hash function
$\|$	Message concatenation
$x \in_R Z_q^*$	Randomly choosing an element x of Z_q^*

2.1   Brief Review of Byun et al.'s Protocol

There are four entities involved in the protocol: $Clients = \{Alice; Bob\}$, and $Server = \{S_A; S_B\}$, where S_A and $Alice$ are in realm A and S_B and Bob are in realm B. S_A helps $Alice$ generate a common session key with Bob in realm B by using a shared key K with S_B. A high-level depiction of the protocol is given in Fig.1, and a more detailed description follows:

(1) Client $Alice$ wishing to initiate a secret communication session by generating a secret session key sk with Bob in a different realm, chooses $x \in Z_q^*$ randomly and computes $E_x = [g^x]_{pwa}$. Then, $Alice$ sends $\langle E_x, ID_A, ID_B \rangle$ to S_A.

(2) S_A obtains g^x by decrypting E_x, chooses $y \in Z_p^*$ randomly and computes $E_y = [g^y]_{pwa}$ and $R = \mathcal{H}(g^{xy})$. It also randomly chooses $k \in Z_q^*$ and computes $E_R = [k, ID_A, ID_B]_R$. It then computes $Ticket_B = [k, ID_A, ID_B, L]_K$, where L is $Ticket_B$'s lifetime. Finally, S_A sends $\langle E_y, E_R, Ticket_B, L \rangle$ to $Alice$.

(3) Upon receiving the message from S_A, $Alice$ computes the ephemeral R and decrypts E_R to obtain k, ID_A, ID_B. It also checks that ID_A and ID_B are valid. Then, $Alice$ sends an additional key confirmation message $[g^y]_R$ to S_A. S_A checks whether g^y is correct or not by correctly decrypting $[g^y]_R$ with the key R.

(4) $Alice$ randomly chooses $a \in Z_q^*$ and computes $E_a = g^a \| MAC_k(g^a)$, and forwards $\langle ID_A, E_a, Ticket_B \rangle$ to Bob.

(5) Bob chooses $x' \in_R Z_q^*$ randomly, and computes $E_{x'} = [g^{x'}]_{pwb}$. Then, he sends $\langle E_{x'}, Ticket_B \rangle$ to S_B.

(6) S_B decrypts $Ticket_B$ using K to obtain k, L, ID_A. It verifies that the lifetime L and ID_A are valid. S_B then randomly chooses $y' \in Z_q^*$ and computes $E_{y'} = [g^{y'}]_{pwb}$ and $E_{R'} = [k, ID_A, ID_B]_{R'}$, where $R' = \mathcal{H}(g^{x'y'})$. It then sends $\langle E_{y'}, E_{R'} \rangle$ to Bob.

(7) Bob decrypts $E_{y'}$ and computes R'. For key confirmation, Bob sends an additional message $[g^{y'}]_{R'}$ to KDC_B. S_B decrypts $[g^{y'}]_{R'}$ and checks whether $g^{y'}$ is correct or not. Then, Bob decrypts $E_{R'}$ to obtain k. Using this, Bob checks the integrity of g^a by verifying the previously received E_a. It then randomly chooses $b \in Z_q^*$, computes $sk = \mathcal{H}(ID_A \| ID_B \| g^a \| g^b \| g^{ab})$ and $E_b = g^b \| MAC_k(g^b)$, and sends E_b to $Alice$.

(8) On receiving E_b, $Alice$ checks the integrity of g^b and also computes sk.

2.2   Password-Compromise Impersonation Attack

Password-compromise impersonation attack represents a serious and subtle threat. Clearly, the discovery by an adversary $\mathcal{A}$ of password of client $Alice$ allows $\mathcal{A}$ to impersonate $Alice$ and establish its own sessions in the name of $Alice$. However, it may be desirable that this loss does not enable $\mathcal{A}$ to impersonate other, uncorrupted clients to $Alice$. For example, $\mathcal{A}$ could impersonate a banking system and cause $Alice$ to accept a predetermined session key and

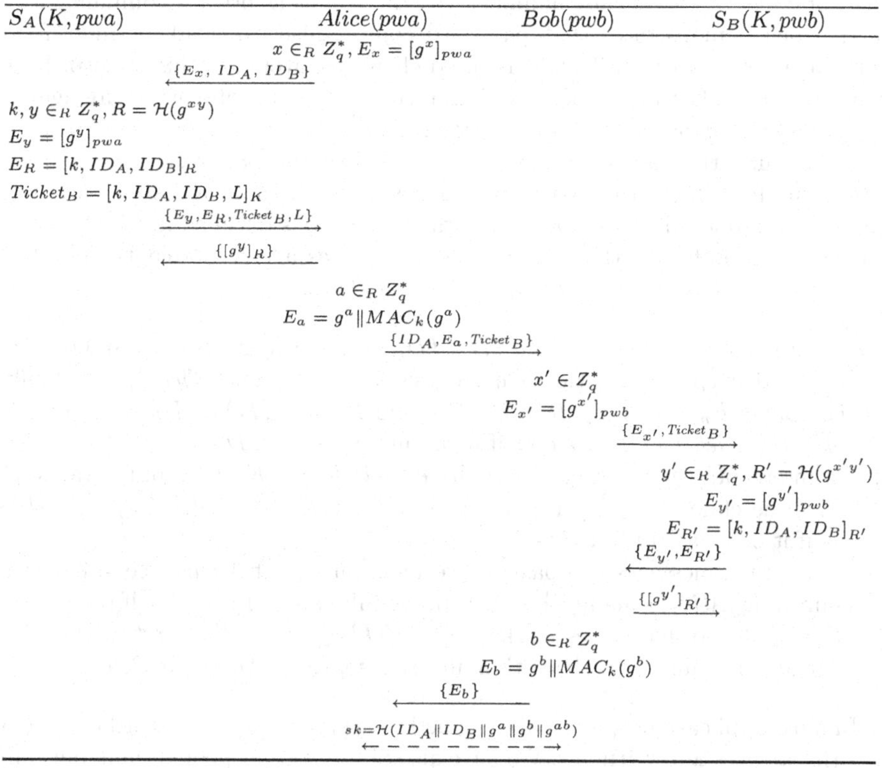

$S_A(K, pwa)$
$Alice(pwa)$
$Bob(pwb)$
$S_B(K, pwb)$
$x \in_R Z_q^*, E_x = [g^x]_{pwa}$
$\{E_x, ID_A, ID_B\}$
$k, y \in_R Z_q^*, R = \mathcal{H}(g^{xy})$
$E_y = [g^y]_{pwa}$
$E_R = [k, ID_A, ID_B]_R$
$Ticket_B = [k, ID_A, ID_B, L]_K$
$\{E_y, E_R, Ticket_B, L\}$
$\{[g^y]_R\}$
$a \in_R Z_q^*$
$E_a = g^a \| MAC_k(g^a)$
$\{ID_A, E_a, Ticket_B\}$
$x' \in Z_q^*$
$E_{x'} = [g^{x'}]_{pwb}$
$\{E_{x'}, Ticket_B\}$
$y' \in_R Z_q^*, R' = \mathcal{H}(g^{x'y'})$
$E_{y'} = [g^{y'}]_{pwb}$
$E_{R'} = [k, ID_A, ID_B]_{R'}$
$\{E_{y'}, E_{R'}\}$
$\{[g^{y'}]_{R'}\}$
$b \in_R Z_q^*$
$E_b = g^b \| MAC_k(g^b)$
$\{E_b\}$
$sk = \mathcal{H}(ID_A \| ID_B \| g^a \| g^b \| g^{ab})$

Fig. 1. Byun et al.'s protocol

$S_A(K, pwa)$
$Alice(pwa)$
$\mathcal{A}(pwa)$
$Bob(pwb)$
$\{E_x = [g^x]_{pwa}, ID_A, ID_B\}$
$k, y \in_R Z_q^*, Ticket_B$
$E_y = [g^y]_{pwa}$
$E_R = [k, ID_A, ID_B]_R$
$\{E_y, E_R, Ticket_B\}$
$\{[g^x]_R\}$
$\{ID_A, E_a = g^a \| MAC_k(g^a), Ticket_B\}$
$\{E_b = g^b \| MAC_k(g^b)\}$
$sk = \mathcal{H}(ID_A \| ID_B \| g^a \| g^b \| g^{ab})$

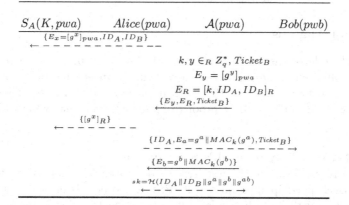

Fig. 2. An attack on Byun et al.'s protocol

then obtain her credit card number over the resulting private communication link, which demonstrates that password compromise can lead to undesirable consequences at least until the corrupted client discovers that his password was compromised. Therefore, it is very significant to design protocols secure against the *password-compromise impersonation* attack.

We assume that an adversary $\mathcal{A}$ compromises the password pwa of client *Alice*. In the attack, the goal of adversary $\mathcal{A}$ is to share a session key with *Alice* by masquerading as *Bob*. The attack scenario is outlined in Fig. 2, where a dashed line indicates that the corresponding message is intercepted by $\mathcal{A}$. A more detailed description of the attack is as follows:

(1) $\mathcal{A}$ intercepts the message $\langle E_x, ID_A, ID_B \rangle$ from *Alice* to KDC_A and obtains g^x by decrypting E_x. He then chooses $k, y \in Z_q^*$ and $Ticket_B$ randomly, computes $E_y = [g^y]_{pwa}$, $R = \mathcal{H}(g^{xy})$, and $E_R = [k, ID_A, ID_B]_R$, and sends $\langle E_y, E_R, Ticket_B \rangle$ to *Alice* as if it originated from KDC_A.

(2) $\mathcal{A}$ intercepts the messages from *Alice* to *Bob* and KDC_A, and obtains g^a. Then he chooses $b \in Z_q^*$ randomly, and sends $E_b = g^b \| MAC_k(g^b)$ to *Alice* as if it originated from *Bob*.

(3) The forged messages will pass the verification test of *Alice* since $\mathcal{A}$ have the same k as *Alice*. Finally, $\mathcal{A}$ would successfully authenticate itself to *Alice* as *Bob*, and also share a session key $sk = \mathcal{H}(ID_A \| ID_B \| g^a \| g^b \| g^{ab})$ with *Alice*, but all the while with *Alice* thinking it is sharing a key with *Bob*.

In most applications, people are inclined to keep the same password in spite of security advises and warnings. We have also observed that people think out only two or three passwords and use them in turn, when they are forced to change their passwords frequently. Thus it is more practical to assume passwords could be compromised by an adversary and *password-compromise impersonation* attack is not only of theoretical importance.

2.3   Proof Analysis

It is worthwhile to discuss why the provable security proof failed to capture *password-compromise impersonation* attack in Byun et al.'s protocol. In this subsection, we first point out the flaws in their security model. Then we investigate Byun et al.'s security proof to figure out what has gone wrong with the proof.

2.3.1   Review of Byun et al.'s Security Model

We briefly recall Byun et al.'s security model based on prior work by Bellare et al.[1]. Each participant U in a C2C-PAKA protocol is either a client C or the trusted server S. Each participant is assumed to be able to execute the protocol multiple times with different partners; this is modeled by allowing each participant to have an unlimited number of *instances* [1] with which to execute the protocol. We denote instance i of participant U as $\prod_U^i$. An adversary, $\mathcal{A}$, is allowed to fully control the communication network by injecting, modifying, blocking, and deleting any messages at will. The capabilities of the adversary are modeled via various oracles to which the adversary is allowed to make queries.

- $Send(\prod_U^i, m)$: This query models an active attack. It outputs the message that client instance $\prod_U^i$ would generate upon receipt of message m.
- $Execute(\prod_U^i, \prod_{U'}^j)$: This oracle query is used to simulate eavesdropping attack of the adversary. The output of this query consists of the messages that were exchanged during the honest execution of the protocol.
- $Reveal(\prod_U^i)$: This query returns to the adversary the session key of client instance $\prod_U^i$, if the latter is defined.
- $Corrupt(U)$: This query models exposure of the password held by the client U or the server U.
- $Test(\prod_U^i)$: This query provides a means of defining security. After querying the oracle, the session key of $\prod_U^i$ or a random number will be returned according to a predefined random bit b. If $b = 1$, the adversary would learn the session key of $\prod_U^i$; otherwise the adversary only learns a random number with the same length. This query can be called only once.

The definition of security is based on the notion of freshness. Intuitively, a fresh instance is an instance which holds a session key about which the adversary should not know. More precisely: An oracle $\prod_U^i$ is *fresh* if neither $\prod_U^i$ nor one of its partners has been asked for a *Reveal* query after oracle $\prod_U^i$ and its partners have computed a session key sk, where two oracles are partners if and only if both oracles accept with a same session key and a same session identifier.

The security of a protocol π against an adversary $\mathcal{A}$ is defined in terms of the probability that $\mathcal{A}$ succeeds in distinguishing a real session key established in an execution of π from a random session key. More precisely, the security is defined in the following context. The adversary $\mathcal{A}$ executes the protocol exploiting as much parallelism as possible and asking any queries allowed in the model. During executions of the protocol, the adversary $\mathcal{A}$, at any time, asks a *Test* query to a fresh instance, gets back a key as the response to this query, and at some later point in time, outputs a bit b' as a guess for the value of the hidden bit b used by the *Test* oracle. Then the advantage of $\mathcal{A}$ in attacking protocol π is denoted by $Adv_\pi^{sk}(\mathcal{A})$ and is defined as

$$Adv_\pi^{sk}(\mathcal{A}) = 2Pr[b = b'] - 1.$$

Protocol π is said to be secure if

$$Adv_\pi^{sk}(\mathcal{A}) \leq \frac{q_S}{|\mathcal{D}|} + \varepsilon(k),$$

for all probabilistic polynomial time adversaries $\mathcal{A}$, where $\varepsilon(k)$ is a negligible function, $|\mathcal{D}|$ is the size of a dictionary, q_S is the number of *Send* queries asked, and k is a security parameter.

2.3.2 Flaw in Byun et al.'s Security Model

We observe the surprising fact that an adversary $\mathcal{A}$ could achieve its goal of breaking the security of protocol π in existing model, even if π is secure in practice. More precisely, $\mathcal{A}$ begins by asking $Corrupt(Alice)$ to obtain the password pwa of client $Alice$. Then $\mathcal{A}$ asks $Send(\prod_B^j, m)$ which prompts instance $\prod_A^i$ to send the message to Bob. It may be easily verified that both $\prod_A^i$ and $\prod_B^j$ are *fresh* because no $Reveal$ query has been made against any instance. Thus, $\mathcal{A}$ may ask the $Test$ query against either of the two instances. Since $\mathcal{A}$ is able to compute the session key sk shared by $Alice$ and Bob, it follows that $Pr[b = b'] = 1$ and hence $Adv_\pi^{sk}(\mathcal{A}) = 1$. Therefore, no protocol is secure in Byun et al.'s security model.

2.3.3 Flaw in Existing Proof

Recall that our attack only makes use of the *Corrupt* query defined in the Byun et al.'s security model. However, in Byun et al.'s proof, *Corrupt* query was not adequately considered in the simulation. This essentially limits the adversary's abilities as he can not reveal the password of any client. Therefore, although the goal that our attack achieves is indeed considered by them, the proof was too specific to catch our attack.

3 Formal Model and Security Notions

3.1 Modification to Security Model

Having identified the flaw in Byun et al.'s security model, we provide a modification to the security model. The original definition of freshness allows an adversary to corrupt the participant running the target session and the participant running the partner session, which is main flaws of security model. Another potential weakness of previous definitions is no formalization of *password-compromise impersonation* resilience. In an attempt to bring these attacks within the scope of analysis we extend the definition of freshness, which shows that as long as an adversary $\mathcal{A}$ is not actively controlling or observing the secret choices made for the generation of the session then, even the knowledge of a client C's password does not allow $\mathcal{A}$ to compromise the session key. In particular, $\mathcal{A}$ cannot impersonate an uncorrupted entity to the client C in a way that allow $\mathcal{A}$ to learn (any information about) the resultant session key.

Definition 3.1 (Freshness). We say an oracle instance Π_U^i is *fresh* if the following conditions hold: (1) It has accepted and generated a valid session key; (2) No $Reveal$ queries have been made to Π_U^i or its partner; (3) U's partner (and the server of U' partner) is not corrupted (not being issued the $Corrupt$ query); and (4) If U (or the server of U) is corrupted, then $Send(\Pi_{\hat{U}}^j, m)$ query and $Send(\Pi_{\hat{S}}^t, m)$ should not be made, where $\hat{U}$ is U's partner, $\hat{S}$ is the server of $\hat{U}$, and m is a message chosen by an adversary.

Another important change we introduce concerns the security definition. A secure *cross-realm* C2C-PAKA protocol should satisfy security requirements: (1)

the session key cannot be distinguished from a random number by an outside malicious adversary; (2) Clients' passwords are not revealed to other clients or servers except for their own servers.

Semantic Security Against Malicious Outside Adversary. For any adversary $\mathcal{A}$, let $Succ(\mathcal{A})$ be the event that $\mathcal{A}$ makes a single *Test* query directed to some fresh instance U^i that has terminated, and eventually outputs a bit b', where $b' = b$ for the bit b that was selected in the *Test* query. Let $\mathcal{D}$ be user's password dictionary. The advantage of $\mathcal{A}$ in violating the semantic security is defined to be

$$Adv_{\mathcal{D}}(\mathcal{A}) \overset{def}{=} 2Pr[Succ(\mathcal{A})] - 1.$$

We say C2C-PAKA protocol is semantically secure against malicious adversary if the advantage $Adv_{\mathcal{D}}(\mathcal{A})$ is only negligibly larger than $O(q_s)/|\mathcal{D}|$, where q_s is the number of active sessions, and $|\mathcal{D}|$ is the size of password dictionary.

Password Protection Against Malicious Client (Malicious Server). For any malicious client $\mathcal{C}$(server $\mathcal{S}$), let $Succ^{pw-mc}(\mathcal{C})(Succ^{pw-ms}(\mathcal{S}))$ be the event that $\mathcal{C}(\mathcal{S})$ successfully learns the password of the client in the other realm. The advantage of $\mathcal{C}$ and $\mathcal{S}$ are defined to be

$$Adv_{\mathcal{D}}^{pw-mc}(\mathcal{C}) \overset{def}{=} Pr[Succ^{pw-mc}(\mathcal{C})] \ \ and \ \ Adv_{\mathcal{D}}^{pw-ms}(\mathcal{S}) \overset{def}{=} Pr[Succ^{pw-ms}(\mathcal{S})],$$

respectively. We say C2C-PAKA protocol satisfies password protection against malicious client (malicious server) if the advantage $Adv_{\mathcal{D}}^{pw-mc}(\mathcal{C})(Adv_{\mathcal{D}}^{pw-ms}(\mathcal{S}))$ is only negligibly larger than $O(q_s)/|\mathcal{D}|$, where q_s is the number of active sessions, and $|\mathcal{D}|$ is the size of password dictionary.

Definition 3.2 (Security). A *cross-realm* C2C-PAKA protocol is said to be secure if it satisfies semantic security against malicious outside adversary and password protection against malicious client (malicious server).

3.2 Cryptographic Assumption

We briefly introduce some assumptions required by our *cross-realm* C2C-PAKA protocol. Let G be a finite cyclic group of prime order q. Let g be a generate element of G. Denote the tuple $\mathbb{G} = (G, g, p)$ as a represented group. Following, DDH assumption is normal and has been heavily used in many works. PCDDH1 and PCDDH2 assumptions are introduced in [14].

Definition 3.3 Decisional Diffie-Hellman Assumption (DDH). Let $\mathbb{G} = (G, g, p)$ be a represented group. DDH assumption means that given (g, g^x, g^y), no probabilistic polynomial time algorithm can distinguish g^{xy} from a random element of G with non-negligible probability.

Definition 3.4 Password-Based Chosen-Basis Decisional Diffie-Hellman Assumption 1 (PCDDH1). PCDDH1 assumption means that given $(g^x \times H(pw), g^{xs}, g^y, g^{xys})$, any adversary cannot distinguish g^{ys} from a random element of G with non-negligible probability, where x, s are two private random number, pw is a password, and the random number y is chosen by the adversary.

Definition 3.5 Password-Based Chosen-Basis Decisional Diffie- Hellman Assumption 2 (PCDDH2). PCDDH2 assumption means that given $(g^x, (g^x/H(pw))^s, g^y)$, any adversary cannot distinguish g^{ys} from a random element of G with non-negligible probability, where s is a private random number, pw is a password, and the other two random numbers x, y are both chosen by the adversary.

The formal security definitions of signature scheme and encryption scheme can be found in [15].

4   The New C2C-PAKA Protocol

The main flaw in Byun et al.'s protocol is that the client *Alice* could not distinguish between interactions with other honest client *Bob* (honest server KDC_A) or with an adversary, if the password of *Alice* is compromised. This means for *Alice* there is no way to verify the identity of *Bob* (KDC_A) except for the corresponding password *pwa*. Therefore, in the *secret-key* setting, it appears infeasible to make any counter measures against *password-compromise impersonation* attack, or else *off-line dictionary* attack would be exploited. However, such attacks may be avoided in the *public-key* setting in which additional security assumptions be required. In this section, we present a new client-to-client password-authenticated key agreement protocol.

4.1   Protocol Description

The protocol proceeds in the following steps (Fig.3).

(1) Client *Alice* sends the request message $\{ID_A, ID_B\}$ to the server S_A. Then, S_A chooses $t \in Z_q^*$ randomly, encrypts g^t with the password *pwa* and sends the ciphertext $M_1 = [g^t]_{pwa}$ to *Alice*.

(2) *Alice* obtains g^t by decrypting M_1, chooses $x \in Z_q^*$ randomly, computes $M_2 = [g^x]_{pwa}$, $R = H(g^{tx})$ and $M_3 = [g^t, ID_A, ID_B]_R$. Finally, *Alice* sends $\{M_2, M_3\}$ to S_A.

(3) Upon receiving the message from *Alice*, S_A obtains g^x and verifies the validity of M_3. Then, S_A chooses $r \in Z_q^*$ randomly, computes $M_4 = \langle g^r \rangle_{Pris_A}$ and $Ticket_B = \{\langle g^r, g^{x \cdot r}, ID_A, ID_B, L \rangle_{Pris_A}\}_{Pubs_B}$, and sends $\{M_4, Ticket_B\}$ to *Alice*.

(4) Upon receiving the message from S_A, *Alice* checks the signature of M_4 is valid. Then A computes $g^{x \cdot r}$ and forwards $\{ID_A, Ticket_B\}$ to *Bob*.

(5) *Bob* chooses $y \in Z_q^*$ randomly, and computes $M_5 = [g^y]_{pwb}$. Then, it sends $\{ID_A, ID_B, M_5, Ticket_B\}$ to S_B.

(6) S_B obtains g^y by decrypting M_5. Then it decrypts $Ticket_B$, verify S_A's signature, and obtains $g^{x \cdot r}$ and g^r. S_B also chooses $r' \in Z_q^*$ randomly, and computes $g^{y \cdot r'}, g^{x \cdot r \cdot r'}, g^{r \cdot r'}, M_6 = [g^{x \cdot r}, g^{x \cdot r \cdot r'}, g^{r \cdot r'}]_{R'}$ and $M_7 = \langle g^{r'} \rangle_{Pris_B}$, where $R' = H(g^{y \cdot r'})$. Finally, S_B sends $\{M_6, M_7\}$ to *Bob*.

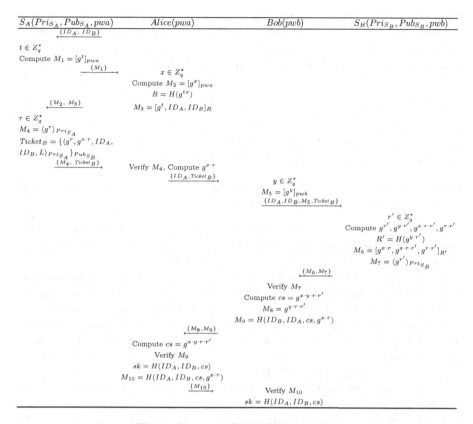

Fig. 3. Proposed C2C-PAKA protocol

(7) Upon receiving the message from S_B, Bob verifies M_7 and computes $R' = H(g^{r' \cdot y})$, and obtains $g^{x \cdot r}, g^{x \cdot r \cdot r'}$, and $g^{r \cdot r'}$ by decrypting M_6. Then B computes $cs = g^{x \cdot y \cdot r \cdot r'}, M_8 = g^{r \cdot r' \cdot y}$, and $M_9 = H(ID_B, ID_A, cs, g^{x \cdot r})$. Finally, Bob sends $\{M_8, M_9\}$ to Alice for session key confirmation.

(8) Upon receiving M_8 and M_9, Alice computes $cs = g^{x \cdot y \cdot r \cdot r'}$. Then, Alice computes $H(ID_B, ID_A, cs, g^{x \cdot r})$ and verifies it with M_9. If it holds, Alice authenticates Bob. Finally, Alice computes $M_{10} = H(ID_A, ID_B, cs, g^{x \cdot r})$, and sends it to Bob for session key confirmation.

(9) Upon receiving M_{10}, Bob computes $H(ID_A, ID_B, cs, g^{x \cdot r})$ and verifies it with M_{10}. If it holds, Bob authenticates Alice. Then, Alice and Bob's common session key is $sk = H(ID_A, ID_B, g^{x \cdot y \cdot r \cdot r'})$.

Informally, we examine the security of our proposed protocol. In the *public-key* setting, clearly an adversary $\mathcal{A}$ that knows the secret key of the server S_A can now impersonate *Alice*. However, it may be desirable that this loss does not enable $\mathcal{A}$ to impersonate other entity *Bob* to *Alice*. In our protocol, $\mathcal{A}$ could forge $Ticket_B$ with the secret key of the server S_A, however, to impersonate *Bob*, $\mathcal{A}$ needs to know $g^{x \cdot r}$. It is impossible for $\mathcal{A}$ to learn $g^{x \cdot r}$, provided that the password pwa is not compromised. Therefore, our protocol is secure against the *key-compromise impersonation* attack. Similarly, our protocol is also immune to the password-compromise impersonation attack, which is mainly due to the signature operation of M_4 and M_7, where $M_4 = \langle g^r \rangle_{Pris_A}$ and $M_7 = \langle g^{r'} \rangle_{Pris_B}$.

4.2   Security

Theorem 4.1. *The proposed C2C-PAKA protocol is secure against malicious adversary, provided that the DDH, PCDDH1, and PCDDH2 assumptions hold, the underlying signature scheme and encryption scheme are secure.*

Proof. (Sketch) First, we show that the proposed C2C-PAKA protocol is semantically secure against malicious adversary.

The security proof defines a sequence of hybrid experiments, starting with the real attack and ending in an experiment in which the adversary has no advantage. Each experiment addresses a different security aspect.

Experiments 1 through 5 show that the adversary gains no information from passive attacks. They do so by showing that keys generated in these sessions can be safely replaced by random ones as long as the DDH assumption holds.

In experiments 6 through 8, we deal with active attacks against malicious server. First, in experiment 6, we show that the output values M_2 and M_5 associated with honest users can be computed using random values and independently of each other as long as the PCDDH1 and PCDDH2 assumptions hold. Then, in the next two experiments, we show that for those cases in which we replaced M_2 and M_5 with random values, the password is no longer used and that the Diffie-Hellman key cs used to compute the session key for these users is indistinguishable from random values. We also show how to bound the probability that the adversary successfully guesses the password of an honest client in the other realm during an active attack against malicious server.

In experiment 9, we consider active attacks against malicious client. More precisely, we show that we can answer all SendClient queries with respect to honest users using random values, without using the password of these users. Moreover, at this moment, we also show how to bound the probability that the adversary successfully guesses the password of an honest client during an active attack against that client.

Next, we show that the malicious client *Bob* (server S_B) cannot learn the client *Alice*'s password pwa. Since pwa is used only for encryption of g^t and g^x, *Bob* (S_B) can not obtain any information about pwa unless fortunately guesses ephemeral values g^t or g^x. From the execution of the protocol, *Bob* (S_B) learns g^{xr}. However, since r is a private random number, it is impossible for *Bob* (S_B) to obtain g^x. Consequently, the proposed protocol has password protection against malicious client (malicious server).

Finally, we conclude that our proposed C2C-PAKA protocol is secure against malicious adversary.

4.3 Efficiency

Now we evaluate the performance of our proposed protocol. It is worth pointing out that our protocol is still efficient even in the *public-key* setting. Our protocol shifts much of the computational burden such as the public key encryption, to the application server equipped with sufficient computational power, which is suitable for low-power clients' device. For the client, some power-consuming operations could also be pre-computed during *off-line* phase.

5 Conclusion

In this paper, we incorporated security against the *password-compromise impersonation* attack directly into the security model of *cross-realm* C2C-PAKA protocol and designed a new C2C-PAKA protocol. We also provided formal security proof under our model and security notions. In addition, we showed that Byun et al.'s C2C-PAKA protocol is vulnerable to the *password-compromise impersonation* attack and conducted a detailed analysis of flaws in the protocol and its security proof.

Acknowledgement

This work was supported by the National Grand Fundamental Research (973) Program of China under Grant No. 2007CB311202 and the National Natural Science Foundation of China (NSFC) under Grants No.60673083 and No.60873197.

References

1. Bellare, M., Pointcheval, D., Rogaway, P.: Authenticated key exchange secure against dictionary attacks. In: Preneel, B. (ed.) EUROCRYPT 2000. LNCS, vol. 1807, pp. 139–155. Springer, Heidelberg (2000)
2. Boyko, V., MacKenzie, P., Patel, S.: Provably secure password-authenticated key exchange using diffie-hellman. In: Preneel, B. (ed.) EUROCRYPT 2000. LNCS, vol. 1807, pp. 156–171. Springer, Heidelberg (2000)
3. Katz, J., Ostrovsky, R., Yung, M.: Efficient password-authenticated key exchange using uuman-memorable passwords. In: Pfitzmann, B. (ed.) EUROCRYPT 2001. LNCS, vol. 2045, pp. 475–494. Springer, Heidelberg (2001)
4. Bellovin, S., Merrit, M.: Encrypted key exchange: password based protocols secure against dictionary attacks. In: Proc. of the Symposium on Security and Privacy, IEE, pp. 72–84 (1992)
5. Byun, J.W., Lee, D.H., Lim, J.: Efficient and Provably Secure Client-to-Client Password-Based Key Exchange Protocol. In: Zhou, X., Li, J., Shen, H.T., Kitsuregawa, M., Zhang, Y. (eds.) APWeb 2006. LNCS, vol. 3841, pp. 830–836. Springer, Heidelberg (2006)
6. Yin, Y., Bao, L.: Secure Cross-Realm C2C-PAKE Protocol. In: Batten, L.M., Safavi-Naini, R. (eds.) ACISP 2006. LNCS, vol. 4058, pp. 395–406. Springer, Heidelberg (2006)

7. Byun, J.W., Jeong, I.R., Lee, D.H., Park, C.: Password-authenticated key exchange between clients with different passwords. In: Deng, R.H., Qing, S., Bao, F., Zhou, J. (eds.) ICICS 2002. LNCS, vol. 2513, pp. 134–146. Springer, Heidelberg (2002)
8. Chen, L.: A weakness of the password-authenticated key agreement between clients with different passwords scheme, ISO/IEC JTC 1/SC27 N3716
9. Wang, S., Wang, J., Xu, M.: Weakness of a password-authenticated key exchange protocol between clients with different passwords. In: Jakobsson, M., Yung, M., Zhou, J. (eds.) ACNS 2004. LNCS, vol. 3089, pp. 414–425. Springer, Heidelberg (2004)
10. Kim, J., Kim, S., Kwak, J., Won, D.: Cryptoanalysis and improvements of password authenticated key exchange scheme between clients with different passwords. In: Laganá, A., Gavrilova, M.L., Kumar, V., Mun, Y., Tan, C.J.K., Gervasi, O. (eds.) ICCSA 2004. LNCS, vol. 3044, pp. 895–902. Springer, Heidelberg (2004)
11. Phan, R.C.-W., Goi, B.: Cryptanalysis of an improved client-to-client password-authenticated key exchange (C2C-PAKE) scheme. In: Ioannidis, J., Keromytis, A.D., Yung, M. (eds.) ACNS 2005. LNCS, vol. 3531, pp. 33–39. Springer, Heidelberg (2005)
12. Yoon, E.J., Yoo, K.Y.: A secure password-authenticated key exchange between clients with different passwords. In: Zhou, X., Li, J., Shen, H.T., Kitsuregawa, M., Zhang, Y. (eds.) APWeb 2006. LNCS, vol. 3841, pp. 659–663. Springer, Heidelberg (2006)
13. Byun, J.W., Lee, D.H., Lim, J.I.: EC2C-PAKA: An efficient client-to-client password-authenticated key agreement. Information Science 177, 3995–4013 (2007)
14. Abdalla, M., Fouque, P.A., Pointcheval, D.: Password-based authenticated key exchange in the three-party setting. In: Vaudenay, S. (ed.) PKC 2005. LNCS, vol. 3386, pp. 65–84. Springer, Heidelberg (2005)
15. Goldreich, O.: Foundation of cryptography, vol. 2. Cambridge University Press, Cambridge (2004)

Elliptic Twin Prime Conjecture

John B. Friedlander[1] and Igor E. Shparlinski[2]

[1] Department of Mathematics, University of Toronto
Toronto, Ontario M5S 3G3, Canada
frdlndr@math.toronto.edu
[2] Department of Computing, Macquarie University
Sydney, NSW 2109, Australia
igor@comp.mq.edu.au

Abstract. Motivated by a recent application to hash functions suggested by O. Chevassut, P.-A. Fouque, P. Gaudry and D. Pointcheval, we study the frequency with which both an elliptic curve over a finite field, and its quadratic twist are cryptographically suitable. Here, we obtain heuristic estimates for the number of such curves for which both the curve and its twist have a number of points which is prime. In a work in progress theoretical extimates are obtained wherein the number of such points on both curves has a prescribed arithmetic structure.

1 Introduction

For a prime power q with $\gcd(q, 6) = 1$ and an elliptic curve $\mathbb{E}$ given by an affine Weierstraß equation

$$\mathbb{E}: \quad y^2 = x^3 + ax + b,$$

over the field $\mathbb{F}_q$ of q elements, we denote by $\overline{\mathbb{E}}$ its quadratic twist given by the Weierstraß equation

$$\overline{\mathbb{E}}: \quad \lambda y^2 = x^3 + ax + b,$$

where $\lambda \in \mathbb{F}_q$ is a quadratic nonresidue (all non-residues lead to the same curve), see [6].

By the celebrated *Hasse bound*, for any curve $\mathbb{E}$ we have

$$\#\overline{\mathbb{E}}(\mathbb{F}_q) = q + 1 - t \qquad \text{and} \qquad \#\overline{\mathbb{E}}(\mathbb{F}_q) = q + 1 + t \tag{1}$$

with some integer $t \in [-2\sqrt{q}, 2\sqrt{q}]$, see [6]. Furthermore, by a result of Deuring [4], see also [2], for any $t \in [-2\sqrt{q}, 2\sqrt{q}]$ there is an elliptic curve $\mathbb{E}$ over $\mathbb{F}_q$ with (1). Moreover, in a narrower interval $t \in [-\sqrt{q}, \sqrt{q}]$, there is almost the expected proportion of such curves, see [5].

It has been shown in [3] that curves $\mathbb{E}$ for which the numbers $\#\overline{\mathbb{E}}(\mathbb{F}_q)$ and $\#\overline{\mathbb{E}}(\mathbb{F}_q)$ of $\mathbb{F}_q$-rational points on both curves are primes can be used for designing cryptographic hash functions.

Accordingly, we call a pair (ℓ, r) of primes *elliptic twins*, if for some prime power q with $\gcd(q, 6) = 1$ and an integer $t \in [-2\sqrt{q}, 2\sqrt{q}]$ we have

$$\ell = q + 1 - t \qquad \text{and} \qquad r = q + 1 + t. \tag{2}$$

C. Xing et al. (Eds.): IWCC 2009, LNCS 5557, pp. 77–81, 2009.

One may certainly guess that there are infinitely many elliptic twins. For example, the *Dickson prime s-tuplets conjecture*, taken with $s = 3$, implies that there are infinitely many prime triples $(q + 1 - t, q, q + 1 + t)$ for any odd t. This conjecture has received substantial computational support, and also theoretical, see for example the work of Balog [1], still, obtaining a rigorous proof seems out of reach at present. Here we present some heuristic arguments supporting the above guess and predicting the the the frequency of occurrence of such elliptic prime twins.

We adopt the convention that implied constants in symbols 'O' and '$\ll$'are absolute (we recall that $A \ll B$ is equivalent to $A = O(B)$).

2   The Conjecture

By a quantitative version of the Dickson prime s-tuplets conjecture, for every fixed t there are asymptotically $C(t)x(\log x)^{-3}$ prime powers $q \leq x$ with both $q + 1 - t$ and $q + 1 + t$ being prime, where

$$C(t) = \prod_p \frac{1 - w_t(p)/p}{(1 - 1/p)^3},$$

where $w_t(p)$ is the number of solutions to the congruence

$$(n + 1 - t)n(n + 1 + t) \equiv 0 \pmod{p}, \qquad 1 \leq n \leq p.$$

Ignoring the issue of uniformity in t one can probably conjecture that the total number $W(x)$ of pairs (q, t) with a prime power $q \leq x$ and an integer $t \in [-2\sqrt{q}, 2\sqrt{q}]$ such that $q + 1 - t$ and $q + 1 + t$ are primes is about

$$
\begin{aligned}
W(x) &\sim 2 \sum_{1 \leq t \leq 2x^{1/2}} C(t) \left(\frac{x}{(\log x)^3} - \frac{t^2}{4(\log(t^2/4))^3} \right) \\
&\sim 2 \frac{x}{(\log x)^3} \sum_{1 \leq t \leq 2x^{1/2}} C(t) - \frac{1}{2} \sum_{1 \leq t \leq 2x^{1/2}} C(t) \frac{t^2}{(\log t^2)^3}.
\end{aligned}
\tag{3}
$$

Hence, we now estimate the mean value:

$$S(T) = \sum_{1 \leq t \leq T} C(t).$$

Theorem 1. *We have*

$$S(T) = (\vartheta + o(1))T,$$

where

$$\vartheta = \frac{9}{4} \prod_{p \geq 5} \left(1 + \frac{1}{(p-1)^3} \right) = 2.300961\ldots.$$

Proof. Clearly if $t = 2m$ is even then $w_{2m}(2) = 2$ thus $C(2m) = 0$.

For an odd $t = 2m - 1$ we have

$$w_{2m-1}(p) = \begin{cases} 1 \text{ if } p = 2, \\ 2 \text{ if } p \geq 3 \text{ and } p \mid m(m-1)(2m-1), \\ 3 \text{ otherwise.} \end{cases}$$

In particular $w_{2m-1}(3) = 2$. Therefore, calculating the contribution from $p = 2$ and $p = 3$ separately, we obtain

$$C(2m - 1) = \frac{9}{2}\beta \prod_{\substack{p \geq 5 \\ p \mid m(m-1)(2m-1)}} \frac{1 - 2/p}{1 - 3/p} = \frac{9}{2}\beta \prod_{\substack{p \geq 5 \\ p \mid m(m-1)(2m-1)}} \left(1 + \frac{1}{p - 3}\right),$$

where

$$\beta = \prod_{p \geq 5} \frac{1 - 3/p}{(1 - 1/p)^3}.$$

Let $f(d)$ be the multiplicative function which is defined as follows:

$$f(d) = \prod_{\substack{p \geq 5 \\ p \mid d}} \frac{1}{p - 3}$$

if d is squarefree and $f(d) = 0$ otherwise. Then

$$\prod_{\substack{p \geq 5 \\ p \mid m(m-1)(2m-1)}} \left(1 + \frac{1}{p - 3}\right) = \sum_{\substack{d \mid m(m-1)(2m-1) \\ \gcd(d,6)=1}} f(d).$$

Therefore, for any integer $M \geq 1$,

$$\sum_{m=1}^{M} C(2m - 1) = \frac{9}{2}\beta \sum_{m=1}^{M} \prod_{\substack{p \geq 5 \\ p \mid m(m-1)(2m-1)}} \left(1 + \frac{1}{p - 3}\right)$$

$$= \frac{9}{2}\beta \sum_{m=1}^{M} \sum_{\substack{d \mid m(m-1)(2m-1) \\ \gcd(d,6)=1}} f(d)$$

$$= \frac{9}{2}\beta \sum_{\substack{d=1 \\ \gcd(d,6)=1}}^{2M^3} f(d) \sum_{\substack{m=1 \\ m(m-1)(2m-1) \equiv 0 \pmod{d}}}^{M} 1$$

$$= \frac{9}{2}\beta \sum_{\substack{d=1 \\ \gcd(d,6)=1}}^{2M^3} f(d) \left(\frac{M}{d}\rho(d) + O(\rho(d))\right),$$

where $\rho(d)$ is the number of solutions to the congruence

$$m(m-1)(2m-1) \equiv 0 \pmod{d}, \qquad 1 \leq m \leq d.$$

By the Chinese Remainder Theorem, we see that if $\gcd(d,6) = 1$ then $\rho(d) = 3^{\nu(d)}$. Therefore

$$\sum_{m=1}^{M} C(2m-1) = \frac{9}{2}\beta M \sum_{\substack{d=1 \\ \gcd(d,6)=1}}^{2M^3} \frac{f(d)3^{\nu(d)}}{d} + O\left(\sum_{d=1}^{2M^3} f(d)3^{\nu(d)}\right). \qquad (4)$$

We note that

$$\sum_{\substack{d=1 \\ \gcd(d,6)=1}}^{2M^3} \frac{f(d)3^{\nu(d)}}{d} = \sum_{\substack{d=1 \\ \gcd(d,6)=1}}^{\infty} \frac{f(d)3^{\nu(d)}}{d} + O(M^{-3}\log^2 M)$$

$$= \prod_{p\geq 5}\left(1 + \frac{3}{p(p-3)}\right) + O(M^{-3}\log^2 M).$$

Also

$$\sum_{d=1}^{2M^3} f(d)3^{\nu(d)} \ll \log^3 M.$$

Hence, using (4), we derive

$$\sum_{m=1}^{M} C(2m-1) = 2\vartheta M + O(\log^3 M),$$

where

$$\vartheta = \frac{9}{4}\beta \prod_{p\geq 5}\left(1 + \frac{3}{p(p-3)}\right).$$

Taking $M = \lfloor T/2 \rfloor$, we conclude the proof.   □

Substituting the result of Theorem 1 in (3) we are led, after partial summation, to expect that:

Conjecture 2. *Let $W(x)$ denote the total number of pairs (q,t), where $q \leq x$ is a prime power and $t \in [-2\sqrt{q}, 2\sqrt{q}]$ is an integer such that $q+1-t$ and $q+1+t$ are primes. Then*

$$W(x) \sim A\frac{x^{\frac{3}{2}}}{(\log x)^3},$$

where

$$A = \frac{8}{3}\vartheta = 6\prod_{p\geq 5}\left(1 + \frac{1}{(p-1)^3}\right) = 6.135897\ldots.$$

Certainly, it would be very desirable to have some numerical evidence confirming Conjecture 2.

Acknowledgements

We are grateful to Andreas Enge for attracting our attention to this problem and to Daniel Sutantyo for his help with the calculation of the numerical value of ϑ in Theorem 1.

During the preparation of this paper, J. F. was supported in part by NSERC Grant A5123, and I. S. by ARC Grant DP0881473.

References

1. Balog, A.: 'The prime k-tuplets conjecture on average', Analytic Number Theory. In: Progress in Mathematics, vol. 85, pp. 47–75. Birkhäuser, Basel (1990)
2. Birch, B.J.: How the number of points of an elliptic curve over a fixed prime field varies. J. Lond. Math. Soc. 43, 57–60 (1968)
3. Chevassut, O., Fouque, P.-A., Gaudry, P., Pointcheval, D.: The twist-AUgmented technique for key exchange. In: Yung, M., Dodis, Y., Kiayias, A., Malkin, T.G. (eds.) PKC 2006. LNCS, vol. 3958, pp. 410–426. Springer, Heidelberg (2006)
4. Deuring, M.: Die Typen der Multiplikatorenringe elliptischer Funktionenkörper'. Abh. Math. Sem. Hansischen Univ. 14, 197–272 (1941)
5. Lenstra, H.W.: Factoring integers with elliptic curves. Ann. Math. 126, 649–673 (1987)
6. Silverman, J.H.: The arithmetic of elliptic curves. Springer, Berlin (1995)

Hunting for Curves with Many Points

Gerard van der Geer

Korteweg-de Vries Instituut, Universiteit van Amsterdam, Plantage Muidergracht 24
1018 TV Amsterdam, The Netherlands
geer@science.uva.nl

Abstract. We construct curves with many points over finite fields by using the class group.

1 Introduction

The question how many rational points a curve of given genus over a finite field of given cardinality can have is an attractive challenge and the table of curves with many points ([10]) has been a reference point for progress for genus ≤ 50 and small fields of characteristic $p = 2$ and $p = 3$. The tables record there for a pair (q, g) an interval $[a, b]$ where b is the best upper bound for the maximum number of points of a curve of genus g over $\mathbb{F}_q$ and a gives a lower bound obtained from an explicit example of a curve C defined over $\mathbb{F}_q$ with a (or at least a) rational points. So $N_q(g) \in [a, b]$ with $N_q(g)$ the maximum number of rational points on a smooth connected projective curve defined over $\mathbb{F}_q$. For progress on the upper bounds we refer to the papers of Howe and Lauter, [1] and to the references given in the tables. It is the purpose of this little paper to record recent improvements of the table and at the same time give constructions for many of the present records of the table. The methods employ the class group and are variations on well-known themes and do not involve new ideas. We hope it inspires people to take up the challenge to improve the tables.

2 Using the Class Group

Let C be a curve of genus g defined over $\mathbb{F}_2$ with 'many' rational points. If f is a rational function not of the form $h^2 + h$ for h in the function field $\mathbb{F}_2(C)$ of C then in the Artin-Schreier cover of C defined by $w^2 + w = f$ all the rational points of C where f vanishes will be split. If the contribution of the poles of f is limited and f vanishes in many points this may yield curves over $\mathbb{F}_2$ with many points. Serre applied this method (cf. [7]) to construct curves over $\mathbb{F}_2$ of genus $g = 11$ with 14 rational points (resp. $g = 13$ with $N = 15$ and $g = 10$ with $N = 13$) thus determining $N_2(10) = 13$, $N_2(11) = 14$ and $N_2(13) = 15$. To find suitable f one uses the class group. Serre used an elliptic curve, but we may of course use higher genus curves as well. We illustrate

C. Xing et al. (Eds.): IWCC 2009, LNCS 5557, pp. 82–96, 2009.
© Springer-Verlag Berlin Heidelberg 2009

the method by constructing many curves that reach or improve the present 'records'.

2.1   Base Curve of Genus 1 Over $\mathbb{F}_2$

Let C be the (projective smooth) curve of genus 1 defined by the affine equation $y^2 + y = x^3 + x$. It has five rational points $P_0 = \infty$, $P_1 = (0,0)$, $P_2 = (1,0)$, $P_3 = (1,1)$ and $P_4 = (0,1)$. Since we have an isomorphism $C(\mathbb{F}_2) \cong \mathbb{Z}/5\mathbb{Z}$ of abelian groups such that under the isomorphism P_i corresponds to $i \pmod 5$, the divisor $\sum a_i P_i$ is of degree 0 and linearly equivalent to 0 if and only if

$$\sum_{i=0}^{4} a_i = 0 \quad \text{and} \quad \sum_{i=0}^{4} a_i i \equiv 0 \pmod 5. \tag{1}$$

So there exists a function $f \in \mathbb{F}_2(C)$ with divisor (f) equal to a given divisor $\sum a_i P_i$ if and only if the 5-tuple $(a_0, a_1, \ldots, a_4) \in \mathbb{Z}^5$ satisfies (1).

Often we shall denote a relation for linear equivalence $\sum_{i=0}^{4} a_i P_i \sim 0$ simply by $[a_0, a_1, \ldots, a_4]$ for shortness. Two examples of such relations for the present curve are $[-3, -1, 2, 1, 1]$ and $[-1, -3, 1, 1, 2]$ with functions f_1 and f_2. This gives curves C_{f_1} and C_{f_2} of genera $g = 4$ with $N = 8$ rational points. Note that $N_2(4) = 8$, so these curves realize the maximum for genus 4. The fibre product $C_{f_1} \times_C C_{f_2}$ has genus $g = 11$ with $N = 14$ rational points. Since the upper bound for $N_2(11)$ is 14, Serre (cf. [7]) thus showed that $N_2(11) = 14$.

Another example is the relation $-7P_0 + 2P_1 + 3P_2 + P_4 + P_5 \sim 0$. If f_3 is a corresponding function then the Artin-Schreier cover is a curve C_{f_3} with genus $g = 5$ and with $N = 9$ rational points as one readily checks. Note that $N_2(5) = 9$. The fibre product $C_{f_1} \times_C C_{f_3}$ is a curve of genus $g = 13$ with $N = 15$ rational points, hence $N_2(13) = 15$, again due to Serre.

Combining f_1 with the function f_4 corresponding to the relation $[-3, 2, 1, -1, 1]$ gives $C_{f_1} \times_C C_{f_4}$ of genus 10 with $N = 13$ rational points. Note that the divisor of $f_1 + f_4$ is $-2P_0 - P_1 + P_2 + 2P_4 + P_5$. Artin-Schreier reduction shows that the conductor of this cover is $2P_0 + 2P_1 + 2P_3$. This determines $N_2(10) = 13$, a result due to Serre.

We list here eight relations that can be used to obtain more good curves;

n	relation	n	relation
1	$[-3, -1, 2, 1, 1]$	5	$[-5, -1, 3, 2, 1]$
2	$[-1, -3, 1, 1, 2]$	6	$[-9, 1, 3, 2, 3]$
3	$[-7, 2, 3, 1, 1]$	7	$[-11, 1, 4, 3, 3]$
4	$[-3, 2, 1, -1, 1]$	8	$[-13, 2, 4, 3, 4]$

and reap the fruits by associating to a tuple $f_{i_1}, \ldots, f_{i_r}$ the corresponding fibre product $C_{f_{i_1}} \times_C C_{f_2} \times \cdots \times_C C_{f_{i_r}}$.

space	g	$\#C(\mathbb{F}_2)$	interval	space	g	$\#$	interval
$\langle f_1 \rangle$	4	8	[8]	$\langle f_1, f_3, f_5 \rangle$	29	25	[25 − 27]
$\langle f_3 \rangle$	5	9	[9]	$\langle f_1, f_3, f_4 \rangle$	30	25	[25 − 27]
$\langle f_1, f_4 \rangle$	10	13	[13]	$\langle f_1, f_2, f_3 \rangle$	32	27	[26 − 29]
$\langle f_1, f_2 \rangle$	11	14	[14]	$\langle f_2, f_3, f_5 \rangle$	34	27	[27 − 30]
$\langle f_1, f_3 \rangle$	13	15	[15]	$\langle f_1, f_3, f_6 \rangle$	35	29	[29 − 31]
$\langle f_3, f_5 \rangle$	14	15	[15 − 16]	$\langle f_3, f_6, f_7 \rangle$	39	33	[33]
$\langle f_3, f_6 \rangle$	15	17	[17]	$\langle f_3, f_6, f_8 \rangle$	43	33	[33 − 36]
$\langle f_1, f_2, f_5 \rangle$	28	25	[25 − 26]	$\langle f_3, f_7, f_8 \rangle$	45	33	[33 − 37]

This reproduces many records of the tables and gives one improvement of the tables for curves over $\mathbb{F}_2$, namely for $g = 32$.

2.2   Base of Genus 2 Over $\mathbb{F}_2$

Let C be the curve of genus 2 defined by the affine equation $y^2 + y = (x^2 + x)/(x^3 + x^2 + 1)$ over $\mathbb{F}_2$. It has six rational points P_i $(i = 1, \ldots, 6)$ and class group $\mathrm{Jac}(C)(\mathbb{F}_2) \cong \mathbb{Z}/19\mathbb{Z}$. Let $\phi : C \to \mathrm{Jac}(C)$ be the Abel-Jacobi map given by $Q \mapsto Q - \deg(Q)\, P_1$. For a suitable numbering of the P_i the images of the P_i in $\mathrm{Jac}(C)(\mathbb{F}_2) \cong \mathbb{Z}/19\mathbb{Z}$ are as follows

i	1	2	3	4	5	6
$\phi(P_i)$	0	1	14	6	16	4

From this table we see that we have the following linear equivalence

$$-3P_1 - 3P_2 + 2P_3 + 2P_4 + P_5 + P_6 \sim 0. \tag{2}$$

If f is the function with the left hand side of (2) as divisor the curve C_f obtained as the double cover of C defined by $w^2 + w = f$ has genus $g = 7$ and 10 rational points. This is optimal. Similarly the relation

$$-9P_1 + 2P_2 + 2P_3 + P_4 + 2P_5 + 2P_6 \sim 0$$

gives rise to a curve with $g = 8$ and 11 rational points, again an optimal curve over $\mathbb{F}_2$. Let Q be the divisor of zeros of the polynomial $x^3 + x^2 + 1$ on C. Then $\phi(Q) = 3$ and $-Q + \sum_{i=1}^{6} P_i \sim 0$ which gives us a curve of genus 9 with 12 points, again an optimal curve.

We list a number of divisors of functions f_n $(n = 1, \ldots, 10)$:

n	relation	n	relation
1	$[-3, -3, 2, 2, 1, 1]$	6	$[-11, 3, 2, 2, 3, 1]$
2	$[-9, 2, 2, 1, 2, 2]$	7	$[1, 0, -5, 2, 1, 1]$
3	$[1, 1, 1, 1, 1, 1] \sim Q$	8	$[3, -3, -3, 1, 1, 1]$
4	$[-1, -5, 1, 2, 2, 1]$	9	$[-7, 0, 2, 2, 2, 1]$
5	$[-5, -1, 2, 1, 1, 2]$	10	$[-13, 3, 1, 3, 3, 3]$

and we find as results:

space	g	$\#C(\mathbb{F}_2)$	interval	space	g	$\#C(\mathbb{F}_2)$	interval
$\langle f_1 \rangle$	7	10	$[10]$	$\langle f_2, f_5 \rangle$	20	19	$[19 - 21]$
$\langle f_2 \rangle$	8	11	$[11]$	$\langle f_2, f_6 \rangle$	22	21	$[21 - 22]$
$\langle f_3 \rangle$	9	12	$[12]$	$\langle f_2, f_{10} \rangle$	24	21	$[21 - 23]$
$\langle f_4, f_8 \rangle$	17	17	$[17 - 18]$	$\langle f_1, f_4, f_5 \rangle$	43	34	$[33 - 36]$
$\langle f_1, f_4 \rangle$	18	18	$[18 - 19]$	$\langle f_2, f_5, f_9 \rangle$	44	33	$[33 - 37]$

Again, we find one improvement, namely for $g = 43$.

We can also employ the class group for making an unramified cover of C of degree 19 in which P_1 splits completely. The genus of this cover is $g = 20$ and it contains 19 rational points. The interval is $[19 - 21]$.

2.3   Base of Genus 3 Over $\mathbb{F}_2$

Consider now the curve C of genus 3 defined over $\mathbb{F}_2$ by the homogeneous equation

$$x^3 y + x^3 z + x^2 y^2 + xz^3 + y^3 z + y^2 z^2 = 0.$$

It has 7 rational points and its class group $\text{Jac}(C)(\mathbb{F}_2)$ is isomorphic to $\mathbb{Z}/71\mathbb{Z}$. We can number the rational points P_i ($i = 1, \ldots, 7$) such that their images $\phi(P_i) = $ class of $(P_i - P_1)$ under the Abel-Jacobi map are as in the following table.

i	1	2	3	4	5	6	7
$\phi(P_i)$	0	1	34	55	10	14	49

There is a divisor Q_3 of degree 3 with $\phi(Q_3 - 3 P_0) = 9$ and one of degree 5, say Q_5, with $\phi(Q_5 - 5 P_0) = 35 \pmod{71}$; there is also a divisor Q_7 of degree 7 with $\phi(Q_7 - 7P_0) = 21$. We list a number of relations:

n	divisor	n	divisor
1	$[-3, 2, 1, 1, 2, 1, -4]$	5	$[-1, 3, 1, 1, 1, 2, -7]$
2	$[-11, 1, 2, 1, 2, 3, 2]$	6	$[2, -1, 1, 2, 2, 2, 1] \sim 3Q_3$
3	$[-13, 2, 1, 2, 2, 3, 3]$	7	$[1, 1, 1, 1, 1, 1, 1] \sim Q_7$
4	$[1, -5, 1, 2, 2, 1, 1] \sim Q_3$	8	$[-15, 4, 1, 3, 4, 1, 2]$

that yield the following results:

curve	g	$\#C(\mathbb{F}_2)$	interval	curve	g	$\#C(\mathbb{F}_2)$	interval
$\langle f_1 \rangle$	9	12	$[12]$	$\langle f_2, f_3 \rangle$	29	25	$[25 - 27]$
$\langle f_7 \rangle$	12	14	$[14 - 15]$	$\langle f_4, f_6 \rangle$	31	25	$[27 - 28]$
$\langle f_1, f_5 \rangle$	24	22	$[21 - 23]$	$\langle f_2, f_3, f_8 \rangle$	69	49	$[49 - 52]$

Note that for f_1 we have to do Artin-Schreier reduction. If the conductor would be $3P_0$ (resp. $3P_0 + P_6$) then C_{f_1} would give $g = 7, N = 11$ (resp. $g = 8, 12$), and these are impossible. So the conductor is $3P_0 + 3P_6$ giving $(g = 9, N = 12)$, an optimal curve. Again we find one improvement (for $g = 24$) for our tables. The interval for genus 69 comes from page 121 of [6].

2.4   Base of Genus 5 Over $\mathbb{F}_2$

Consider the fibre product C of C_1 given by $y^2 + y = x^3 + x$ and C_2 given by $y^2 + y = x^5 + x^3$ over the x-line. This is an optimal curve of genus 5 with 9 points. Its Jacobian is isogenous with the product of the three Jacobians of the curves C_1 and C_2 and the curve C_2' defined by $y^2 + y = x^5 + x$. The corresponding L-polynomials are $2t^2 + 2t + 1$, $4t^4 + 4t^3 + 2t^2 + 2t + 1$ and $4t^4 + 4t^3 + 4t^2 + 2t + 1$. The class groups are $\mathbb{Z}/5\mathbb{Z}$, $\mathbb{Z}/13\mathbb{Z}$ and $\mathbb{Z}/15\mathbb{Z}$. The nine rational points are P_∞ and eight points that can be identified by their (x, y_1, y_2, y_3)-coordinates. Writing the class group of C as $\mathbb{Z}/65\mathbb{Z} \times \mathbb{Z}/15\mathbb{Z}$ we have the following table:

i	1	2	3	4	5	6	7	8	9
$\phi(P_i)$	$(0,0)$	$(1,1)$	$(51,14)$	$(64,1)$	$(14,14)$	$(57,11)$	$(47,4)$	$(8,11)$	$(18,4)$

The relation $[1, 2, -13, 1, 2, 2, 2, 2, 1]$ gives a curve with $g = 16$ and 17 points (interval $[17 - 18]$). The relation $[1, 1, -11, 1, -1, 3, 2, 1, 3]$ gives a curve with $g = 16$ and 16 points. The fibre product has genus $g = 39$ with 31 points. The combination of the relations $[1, 2, -13, 1, 2, 2, 2, 2, 1]$ and $[3, 3, -17, 1, 2, 1, 3, 1, 3]$ gives a curve with $g = 42$ and 33 points (interval $[33 - 35]$).

2.5   Genus 1 Over $\mathbb{F}_3$

Consider the elliptic curve defined by the equation $y^2 = x^3 + 2x + 1$ over $\mathbb{F}_3$. It has 7 points P_i with $i = 0, \ldots, 7$ such that $P_i \mapsto i \pmod 7$ defines an isomorphism $C(\mathbb{F}_3) \cong \mathbb{Z}/7\mathbb{Z}$. The relations $[-4, -1, 0, 1, 2, 1, 1]$ and $[-4, -2, 1, 2, 1, 1, 1]$ give a fibre product of genus $g = 30$ with 38 points that improves the interval $[37 - 46]$ slightly.

2.6   Curves of Genus 2 Over $\mathbb{F}_3$

The curve C of genus 2 over $\mathbb{F}_3$ given by $y^2 = x^5 + x^3 + x + 1$ has zeta function $(9t^4 + 9t^3 + 5t^2 + 3t + 1)/(1 - t)(1 - 3t)$. The curve has 7 rational points and class group $\mathbb{Z}/27\mathbb{Z}$. The 7 rational points map to $0, 1, 26, 17, 10, 23, 4$ in the group. We have the relations $[1, 1, 1, 1, -2, 2, -4]$ and $[1, 1, 1, 2, -1, 1, -5]$ and the corresponding fibre product has $g = 44$ with 47 points. This gives a new entry for the table, albeit not a very strong one (resulting interval $[47 - 61]$).

2.7 Base Curve of Genus 3 Over $\mathbb{F}_3$

The curve C of genus 3 given by

$$2\,x^4 + x^3z + 2\,x^2y^2 + x^2yz + x^2z^2 + 2\,xz^3 + 2\,y^4 + 2\,y^3z + 2y^2z^2 = 0$$

has 10 rational points and class group $\mathbb{Z}/204$. The 10 points map to

$$0, 72, 129, 59, 182, 121, 172, 45, 47, 26$$

in $\mathbb{Z}/204$. We have the relation $[1, 2, 1, 1, 1, 1, 1, 1, 1, -10, 1]$ leading to an Artin-Schreier cover with genus 18 with 28 points improving the interval $[26 - 31]$ to $[28 - 31]$.

3 Subgroups of the Class Group

Let C be a smooth complete irreducible curve over $\mathbb{F}_q$ with Jacobian $\mathrm{Jac}(C)$ and class group $G = \mathrm{Jac}(C)(\mathbb{F}_q)$. We choose a rational point P, provided there is one, and consider the morphism $\phi : C \to \mathrm{Jac}(C)$ given by $Q \mapsto Q - \deg(Q)P$. If H is a subgroup of G of index d containing the images $\{\phi(P_i) : i \in I\}$ then there exists an unramified degree d cover $\tilde{C}$ of C in which the points P_i with $i \in I$ split completely. Choosing C and I appropriately can produce curves with many points.

3.1 Base of Genus 4 Over $\mathbb{F}_2$

The reader might think that it is necessary to start with a curve with many points. To dispel this idea start with the genus 4 curve given by the equation

$$y^2 + y = (x^7 + x^5 + 1)/(x^2 + x)$$

over $\mathbb{F}_2$ which has three rational points P_1, P_2, P_3. The L-polynomial is $16\,t^8 + 4\,t^6 + 4\,t^5 + 4\,t^4 + 2\,t^3 + t^2 + 1$, hence the class number is 32. In fact, the class group is $\mathbb{Z}/2\mathbb{Z} \times \mathbb{Z}/16\mathbb{Z}$ and the differences $P_i - P_1$ map to $(0,0)$, $(1,0)$ and $(1,8)$. Hence there exists an étale cover of degree 8 in which the three points split completely. This gives a curve of genus 25 with 24 points. This is optimal. Or start with the hyperelliptic curve of genus 5 given by

$$y^2 + y = (x^9 + x^7 + x^3 + x + 1)/(x^2 + x)$$

again with three rational points and class group $\mathbb{Z}/2\mathbb{Z} \times \mathbb{Z}/30\mathbb{Z}$ with the points now mapping to $(0,0)$, $(1,0)$ and $(0,15)$. This gives an étale cover of degree 15 with genus $g = 61$ with 45 points, very close to the best upper bound 47 and improving the interval $[41 - 47]$ of [6], p. 121.

3.2   Genus 2 Over $\mathbb{F}_3$

We consider the curve C of genus 2 over $\mathbb{F}_3$ given by the affine equation $y^3 - y = x - 1/x$. The zeta function is $(9t^4 + 12t^3 + 10t^2 + 4t + 1)/(1-t)(1-3t)$ and its class group is isomorphic to $\mathbb{Z}/6\mathbb{Z} \times \mathbb{Z}/6\mathbb{Z}$. The images of the 8 rational points under the Abel-Jacobi map are

$$[0,0], [2,4], [1,0], [1,4], [0,1], [2,3], [5,5], [3,5]$$

in $(\mathbb{Z}/6\mathbb{Z})^2$. The subgroup of the class group of index 12 given by $x_1 + x_2 \equiv 0 \,(\mathrm{mod}\,3)$ and $x_1 \equiv x_2 \equiv 0 \,(\mathrm{mod}\,2)$ contains $[0,0]$ and $[2,4]$ and thus leads to a cover of degree 12 of genus $g = 13$ and with 24 rational points (interval $[24 - 25]$). There is a place Q_2 of degree 2 which maps to $[2,1]$ in the class group. The relation $\sum_{i=1}^{8} P_i \sim 4Q_2$ leads to an Artin-Schreier cover with $g = 14$ with 24 points (interval $[24 - 25]$).

An alternative case: consider $y^2 = x(x^2+1)(x^2-x-1)$ over $\mathbb{F}_3$ of genus 2 with 6 points and class group $\mathbb{Z}/2\mathbb{Z} \times \mathbb{Z}/10\mathbb{Z}$. The points map under an Abel-Jacobi map to

$$[0,0], [1,0], [0,9], [0,1], [1,7], [1,3].$$

So the index 10 subgroup defined by $x_2 \equiv 0 \,(\mathrm{mod}\,10)$ gives a cover of degree 10 of genus 11 with 20 points. (Interval $[20 - 22]$.)

Finally, consider the curve C defined over $\mathbb{F}_3$ by the equation $y^2 = 2x^5 + x^4 + x$. It has 6 rational points and zeta function $(9t^4 + 6t^3 + 4t^2 + 2t + 1)/(1-t)(1-3t)$ and class group $\mathbb{Z}/22\mathbb{Z}$. The six points map to $0, 11, 12, 10, 3, 19 \,(\mathrm{mod}\,22)$. We thus see that there is an étale cover of degree 11 with genus 12 and 22 rational points (interval $[22 - 23]$).

Let C be the curve of genus 2 over $\mathbb{F}_3$ defined by the equation $y^3 - y = x + 1/x$. The zeta function of this curve is

$$(81t^4 + 90t^3 + 43t^2 + 10t + 1)/(1-t)(1-9t)$$

and its class group is $\mathbb{Z}/15\mathbb{Z} \times \mathbb{Z}/15\mathbb{Z}$. The curve possesses 20 rational points over $\mathbb{F}_9$. The images of these points under a suitable Abel-Jacobi map are given by

$$[0,13], [0,8], [7,9], [8,12], [14,13], [1,8], [6,12], [9,9], [14,2], [1,4],$$
$$[2,9], [13,12], [4,7], [11,14], [9,6], [6,0], [4,8], [11,13], [0,6], [0,0].$$

The equation $5a + 5b \equiv 0 \,(\mathrm{mod}\,15)$ defines a subgroup of index 3 in $\mathbb{Z}/15\mathbb{Z} \times \mathbb{Z}/15\mathbb{Z}$ containing the images of 10 points, hence we get a curve $\tilde{C}$, an unramified cover of C of degree 3 with genus 4 with 30 points, the maximum possible.

Similarly, the six points P_i corresponding to pairs $[a,b]$ with $a \equiv b \equiv 0\,(\,\mathrm{mod}\,3)$ lie in a subgroup of index 9. So there is an unramified cover $\tilde{C}$ of degree 9 and genus 10 with $9 \times 6 = 54$ rational points. The interval in the tables is $[54 - 55]$.

3.3   Base Curve of Genus 3 Over $\mathbb{F}_3$

Consider the plane curve of degree 4 over $\mathbb{F}_3$ given by

$$2\,x^3y + 2\,x^3z + x^2y^2 + xz^3 + 2\,y^3z + yz^3 = 0.$$

This curve has 10 points over $\mathbb{F}_3$ and has class group isomorphic to $(\mathbb{Z}/14\mathbb{Z})^2$ and the 10 points go to

$$[0,0],[1,0],[6,13],[7,3],[13,7],[4,12],[11,11],[6,2],[4,7]$$

The subgroup defined by $x_2 \equiv 0 \,(\mathrm{mod}\ 7)$ has index 7 and contains the images of 4 points. The corresponding curve over $\mathbb{F}_3$ has genus 15 with 28 rational points; this is optimal.

Or consider the hyperelliptic curve given by the equation

$$y^2 + (x^3 - x)y = x^7 - x^2 + x$$

over $\mathbb{F}_3$. It is of genus 3, has five rational points mapping in the class group $\mathbb{Z}/2\mathbb{Z} \times \mathbb{Z}/2\mathbb{Z} \times \mathbb{Z}/12\mathbb{Z}$ to $(0,0,0,),(0,0,6),(0,1,0),(0,0,11),(0,0,1)$. The index 12 subgroup containing the first three points gives rise to a cover of genus $g = 25$ with 36 points (the interval being $[36 - 40]$).

The hyperelliptic curve $y^2 = x^7 - x^2 + x$ over $\mathbb{F}_3$ has similarly a class group $\mathbb{Z}/2\mathbb{Z} \times \mathbb{Z}/2\mathbb{Z} \times \mathbb{Z}/14\mathbb{Z}$ and an index 14 subgroup containing three points giving rise to a curve of genus 29 with 42 points (interval $[42 - 44]$).

The hyperelliptic curve of genus 4 over $\mathbb{F}_3$ given by $y^2 + xy = x^9 - x$ has class group $\mathbb{Z}/2\mathbb{Z} \times \mathbb{Z}/126\mathbb{Z}$ and the six rational points map to

$$(0,0),(1,0),(0,125),(0,1),(0,14),(0,112)$$

and so the index 14 subgroup containing four of these points gives rise to a cover of genus 43 with 56 points improving the interval $[55 - 60]$.

3.4   Examples Over $\mathbb{F}_4$

Let C be the hyperelliptic curve of genus 2 over $\mathbb{F}_4$ given by $y^2 + y = x^5 + x^3 + x$ with class group $\mathbb{Z}/7\mathbb{Z} \times \mathbb{Z}/7\mathbb{Z}$. It has a cover of degree 7 in which three points split completely giving a curve of genus 8 with 21 rational points (interval $[21 - 24]$).

3.5   Genus 3 Over $\mathbb{F}_9$

Let α be a generator of the multiplicative group $\mathbb{F}_9^*$ and consider the curve C of genus 3 given by $y^3 - y = \alpha^2 x^4$ over the field $\mathbb{F}_9$. This curve has 28 rational points, the bitangent points of a plane model. The class group is of the form

$(\mathbb{Z}/4\mathbb{Z})^6$. The 28 points map under an Abel-Jacobi map to the following elements in $(\mathbb{Z}/4\mathbb{Z})^6$.

$$\begin{aligned}
&[0,0,0,0,0,0],[3,1,3,3,0,1],[3,2,1,0,3,0],[2,0,3,1,1,3],[2,1,1,1,1,3],\\
&[2,2,0,1,1,3],[3,1,0,1,1,3],[1,1,0,0,1,3],[2,1,0,2,1,3],[1,3,3,2,0,3],\\
&[1,2,2,3,2,0],[0,2,3,2,1,2],[2,0,0,0,1,2],[1,1,3,1,2,0],[3,2,1,2,0,3],\\
&[0,0,1,3,2,1],[0,3,1,1,2,2],[2,0,2,3,3,2],[3,3,0,3,2,3],[3,1,2,0,1,0],\\
&[0,3,2,0,0,2],[1,1,1,2,1,1],[1,2,0,3,3,3],[0,0,3,2,3,1],[2,1,0,1,1,0],\\
&\qquad\qquad [2,1,0,1,0,2],[2,1,0,1,2,3],[2,1,0,1,1,3]
\end{aligned}$$

The index 2 subgroup of the class group defined by the equation $2x_1 \equiv 0\,(\mathrm{mod}\ 4)$ contains the images of 16 points. The corresponding étale covering of C of degree 2 is of genus 5 and has 32 points (interval $[32-35]$).

The index 4 subgroup of the class group defined by the equation $x_2 + x_3 + x_4 + x_5 \equiv 0\,(\mathrm{mod}\,4)$ contains the images of 12 points. We thus find a degree 4 étale cover of C of genus 9 with 48 points (interval $[48-50]$).

4   Using the Class Group Over Extension Fields

Let C a smooth projective curve defined over $\mathbb{F}_q$ of genus g and with m rational points. Let J be its Jacobian variety. It is easy to see that if $Z(C,t)$ is the zeta function of $C/\mathbb{F}_q$ and $Z_n(C,t)$ the zeta function of C considered as a curve over $\mathbb{F}_{q^n}$ then we have $Z_n(C,t^n) = \prod_\zeta Z(C,\zeta t)$, where the product is taken over the nth roots ζ of 1. If we write $Z(C,t) = L(C,t)/(1-t)(1-qt)$ and $Z_n(C,t) = L_n(C,t)/(1-t)(1-q^n t)$ we get $L_n(C,1) = \prod_\zeta L(C,\zeta)$. Moreover, we know that $\#J(\mathbb{F}_q) = L(C,1)$ and $\#J(\mathbb{F}_{q^n}) = L_n(C,1)$ and we thus get

$$[J(\mathbb{F}_{q^n}) : J(\mathbb{F}_q)] = \prod_{\zeta^n=1,\zeta\neq 1} L(C,\zeta).$$

Since under the Abel-Jacobi map the $\mathbb{F}_q$-rational points of C map to $J(\mathbb{F}_q)$ we conclude that there exists an unramified cover of C defined over $\mathbb{F}_{q^n}$ of degree $d = \prod_{\zeta^n=1,\zeta\neq 1} L(C,\zeta)$ in which all the $\mathbb{F}_q$-rational points split completely. Thus we find a curve $\tilde{C}$ of genus $d(g-1)+1$ with at least dm rational points. This idea was exploited very succesfully by Niederreiter and Xing in their papers [2] up to [6]. Here we employ it more systematically by going through all isomorphism classes of curves of low genera. This improves the tables at many places.

We now list all possible L-functions of genus 2 curves over $\mathbb{F}_2$. By N_i we mean $\#C(\mathbb{F}_{2^i})$.

f	$[N_1, N_2]$	Cl	L
$(x^2 + x)/(x^3 + x^2 + 1)$	$[6, 6]$	$\mathbb{Z}/19\mathbb{Z}$	$4t^4 + 6t^3 + 5t^2 + 3t + 1$
$(x^3 + x + 1)/(x^3 + x^2 + 1)$	$[0, 6]$	(0)	$4t^4 - 6t^3 + 5t^2 - 3t + 1$
$1/(x^3 + x^2 + 1)$	$[2, 6]$	$\mathbb{Z}/3\mathbb{Z}$	$4t^4 - 2t^3 + t^2 - t + 1$
$x/(x^3 + x^2 + 1)$	$[4, 6]$	$\mathbb{Z}/9\mathbb{Z}$	$4t^4 + 2t^3 + t^2 + t + 1$
$x^2/(x^3 + x^2 + 1)$	$[4, 10]$	$\mathbb{Z}/11\mathbb{Z}$	$4t^4 + 2t^3 + 3t^2 + t + 1$
$(x^3 + 1)/(x^3 + x^2 + 1)$	$[2, 10]$	$\mathbb{Z}/5\mathbb{Z}$	$4t^4 - 2t^3 + 3t^2 + t + 1$
$1/x(x^2 + x + 1)$	$[3, 7]$	$\mathbb{Z}/6\mathbb{Z}$	$4t^4 + t^2 + 1$
$(x + 1)/x(x^2 + x + 1)$	$[5, 7]$	$\mathbb{Z}/11\mathbb{Z}$	$4t^4 + 4t^3 + 3t^2 + 2t + 1$
$(x^3 + x^2 + 1)/(x(x^2 + x + 1))$	$[1, 7]$	$\mathbb{Z}/2\mathbb{Z}$	$4t^4 - 4t^3 + 3t^2 - 2t + 1$
$(x^3 + x^2 + 1)/x(x + 1)$	$[3, 3]$	$\mathbb{Z}/2\mathbb{Z} \times \mathbb{Z}/2\mathbb{Z}$	$4t^4 - t^2 + 1$
$1/x + x^3$	$[4, 4]$	$\mathbb{Z}/8\mathbb{Z}$	$4t^4 + 2t^3 + t + 1$
$1/x + x^2 + x^3$	$[2, 8]$	$\mathbb{Z}/4\mathbb{Z}$	$4t^4 - 2t^3 + 2t^2 - t + 1$
$1/x + 1 + x^3$	$[2, 4]$	$\mathbb{Z}/2\mathbb{Z}$	$t^4 - 2t^3 - t + 1$
$1/x + 1 + x^2 + x^3$	$[4, 8]$	$\mathbb{Z}/10\mathbb{Z}$	$4t^4 + 2t^3 + 2t^2 + t + 1$
x^5	$[3, 5]$	$\mathbb{Z}/5\mathbb{Z}$	$4t^4 + 1$
$x^5 + x^3 + x$	$[3, 9]$	$\mathbb{Z}/7\mathbb{Z}$	$4t^4 + 2t^2 + 1$
$x^5 + x$	$[5, 9]$	$\mathbb{Z}/15\mathbb{Z}$	$4t^4 + 4t^3 + 4t^2 + 2t + 1$
$x^5 + x + 1$	$[1, 5]$	$\mathbb{Z}/3\mathbb{Z}$	$4t^4 - 4t^3 + 4t^2 - 2t + 1$
$x^5 + x^3$	$[5, 5]$	$\mathbb{Z}/13\mathbb{Z}$	$4t^4 + 4t^3 + 2t^2 + 2t + 1$
$x^5 + x^3 + 1$	$[1, 9]$	(0)	$4t^4 - 4t^3 + 2t^2 - 2t + 1$

If we now take $n = 2$ we get curves defined over $\mathbb{F}_4$; for example we find the following cases that do not give improvements of the table for $\mathbb{F}_4$ but good realizations of the present records.

$[N_1, N_2]$	g	$\#C(\mathbb{F}_4)$	interval
$[5, 9]$	4	15	$[15]$
$[4, 10]$	6	20	$[20]$
$[3, 9]$	8	21	$[21 - 24]$

With $n = 3$ we get the following cases

$[N_1, N_2]$	g	$\#C(\mathbb{F}_8)$	interval
$[5, 7]$	8	35	$[35 - 42]$
$[5, 5]$	14	65	$[65]$
$[4, 4]$	20	76	$[68 - 83]$

Here the case $g = 14$ with $\#C(\mathbb{F}_4) = 65$ is optimal; the case $g = 8$ with 35 rational points equals the lower bound of the interval $[35 - 42]$ of the tables while $g = 20$ with 76 points improves the interval $[68 - 83]$ of the tables.

Starting with the curves of genus 2 over $\mathbb{F}_4$ gives the following results:

$[N_1, N_2]$	g	$\#C(\mathbb{F}_{16})$	interval
$[9, 24]$	9	72	$[72 - 81]$
$[9, 25]$	10	81	$[81 - 87]$
$[8, 24]$	11	80	$[80 - 91]$
$[8, 26]$	12	88	$[83 - 97]$
$[7, 27]$	15	98	$[98 - 113]$
$[7, 31]$	17	112	$[112 - 123]$

And the same can be done starting with the curves of genus 2 over $\mathbb{F}_8$. Here there are still many empty places in the tables due to the lack of good examples.

$[N_1, N_2]$	g	$\#C(\mathbb{F}_{64})$	interval
$[18, 54]$	20	342	
$[17, 63]$	25	408	
$[17, 65]$	26	425	
$[16, 64]$	27	416	
$[16, 70]$	30	464	
$[15, 71]$	33	480	
$[14, 66]$	34	462	
$[15, 75]$	35	510	
$[14, 70]$	36	490	
$[15, 79]$	37	540	
$[14, 74]$	38	518	
$[13, 67]$	39	494	$[489 - 650]$
$[14, 78]$	40	546	
$[14, 80]$	41	560	
$[14, 82]$	42	574	
$[12, 66]$	44	516	
$[13, 79]$	45	572	
$[13, 81]$	46	585	
$[13, 83]$	47	598	
$[12, 74]$	48	564	
$[13, 87]$	49	624	
$[12, 78]$	50	588	

We can use curves of higher genus too. Going through all non-hyperelliptic curves of genus 3 over $\mathbb{F}_2$ gives two improvements for the table over $\mathbb{F}_4$:

$[N_1, N_2, N_3]$	g	$\#C(\mathbb{F}_4)$	interval
$[5, 9, 5]$	15	35	$[33 - 37]$
$[5, 11, 5]$	17	40	$[40]$
$[4, 12, 7]$	27	52	$[50 - 56]$

and considering these curves over $\mathbb{F}_4$ gives three improvements for the table over $\mathbb{F}_{16}$:

$[N_1, N_2, N_3]$	g	$\#C(\mathbb{F}_{16})$	interval
$[4, 12, 7]$	27	156	$[145 - 176]$
$[3, 11, 6]$	35	187	
$[5, 11, 5]$	41	220	$[216 - 249]$

Going through the hyperelliptic curves of genus 4 over $\mathbb{F}_2$ yields one improvement of the table over $\mathbb{F}_4$: the curve with $[N_1, N_2, N_3, N_4] = [6, 10, 6, 26]$ yields a curve of genus 28 with 54 points over $\mathbb{F}_4$; the interval was $[53 - 58]$.

Taking the curve of genus 4 over $\mathbb{F}_2$ with $[N_1, N_2, N_3, N_4] = [8, 8, 8, 16]$ over $\mathbb{F}_2$ and L-polynomial $16t^8 + 40t^7 + 56t^6 + 56t^5 + 44t^4 + 28t^3 + 14t^2 + 5t + 1$ gives a degree 13 cover with genus 40 and 104 points, improving the interval $[103 - 141]$.

Similarly, we can list all pairs $[N_1, N_2]$ with $N_i = \#C(\mathbb{F}_{3^i})$ occuring for curves of genus 2 over $\mathbb{F}_3$. This yields the following harvest.

$[N_1, N_2]$	g	$\#C(\mathbb{F}_9)$	interval
$[8, 14]$	5	32	$[32 - 35]$
$[7, 15]$	6	35	$[35 - 40]$
$[6, 16]$	8	42	$[40 - 47]$
$[6, 18]$	9	48	$[48 - 50]$
$[6, 20]$	10	54	$[54]$

Doing the same for curves of genus 2 over $\mathbb{F}_9$ gives the following.

$[N_1, N_2]$	g	$\#C(\mathbb{F}_{81})$	interval
$[20 - 68]$	26	500	
$[19 - 75]$	30	551	
$[18 - 78]$	33	576	
$[18 - 80]$	34	594	$[494 - 689]$
$[18 - 82]$	35	612	
$[18 - 86]$	37	648	$[568 - 742]$
$[17 - 83]$	38	629	
$[17 - 85]$	39	646	
$[17 - 87]$	40	663	
$[17 - 89]$	41	680	
$[17 - 91]$	42	697	
$[16 - 86]$	43	672	
$[16 - 90]$	45	704	
$[16 - 92]$	46	720	
$[15 - 85]$	47	690	
$[16 - 96]$	48	752	$[676 - 885]$
$[16 - 86]$	49	672	$[656 - 898]$
$[16 - 100]$	50	784	

Similarly, the genus 2 curves over $\mathbb{F}_3$ give with $n = 3$:

$[N_1, N_2]$	g	$\#C(\mathbb{F}_{27})$	interval
$[8, 10]$	26	200	
$[7, 13]$	29	196	
$[7, 11]$	40	273	$[244 - 346]$

For the case of non-hyperelliptic curves of genus 3 over $\mathbb{F}_3$ we find the following cases:

m	$\#C(\mathbb{F}_3)$	$L(-1)$	g	$\#\tilde{C}(\mathbb{F}_9)$	interval
687439	8	12	25	96	$[82 - 108]$
787452	8	13	27	104	$[91 - 114]$
687411	7	17	35	119	$[116 - 139]$
787567	7	18	37	126	$[120 - 145]$
884286	7	20	41	140	$[128 - 158]$
687541	7	21	43	147	$[120 - 164]$

Here we use the following notation for plane curves of degree 3 over $\mathbb{F}_3$. We write the polynomials in x, y, z in pure lexicographic order and use then the coefficients c_i $(i = 1, \ldots, 15)$ to associate the natural number $m = \sum_{i=1}^{15} c_i 3^{i-1}$ to the curve. So x^4 corresponds to 1 and the Klein curve $x^3 y + y^3 z + z^3 x = 0$ to 196833.

5 Tables

Table p=2.

$g\backslash q$	2	4	8	16	32	64	128
1	5	9	14	25	44	81	150
2	6	10	18	33	53	97	172
3	7	14	24	38	64	113	192
4	8	15	25	45	71–74	129	215
5	9	17	29–30	49–53	83–85	132–145	227–234
6	10	20	33–35	65	86–96	161	243–258
7	10	21–22	34–38	63–69	98–107	177	262–283
8	11	21–24	35–42	62–75	97–118	169–193	276–302
9	12	26	45	72–81	108–128	209	288–322
10	13	27	42–49	81–87	113–139	225	296–345
11	14	26–29	48–53	80–91	120–150	201–236	294–366
12	14–15	29–31	49–57	88–97	129–161	257	321–388
13	15	33	56–61	97–102	129–172	225–268	
14	15–16	32–35	65	97–107	146–183	241–284	353–437
15	17	35–37	57–67	98–113	158–194	258–300	386–455
16	17–18	36–38	56–71	95–118	147–204	267–316	
17	17–18	40	63–74	112–123	154–212		
18	18–19	41–42	65–77	113–129	161–220	281–348	
19	20	37–43	60–80	129–134	172–228	315–364	
20	19–21	40–45	76–83	127–139	177–236	342–380	
21	21	44–47	72–86	129–145	185–243	281–396	
22	21–22	42–48	74–89	129–150		321–412	
23	22–23	45–50	68–92	126–155			
24	22–23	49–52	81–95	129–161	225–267	337–444	513–653
25	24	51–53	86–97	144–165		408–460	
26	24–25	55	82–100	150–171		425–476	
27	24–25	52–56	96–103	156–176	213–290	416–492	
28	25–26	54–58	97–106	145–181	257–298	513	577–745
29	25–27	52–60	97–109	161–186	227–305		
30	25–27	53–61	96–112	162–191	273–313	464–535	609–784
31	27–28	60–63	89–115	165–196		450–547	578–807
32	27–29	57–65	90–118				
33	28–29	65–66	97-121	193–207		480–570	
34	27–30	65–68	98–124	183–212		462–581	
35	29–31	64–69	112–127	187–217	253–351	510–593	
36	30–31	64–71	112–130	185–222		490–604	705–917
37	30–32	66–72	121–132	208–227		540–616	
38	30–33	64–74	129–135	193–233	291–375	518–627	
39	33	65–75	120–138	194–238		494–638	
40	32–34	75–77	103–141	225–243	293–390	546–649	
41	33–35	65–78	118–144	220–249	308–398	560–661	
42	33–35	75–80	129–147	209–254	307–405	574–672	
43	34–36	72–81	116–150	226–259	306–413	546–684	
44	33–37	68–83	130–153	226–264	325–420	516–695	
45	33–37	80–84	144–156	242–268	313–428	572–706	
46	34–38	81–86	129–158	243–273		585–717	
47	36–38	73–87	126–161			598–729	
48	34–39	80–89	128–164	243–282		564–740	
49	36–40	81–90	130–167	213–286		624–751	913–1207
50	40	91–92	130–170	255–291		588–762	

Table p=3.

$g\backslash q$	3	9	27	81
1	7	16	38	100
2	8	20	48	118
3	10	28	56	136
4	12	30	64	154
5	13	32–35	72–75	160–172
6	14	35–40	76–85	190
7	16	40–43	82–95	180–208
8	17–18	42–47	92–105	226
9	19	48–50	99–113	244
10	20–21	54	94–123	226–262
11	20–22	55–58	100–133	220–280
12	22–23	56–62	109–143	298
13	24–25	64–65	136–153	256–312
14	24–26	56–69		278–330
15	28	64–73	136–170	292–348
16	27–29	74–77	144–178	370
17	25–30	74–81	128–185	288–384
18	28–31	67–84	148–192	306–401
19	32	84–88	145–199	
20	30–34	70–91		
21	32–35	88–95	163–213	352–455
22	30–36	78–98		
23	32–37	92–101		
24	31–38	91–104	208–234	
25	36–40	96–108	196–241	392–527
26	36–41	110–111	200–248	500–545
27	39–42	104–114		
28	37–43	105–117		
29	42–44	104–120	196–269	
30	38–46	91–123	196–276	551–617
31	40–47	120–127		460–635
32	40–48	92–130		
33	46–49	128–133	220–297	576–671
34	46–50	111–136		594–689
35	47–51	119–139		612–707
36	48–52	118–142	244–318	730
37	52–54	126–145	236–325	648–742
38		105–149		629–755
39	48–56	140–152	271–340	646–768
40	56–57	118–155	273–346	663–781
41	50–58	140–158		680–795
42	52–59	122–161	280–360	697–808
43	56–60	147–164		672–821
44	47–61	119–167	278–374	
45	54–62	136–170		704–847
46	55–63	162–173		720–859
47	54–65	154–177	299–395	690–872
48	55–66	163–180	325–402	752–885
49	64–67	168–183	316–409	768–898
50	63–68	182–186	312–416	784–911

For exhaustive references we refer to the bibliography of the tables, [10]. For general background on curves over finite fields we refer to the book by Stichtenoth [9] and for general background on curves over finite fields with many points to Serre's notes [8] and for an overview of the methods of Niederreiter and Xing to [6].

References

1. Howe, E., Lauter, K.: Improved upper bounds for the number of points on curves over finite fields. Ann. Inst. Fourier 53, 1677–1737 (2003); corrigendum to: Improved upper bounds for the number of points on curves over finite fields. Ann. Inst. Fourier (Grenoble) 57(3), 1019–1021 (2007)
2. Niederreiter, H., Xing, C.P.: Explicit global function fields over the binary field with many rational places. Acta Arithm. 75, 383–396 (1996)
3. Niederreiter, H., Xing, C.P.: Cyclotomic function fields, Hilbert class fields and global function fields with many rational places. Acta Arithm. 79, 59–76 (1997)
4. Niederreiter, H., Xing, C.P.: Global function fields fields with many rational points over the ternary field. Acta Arithm. 83, 65–86 (1998)
5. Niederreiter, H., Xing, C.P.: Algebraic curves with many rational points over finite fields of characteristic 2. In: Proc. Number Theory Conference (Zakopane 1997), pp. 359–380. de Gruyter, Berlin (1999)
6. Niederreiter, H., Xing, C.P.: Rational Points on Curves over Finite Fields—Theory and Applications. London Math. Soc. Lecture Note Ser., vol. 285. Cambridge University Press, Cambridge (2002)
7. Serre, J.-P.: Letter to G. van der Geer, September 1 (1997)
8. Serre, J.-P.: Rational points on curves over finite fields. Notes of lectures at Harvard University (1985)
9. Stichtenoth, H.: Algebraic Function Fields and Codes. Universitext. Springer, Heidelberg (1993)
10. van der Geer, G., van der Vlugt, M.: Tables of curves with many points. Math. Comp. 69(230), 797–810 (2000),
    http://www.science.uva.nl/~geer/tables-mathcomp20.pdf

List Decoding of Binary Codes–A Brief Survey of Some Recent Results

Venkatesan Guruswami[*]

Department of Computer Science & Engineering
University of Washington
Seattle, WA 98195

Abstract. We briefly survey some recent progress on list decoding algorithms for *binary* codes. The results discussed include:

- Algorithms to list decode binary Reed-Muller codes of any order up to the minimum distance, generalizing the classical Goldreich-Levin algorithm for RM codes of order 1 (Hadamard codes). These algorithms are "local" and run in time polynomial in the *message* length.
- Construction of binary codes efficiently list-decodable up to the Zyablov (and Blokh-Zyablov) radius. This gives a factor two improvement over the error-correction radius of traditional "unique decoding" algorithms.
- The existence of binary linear *concatenated* codes that achieve list decoding capacity, i.e., the optimal trade-off between rate and fraction of worst-case errors one can hope to correct.
- Explicit binary codes mapping k bits to $n \leqslant \text{poly}(k/\varepsilon)$ bits that can be list decoded from a fraction $(1/2 - \varepsilon)$ of errors (even for $\varepsilon = o(1)$) in $\text{poly}(k/\varepsilon)$ time. A construction based on concatenating a variant of the Reed-Solomon code with dual BCH codes achieves the best known (cubic) dependence on $1/\varepsilon$, whereas the existential bound is $n = O(k/\varepsilon^2)$. (The above-mentioned result decoding up to Zyablov radius achieves a rate of $\Omega(\varepsilon^3)$ for the case of *constant* ε.)

We will only sketch the high level ideas behind these developments, pointing to the original papers for technical details and precise theorem statements.

1   Introduction

Shannon's capacity theorem states that for the binary symmetric channel BSC_p with crossover probability p,[1] there exist codes using which one can reliably communicate at any rate less than $1 - H(p)$, and conversely, no larger rate is possible.

---

[*] Currently visiting the Computer Science Department, Carnegie Mellon University, Pittsburgh, PA. Research supported by NSF grants CCF-0343672 and CCF-0835814, and a Packard Fellowship.
[1] The BSC_p is a communication channel that transmits bits, and flips each bit independently with probability p.

(Here $H(\cdot)$ is the binary entropy function.) But what if the errors are *worst-case* and not random and independent? Specifically, if the channel is modeled as a "jammer" than can corrupt up to a fraction $p < 1/2$ of symbols in an arbitrary manner, what is the largest rate possible for error-free communication?

If we want a guarantee that the original message can be correctly and uniquely recovered, then this is just the question of the largest rate of a binary code every pair of whose codewords have different bits in at least a fraction $2p$ of the positions. This remains one of the biggest open questions in combinatorial coding theory. It is known, however, that in this case the rate has to be much smaller than $1 - H(p)$. The best rate known to be possible (even by non-constructive random coding methods) is the much smaller $1 - H(2p)$. (Note, in particular, that the rate is 0 already for $p > 1/4$.)

The above limitation arises due to the fact that, for rates close to Shannon capacity, there will be two codewords that both differ from some string in less than a fraction p of bits. However, the closest codeword is usually unique, and in those cases, it makes sense to decode up to a radius p. A clean way to model this algorithmic task is to relax the requirement on the error-correction algorithm and allow it to output a *small list* of messages in the worst-case (should there be multiple close-by codewords). This model is called *list decoding*. Perhaps surprisingly (and fortunately), using list decoding, one achieve a rate approaching the Shannon capacity $1 - H(p)$, even if the errors are worst-case. Formally, there *exist* binary codes $C \subseteq \{0,1\}^n$ of rate $1 - H(p) - \frac{1}{L}$ which are (p, L)-list-decodable, i.e., every Hamming ball of radius pn has at most L codewords of C [19,1]. In fact, such binary *linear* codes exist [6]. If a codeword from such a code is transmitted and corrupted by at most a fraction p of errors, there will be at most L possible codewords that could have resulted in the received word. Thus it can be used for error recovery with an ambiguity of at most L. By allowing the worst-case list size L to grow, one can approach the best possible rate of $1 - H(p)$, which we call the *list decoding capacity*.

Thus, list decoding offers the potential of realizing the analog of Shannon's result for worst-case errors. However, the above is a *non-constructive* result. The codes achieving this trade-off are shown to exist via a random coding argument and are not explicitly specified. Further, for a code to be useful, the decoding algorithm must be efficient, and for a random, unstructured code only brute-force decoders running in exponential time are known.

Therefore, the grand challenge in the subject of list decoding binary codes is to give an explicit (polynomial time) construction of binary codes approaching list decoding capacity, together with an efficient list decoding algorithm. This remains a challenging long term goal that seems out of the reach of currently known techniques. For *large* alphabets, recent progress in algebraic list decoding algorithms [16,8,5] has led to the construction of explicit codes that achieve list decoding capacity — namely, they admit efficient algorithms to correct close to the optimal fraction $1 - R$ of errors with rate R. This in turn has spurred some (admittedly modest) progress on list decoding of binary codes, using code concatenation, soft decoding, and other techniques.

We give an informal discussion of some of this progress in this paper. The specific problems we discuss are those mentioned in the abstract of the paper, and are based on results in [3,8,10,7,9]. The second and third results (discussed in Sections 3 and 4) have straightforward extensions to codes over the field $\mathbb{F}_q$ with q elements for any fixed prime power q. The exact list decodability of Reed Muller codes over non-binary fields $\mathbb{F}_q$ remains an intriguing open problem (some progress is made in [3] but the bounds are presumably not tight). Our technical discussion below assumes that the reader is familiar with the basic background material and terminology of coding theory.

2   List Decoding Reed-Muller Codes

The first non-trivial algorithm for list decoding was the Goldreich-Levin algorithm for Hadamard codes (or first order Reed-Muller codes) [2]. The messages of the (binary) Reed-Muller code $\mathsf{RM}(m, r)$ of order r consist of m-variate multilinear polynomials of degree at most r over the binary field $\mathbb{F}_2$. The encoding of such a polynomial f consists of the evaluations $f(\mathbf{x})$ at all $\mathbf{x} \in \mathbb{F}_2^m$. The length of the encoding is thus 2^m. The minimum distance of $\mathsf{RM}(m, r)$ is 2^{m-r}.

The order 1 RM code corresponds to evaluations of *linear* polynomials on $\mathbb{F}_2^m$ and is often called the Hadamard code.[2] The Goldreich-Levin algorithm list decodes the Hadamard code up to a fraction $(1/2 - \varepsilon)$ of errors in time $\mathrm{poly}(m/\varepsilon)$, outputting a list of size $O(1/\varepsilon^2)$. Note that the decoding radius approaches the relative distance, and further the runtime of the decoder is polynomial in the message length and *sub-linear* (in fact just polylogarithmic) in the length of the code (which is 2^m). The decoder is a "local" algorithm that only randomly probes a small portion of the received word in order to recover the messages corresponding to the close-by codewords.

An extension of the Goldreich-Levin algorithm to higher order RM codes was open for a long time. In recent work, Gopalan, Klivans, and Zuckerman [3] solved this problem, giving a local list decoding algorithm to correct a fraction $(2^{-r} - \varepsilon)$ errors for $\mathsf{RM}(m, r)$, i.e., arbitrarily close to the relative distance 2^{-r}. The algorithm runs in time $(m/\varepsilon)^{O(r)}$ and outputs a list of size at most $(1/\varepsilon)^{O(r)}$. This list size bound is also shown to be tight (up to constant factors in the exponent $O(r)$). Reed-Muller codes of constant order have only polynomially many codewords and thus have vanishing rate. Thus, this result is not directly related to the program of constructing binary codes with good trade-off between rate and list decoding radius. It is nevertheless an exciting development since Reed-Muller codes are one of the classical and most well-studied binary code constructions.

We now describe the approach behind the algorithm at a very informal level. Suppose we are given oracle access to a received word R which we think of as a function $R : \mathbb{F}_2^m \to \mathbb{F}_2$. The goal is to find all degree r polynomials f which differ from R on at most a fraction $(2^{-r} - \varepsilon)$ of points. The algorithm picks

---

[2] To be accurate, the Hadamard code only encodes linear polynomials with no constant term but this is a minor difference.

a random subspace A of $\mathbb{F}_2^m$ of size $a = 2^{O(r)}/\varepsilon^2$. Then it guesses the correct value of $f_{|A}$, the target polynomial f restricted to A (we remark on the number of such guesses shortly). Then given a point $\mathbf{b} \in \mathbb{F}_2^m$, the algorithm determines $f(\mathbf{b})$ as follows: Consider the subspace $B = A \cup (A + \mathbf{b})$. Run a unique decoding algorithm for a Reed-Muller code of order r restricted to B (correcting up to a fraction $\frac{1}{2} \cdot 2^{-r}$ of errors), to find a degree r polynomial $g_{|B}$, if one exists. Note that if the error rate on $A + \mathbf{b}$ w.r.t f is less than 2^{-r}, the error rate on B will be less than $2^{-(r+1)}$ and thus $g_{|B}$ must be $f_{|B}$. We will recover $f(\mathbf{b})$ correctly in this case from the corresponding value of g.

Since A is a random subspace of size a, by pairwise independence, with high probability (at least $1 - \frac{1}{a\varepsilon^2} = 1 - 2^{-\Omega(r)}$), the error rate on $A + \mathbf{b}$ is within ε of the original error rate, and therefore less than 2^{-r}. Therefore, with high probability over the choice of the subspace A, when the algorithm guesses $f_{|A}$ correctly, it will correctly compute $f(\mathbf{b})$ for all but a $2^{-\Omega(r)}$ fraction of points $\mathbf{b}$. The function computed by the algorithm is thus within fractional distance at most $2^{-(r+1)}$ from the encoding of f. Since the unique decoding radius of $\mathrm{RM}(m, r)$ is $2^{-(r+1)}$, running a local unique decoder on the function computed by the algorithm then returns f.

The list size is governed by the number of guesses for $f_{|A}$. When $r = 1$, since f is a linear polynomial, it suffices to guess the value on a basis for the subspace A which consists of $\log a = 2\log(1/\varepsilon) + O(1)$ points. This leads to the $O(1/\varepsilon^2)$ list-size bound for Hadamard codes. For $r > 1$, the total number of guesses becomes quasi-polynomial in $1/\varepsilon$. Using additional ideas, it is possible to improve this to $(1/\varepsilon)^{O(r)}$. The above description was meant to only give the flavor of the algorithm, and hides many subtleties. The reader can find the formal details and the arguments for the improved list size bound in [3]. List size bounds for decoding binary Reed-Muller codes *beyond* the minimum distance are established in [14].

3   List Decoding Concatenated Codes Up to Zyablov Radius

As mentioned in the introduction, the problem of constructing explicit binary codes achieving list decoding capacity, i.e., binary codes of rate R list-decodable up to radius $H^{-1}(1 - R)$, remains wide open. In this section, we report on recent constructions that achieve the best known trade-offs between rate and list decoding radius.

Guruswami and Rudra [8] construct a variant of Reed-Solomon codes, called folded Reed-Solomon codes, over an alphabet of size polynomial in the block length, that can list decode a fraction $(1 - R_0 - \varepsilon)$ of errors with rate R_0 for any desired $0 < R_0 < 1$ and any constant $\varepsilon > 0$. This achieves the list decoding capacity over large alphabets. A natural approach to construct binary codes is to concatenate these codes with optimal binary list-decodable codes at the inner level. Since the inner codes have only polynomially codewords, one can find by a greedy "brute-force" search such codes close to list decoding capacity, i.e., with rate r which are list-decodable up to a fraction $H^{-1}(1 - r - \varepsilon)$ of errors (with list-size $O(1/\varepsilon)$). The search for the inner code is not based on the most obvious brute-force algorithm (which would take quasi-polynomial time in the block length of the outer code), but rather a greedy derandomization of the probabilistic method; see [6] for details.

The resulting concatenated code has rate rR_0, and an algorithm to list decode the code up to a fraction $(1 - R_0)H^{-1}(1 - r) - \varepsilon$ of errors is given in [8]. This leads to binary codes of rate R list-decodable up to the so-called *Zyablov radius* given by

$$\text{Zyablov}(R) = \max_{\substack{0 < R_0, r \leqslant 1 \\ R_0 r = R}} (1 - R_0)H^{-1}(1 - r) .$$

We note that the Zyablov radius is also the standard product bound on the relative distance of concatenated codes with an outer Maximum Distance Separable code and an inner binary code meeting the Gilbert-Varshamov bound. The above result is able to *decode* up to this radius. In comparison, traditional algorithms based on Generalized Minimum Distance decoding are able to (unique) decode up to *half* the Zyablov radius.

The idea behind the above algorithm is very natural. The various inner blocks are first decoded using a brute-force search algorithm up to a radius of $H^{-1}(1 - r - \varepsilon)$. By the assumed list decoding properties of the inner code, this step returns a set of at most $\ell = O(1/\varepsilon)$ candidate symbols for each possible symbol of the folded Reed-Solomon codeword. If the fraction of errors is at most $(1 - R_0)H^{-1}(1 - r) - O(\varepsilon)$, then at most a $(1 - R_0 - \varepsilon)$ fraction of the sets returned by the inner decodings will fail to contain the correct outer symbol. Now comes the part where a crucial, powerful feature of the Guruswami-Rudra list decoder comes in handy — a fraction $(1 - R_0 - \varepsilon)$ of errors can be list decoded even if the input is not a received word but a collection of sets of possible symbols (of some bounded size, such as ℓ), one for each codeword position, and where a codeword position is counted as an error if the correct codeword symbol does not belong to the set of candidates corresponding to that position. This generalization of list decoding is called *list recovery* in the literature. The details of list recovering folded Reed-Solomon codes and formal details about the above algorithm can be found in [8].

In [10], the authors use multilevel concatenated codes to improve the above trade-off and construct codes list-decodable up to the *Blokh-Zyablov* radius. The outer codes are folded Reed-Solomon codes, and the inner codes are picked (via a careful brute-force search, guided by the derandomization of a probabilistic argument) to satisfy a certain "nested" list-decodability property.

Folded codes based on cyclotomic function fields are constructed in [5] with list decoding properties similar to folded Reed-Solomon codes but over an alphabet of size polylogarithmic (instead of polynomial) in the block length. This is useful to give a *Justesen-style* explicit binary concatenated code list-decodable up to the Zyablov radius *without a brute-force search* for a good inner code, by using *all* possible linear codes at the inner level.

4   Concatenated Codes Can Achieve List Decoding Capacity

Despite achieving some good trade-offs, the above concatenated code constructions fall well short of achieving the list-decoding capacity for binary codes.
Given the almost exclusive stronghold of concatenated codes on progress in explicit constructions of list-decodable codes over small alphabets, a natural question that arises is the following: Do there exist binary *concatenated* codes that
achieve list-decoding capacity, or does the stringent structural restriction imposed on the code by concatenation preclude such codes achieving list-decoding
capacity?

In [7], the authors prove that there *do* exist binary linear concatenated codes
that achieve list-decoding capacity for any desired rate. In fact, it is shown that a
random concatenated code drawn from a certain ensemble achieves capacity with
overwhelming probability. This is somewhat encouraging news for the eventual
goal of achieving list-decoding capacity (or at least, going beyond the Blokh-
Zyablov radius) for binary codes with polynomial time decodable codes, since
code concatenation has been the preeminent method for constructing good codes
over small alphabets.

The outer codes in this construction are folded Reed-Solomon codes of rate R_0
over an extension field $\mathbb{F}_{2^m}$ with near-optimal list-recovering properties (these
were constructed in [8]). The inner codes for the various positions are random
binary linear codes of dimension m and rate r (which can even be chosen to
equal 1), with a completely *independent* random choice for each outer codeword position. This gives independence across coordinates of the outer code,
which is crucially exploited in the analysis. To prove the desired list decoding
property, the goal is to show that a large number of codewords of the concatenated code are unlikely to lie in some Hamming ball of fractional radius
$H^{-1}(1 - rR_0) - \varepsilon$.

Using the list recovering properties of folded Reed-Solomon codes, it is shown
that for every integer J, any large enough collection (compared to J) of outer
codewords has a "good" subset, say $c_1, \ldots, c_J$, of size J with the property that
each c_i has at least a fraction $(1 - R_0 - \varepsilon)$ of symbols which are linearly independent (over $\mathbb{F}_2$) of the corresponding symbols of $c_1, c_2, \ldots, c_{i-1}$. Since the
inner encoding at each position is a random linear code, each such linearly independent symbol of c_i is mapped to a random binary string that is independent
of where the corresponding symbols of $c_1, c_2, \ldots, c_{i-1}$ were mapped. This is to
used to upper bound the probability that c_i also falls inside a fixed Hamming
ball, even conditioned on $c_1, \ldots, c_{i-1}$ belonging to that ball.

Finally, a union ball over all centers for the Hamming ball and all choices of
"good" (in the above sense) J tuples of outer codewords is used to show that
with high probability (over the choice of the independent inner encodings) the
concatenated code will not have too many codewords in *any* Hamming ball of
fractional radius $H^{-1}(1 - rR_0) - \varepsilon$. We refer the reader to [7] for the formal
details and the precise probability calculations.

5   List Decoding Up to $1/2 - o(1)$ Radius

In this section, we consider the problem of binary codes that can correct close to the information-theoretically maximum possible fraction $1/2$ of errors. Consider the task of communicating k bits of information over a channel that can flip an arbitrary subset of up to a fraction $1/2 - \varepsilon$ of the transmitted bits. Here we think of $\varepsilon \rightarrow 0$ as very small, and even allow $\varepsilon = o(1)$ (as k grows). We are interested in the following question: What is the fewest (asymptotic) number $n = n(k, \varepsilon)$ of bits we need to communicate so that no matter which subset of at most $(1/2 - \varepsilon)n$ bits are corrupted, we can recover the k message bits *efficiently* (in time polynomial in k and $1/\varepsilon$)? This question arises in many applications of list decoding in complexity theory and cryptography such as the construction of hardcore predicates from one-way functions, constructions of randomness extractors and pseudorandom generators, approximability of the VC dimension, membership comparability of NP-complete sets, approximating NP-witnesses, etc. We refer the reader to [18] and [4, Chap. 12] for a survey of some of these applications of list decoding.

It can be shown by a random coding argument that there exist codes with $n = O(k/\varepsilon^2)$ such that every Hamming ball of radius $(1/2 - \varepsilon)n$ has at most poly(k/ε) codewords. Further, this bound is tight — for any code with $n < k/\varepsilon^a$ for $a < 2$, there must exist some error pattern for which a super-polynomial number of codewords need to be output. The upper bound of $n = O(k/\varepsilon^2)$ above is non-constructive, and all known explicit constructions achieve weaker bounds on n. In the sequel, we focus on codes that can be constructed in time polynomial in k/ε, as well as list decoded from a fraction $(1/2 - \varepsilon)$ of errors in poly(k/ε) time. (The above mentioned applications in complexity theory demand such efficiency of the construction and the decoding algorithms.) In particular, brute-force search that takes time exponential in $1/\varepsilon$ is not permitted. The construction of codes list-decodable up to the Zyablov radius due to [8] that we discussed in Section 3 achieves an encoding length $n = O(k/\varepsilon^3)$; however, it has construction and decoding complexity exponential in $1/\varepsilon$. Prior to [8], an encoding length of $n = O(k/\varepsilon^4)$ was obtained in [6], but the construction time was once again exponential in $1/\varepsilon$.

We should remark that in the complexity-theoretic applications, the exact dependence of n on k often does not matter as long as the exponent of k is some constant. In the "traditional" setting of coding theory, the most interesting regime is when the exponent of k equals one as this corresponds to constant rate codes (when ε is thought of as a constant). On the other hand, in some complexity theory settings, one would like ε to be as small a function of n as possible. This makes the goal of approaching the optimal $1/\varepsilon^2$ dependence of the block length n on ε important.

The approach to construct such binary codes is again to concatenate an outer algebraic code with good distance/list decoding properties with some special low-rate inner code. By concatenating an outer Reed-Solomon code with an inner Hadamard code, a bound of $n = O(k^2/\varepsilon^4)$ was achieved in [12]. This quartic dependence on ε was the best known bound, till a recent improvement by Guruswami and Rudra [9] who constructed codes with encoding length $n = O\left(\frac{k^3}{\varepsilon^{3+\gamma}}\right)$ for any constant $\gamma > 0$. Their construction was based on concatenation of an outer folded Reed-Solomon code (or their precursor, the Parvaresh-Vardy codes [16]), with a dual BCH code as inner code.

We remark that the Parvaresh-Vardy codes are a generalization of Reed-Solomon codes and the dual BCH codes are generalizations of Hadamard codes. Thus, the above result from [9] generalizes both the outer and inner codes in the Reed-Solomon concatenated with Hadamard construction. This seems necessary for reasons we will sketch shortly.

Consider a Reed-Solomon code of block length $n = 2^m$ over an extension field $\mathbb{F}_{2^m}$ concatenated with a binary Hadamard code of dimension m and block length 2^m. The block length of the concatenated code is thus $N = 2^{2m} = n^2$. Let k' denote the dimension of the Reed-Solomon code, and let $k = k'm$ denote the dimension of the binary concatenated code. If the total fraction of errors w.r.t a codeword is at most $(1/2 - \varepsilon)$, at least $\varepsilon/2$ of the inner Hadamard blocks have at most a fraction $(1/2 - \varepsilon/2)$ of errors. List decoding the Hadamard code up to radius $(1/2 - \varepsilon/2)$ will return a list of at most $1/\varepsilon^2$ candidate field elements for each position of the Reed-Solomon code. At the outer level, it suffices to output all polynomials which agree with one of these candidate symbols for at least $\varepsilon n/2$ positions. This can be accomplished using Sudan's list decoding algorithm for Reed-Solomon codes [17] provided the rate k'/n of the Reed-Solomon code is at most $\varepsilon^4/16$. For $k' = \Omega(\varepsilon^4 n)$, the block length of the concatenated code satisfies $N = n^2 \leqslant O(k'^2/\varepsilon^8) \leqslant O(k^2/\varepsilon^8)$.

To improve this to the $O(k^2/\varepsilon^4)$ bound attained in [12], one must pass more sophisticated information from the Hadamard decoding stage to the outer Reed-Solomon decoder, and use a *soft* list decoding algorithm for Reed-Solomon codes [11,15]. Informally, for each inner block i the Hadamard decoder returns every field symbol α as a candidate symbol along with a confidence estimate $w_{i,\alpha}$ which is a (decreasing) linear function of the distance of that inner block to the Hadamard encoding of α. The second moment of these confidence estimates is at most $O(1)$ (this follows from the Parseval's identity for Fourier coefficients, and was the key to the result in [12]). When plugged into the soft decoding bound [11] for Reed-Solomon codes, this implies that the outer decoding succeeds for the larger rate $k'/n = \Omega(\varepsilon^2)$, leading to a block length $N \leqslant O(k^2/\varepsilon^4)$.

The improvement in [9] is based on using the Parvaresh-Vardy codes, which have better list-decoding guarantees, for the outer encoding. For this construction, the Hadamard code is too wasteful to be used at the inner level to encode the PV codeword symbols, since PV codes have a much larger alphabet size than Reed-Solomon codes. Dual BCH codes are much more efficient in encoding length. On the other hand, they necessarily provide weaker guarantees. In particular, the second moment of a certain coset weight distribution is bounded in the case of Hadamard codes, and this found use as appropriate confidence estimates to pass to the Reed-Solomon soft decoder. For duals of BCH codes with distance $(2t+1)$, an analogous bound holds only for the $2t$'th moment (this bound also arose in the work of Kaufman and Litsyn [13] on property testing of dual BCH codes). But quite remarkably, there is a *soft* decoding algorithm for PV codes that works under the weaker guarantee of bounded higher order moments, and this enables obtaining a near-cubic dependence on $1/\varepsilon$ in [9]. We refer the reader to [9] for further details and the precise calculations. We stress that the power of the soft decoding algorithm for the newly discovered algebraic codes [16,8], and the ability to exploit it via meaningful weights passed from the dual BCH decoder, is crucial for this result.

References

1. Elias, P.: Error-correcting codes for list decoding. IEEE Transactions on Information Theory 37, 5–12 (1991)
2. Goldreich, O., Levin, L.: A hard-core predicate for all one-way functions. In: Proc. of the 21st ACM Symp. on Theory of Computing, pp. 25–32 (1989)
3. Gopalan, P., Klivans, A., Zuckerman, D.: List-decoding Reed-Muller codes over small fields. In: Proc. 40th ACM Symposium on Theory of Computing (STOC 2008), pp. 265–274 (2008)
4. Guruswami, V.: List Decoding of Error-Correcting Codes. LNCS, vol. 3282. Springer, Heidelberg (2004)
5. Guruswami, V.: Artin automorphisms, cyclotomic function fields, and folded list-decodable codes. arXiv:0811.4139 (November 2008)
6. Guruswami, V., Håstad, J., Sudan, M., Zuckerman, D.: Combinatorial bounds for list decoding. IEEE Transactions on Information Theory 48(5), 1021–1035 (2002)
7. Guruswami, V., Rudra, A.: Concatenated codes can achieve list decoding capacity. In: Proceedings of 19th ACM-SIAM Symposium on Discrete Algorithms, pp. 258–267 (January 2008)
8. Guruswami, V., Rudra, A.: Explicit codes achieving list decoding capacity: Error-correction with optimal redundancy. IEEE Transactions on Information Theory 54(1), 135–150 (2008); Preliminary version in STOC 2006
9. Guruswami, V., Rudra, A.: Soft decoding, dual BCH codes, and better list-decodable ε-biased codes. In: Proceedings of the 23rd Annual IEEE Conference on Computational Complexity, pp. 163–174 (2008)
10. Guruswami, V., Rudra, A.: Better binary list-decodable codes via multilevel concatenation. IEEE Transactions on Information Theory 55(1), 19–26 (2009)
11. Guruswami, V., Sudan, M.: Improved decoding of Reed-Solomon and algebraic-geometric codes. IEEE Transactions on Information Theory 45, 1757–1767 (1999)
12. Guruswami, V., Sudan, M.: List decoding algorithms for certain concatenated codes. In: Proceedings of the 32nd Annual ACM Symposium on Theory of Computing (STOC), pp. 181–190 (2000)
13. Kaufman, T., Litsyn, S.: Almost orthogonal linear codes are locally testable. In: Proceedings of the 46th Annual IEEE Symposium on Foundations of Computer Science (FOCS), pp. 317–326 (2005)
14. Kaufman, T., Lovett, S.: The list-decoding size of Reed-Muller codes. arXiv:0811.2356 (November 2008)
15. Koetter, R., Vardy, A.: Algebraic soft-decision decoding of Reed-Solomon codes. IEEE Transactions on Information Theory 49(11), 2809–2825 (2003)
16. Parvaresh, F., Vardy, A.: Correcting errors beyond the Guruswami-Sudan radius in polynomial time. In: Proc. of the 46th IEEE Symposium on Foundations of Computer Science, pp. 285–294 (2005)

17. Sudan, M.: Decoding of Reed-Solomon codes beyond the error-correction bound. Journal of Complexity 13(1), 180–193 (1997)
18. Sudan, M.: List decoding: Algorithms and applications. SIGACT News 31, 16–27 (2000)
19. Zyablov, V.V., Pinsker, M.S.: List cascade decoding. Problems of Information Transmission 17(4), 29–34 (1981) (in Russian); pp. 236–240 (1982) (in English)

Recent Developments in Low-Density Parity-Check Codes

Wen-Ching Winnie Li*, Min Lu, and Chenying Wang

Department of Mathematics, The Pennsylvania State University,
University Park PA 16802, USA

Abstract. In this paper we prove two results related to low-density parity-check (LDPC) codes. The first is to show that the generating function attached to the pseudo-codewords of an LDPC code is a rational function, answering a question raised in [6]. The combinatorial information of its numerator and denominator is also discussed.

The second concerns an infinite family of q-regular bipartite graphs with large girth constructed in [8]. The LDPC codes based on these graphs have attracted much attention. We show that the first few of these graphs are Ramanujan graphs.

1   Introduction

Introduced by Gallager [3], a low-density parity-check (LDPC) code C is a binary linear code equipped with a sparse parity-check matrix H. The check equations included in H are graphically described by a bipartite Tanner graph $T(H)$ whose two vertex sets consist of the bit nodes and check nodes, corresponding to the columns and rows of H, respectively. A bit node v_j is adjacent to a check node f_i if and only if the ijth entry of H is 1, in other words, the jth variable occurs in the ith check equation. Thus a codeword is an assignment of 0 and 1 of the bit nodes such that the neighbors of each check node sum to zero.

The vertices in $T(H)$ have low degree because the matrix H is sparse. To capitalize on this feature, an efficient decoding scheme is designed, first by Gallager [3], to send information back and forth between bit nodes and check nodes across the Tanner graph as follows [9]. The bit nodes are initialized by the received word, which is first screened to see if it is already a codeword. If it is, then stop. If not, the information is passed to the neighboring check nodes, which compute new estimates based on the received information and error probability, and then send the new information to their neighboring bit nodes. After one round of message passing, the algorithm computes the error probability and outputs an assignment of 0 and 1 at each bit node which is again checked to see if it is a codeword. If yes, then stop; if not, repeat the above procedure. To date there are several message-passing iterative decoding (MPID) algorithms, such as min-sum and sum-product algorithms [16,17]. Some questions arise naturally: how does

---

* The research of the first author is supported in part by the NSF grants DMS-0457574 and DMS-0801096 and by the DARPA grant HR0011-06-1-001.

C. Xing et al. (Eds.): IWCC 2009, LNCS 5557, pp. 107–123, 2009.

MPID compare with the traditional maximum likelihood (ML) decoding? how to avoid mis-correction? and what will happen if MPID is repeatedly applied to a codeword? These questions are addressed in §2.

Since the MPID algorithms operate locally, they converge very fast. On the other hand, an algorithm acting locally can not distinguish the Tanner graph $T(H)$ from its finite unramified covers, which are the same locally. Let T_m be an m-fold unramified cover of $T(H)$ and C_m the code whose Tanner graph is T_m. Since each vertex of $T(H)$ is covered by m vertices in T_m, the length of C_m is m times the length of C, which is the number, n, of the bit nodes in $T(H)$. Given a codeword $\mathbf{c}$ of C_m, the unscaled pseudo-codeword $\mathbf{p}(\mathbf{c}) = (c_1, ..., c_n)$ associated to $\mathbf{c}$ is a vector in $\mathbb{Z}^n$ such that for each $1 \leq j \leq n$, c_j counts the number of 1's in $\mathbf{c}$ occurring at the bit nodes of T_m above the bit node v_j of $T(H)$. Notice that $\mathbf{p}(\mathbf{c})$ comes from a codeword in C if and only if its normalization $\frac{1}{m}\mathbf{p}(\mathbf{c})$ lies in C. If the Tanner graph $T(H)$ is a tree, then all normalized pseudo-codewords are codewords. Otherwise, the relation between pseudo-codewords and MPID algorithms is seen through the "computation tree", as explained in [16,17]. Suffice it to say that the pseudo-codewords record all decoding errors, hence it is of utmost importance to understand them.

The smallest cone containing all pseudo-codewords, called the fundamental cone $\mathcal{K}(H)$ attached to the LDPC code C, is determined by Koetter and Vontobel [7] in terms of inequalities derived from the check equations contained in H. Two characterizations for pseudo-codewords are given by Koetter-Li-Vontobel-Walker in [5,6]. The first one says that they are the integral points in the cone which, modulo 2, reduce to codewords. The second is in terms of the monomials occurring in the edge zeta function of a graph. When the code C is a cycle code on a graph X, the pseudo-codewords are described by the monomials in the edge zeta function of X. But for a general C, they correspond to monomials in the edge zeta function of the Tanner graph of C satisfying a special property. In view of this, a question was raised in [6] to find a zeta function attached to the code C whose monomials describe the pseudo-codewords of C, and the zeta function itself is a rational function providing interesting combinatorial information.

The first topic of this paper is to present an answer to this question. A full discussion is contained in Lu's thesis [10].

Given a vector of n variables $\mathbf{u} = (u_1, ..., u_n)$ and a vector $\mathbf{w} = (w_1, ..., w_n)$ in $\mathbb{R}^n$, write $\mathbf{u}^{\mathbf{w}}$ for the monomial $u_1^{w_1} \cdots u_n^{w_n}$. The zeta function attached to an LDPC code C of length n is the zeta function attached to its fundamental cone $\mathcal{K}(H)$, defined as the generating function of its pseudo-codewords:

$$Z_C(u_1, ..., u_n) = \sum_{\mathbf{p} \text{ pseudo-codeword of } C} \mathbf{u}^{\mathbf{p}}.$$

Using the aforementioned first criterion of pseudo-codewords, we shall prove

Theorem 1. *Let C be an LDPC code of length n with parity check matrix H and fundamental cone $\mathcal{K}(H)$. Then its zeta function is a rational function of the form*

$$Z_C(u_1, ..., u_n) = \frac{\sigma(\mathbf{u})}{(1 - \mathbf{u}^{\mathbf{w}_1}) \cdots (1 - \mathbf{u}^{\mathbf{w}_d})},$$

where $\mathbf{w}_1, ..., \mathbf{w}_d$ *are pseudo-codewords with even components which generate the cone, namely*

$$\mathcal{K}(H) = \{\lambda_1\mathbf{w}_1 + \cdots + \lambda_d\mathbf{w}_d \; : \; \lambda_i \geq 0 \text{ for } i = 1, ..., d\},$$

and $\sigma(\mathbf{u})$ *is a polynomial.*

To describe the numerator, more information about the cone $\mathcal{K}(H)$ is needed. First assume that the cone $\mathcal{K}(H)$ is d-dimensional generated by d linearly independent pseudo-codewords $\mathbf{w}_1, ..., \mathbf{w}_d$, called a simplicial d-cone. In this case $\sigma(\mathbf{u}) = \sigma_\Pi(\mathbf{u}) := \sum_{\mathbf{p} \text{ pseudo-codeword in } \Pi} \mathbf{u}^\mathbf{p}$, where Π is the half-open fundamental parallelopiped

$$\Pi = \{\lambda_1\mathbf{w}_1 + \cdots + \lambda_d\mathbf{w}_d : 0 \leq \lambda_i < 1, i = 1, ..., d\}.$$

In general, the cone is triangulated into a finite union of simplicial cones, each generated by a subset of $\mathbf{w}_1, ..., \mathbf{w}_d$. The generating function of the pseudo-codewords contained in each simplicial cone is a rational function, and Z_C is a finite sum and difference of these rational functions. The least common denominator of these rational functions is the denominator of Z_C as described in Theorem 1.

Thus the first step to explicitly determine the zeta function of the code is to find generators of the fundamental cone. To avoid the subsequent time-consuming steps of triangulation and computing the generating function of each simplicial cone, we propose a much simpler alternative

$$Z'_C(u_1, ..., u_n) = \frac{\sigma_\Pi(\mathbf{u})}{(1 - \mathbf{u}^{\mathbf{w}_1}) \cdots (1 - \mathbf{u}^{\mathbf{w}_d})}$$

using the same notation as above. Its power series expansion is the sum over the pseudo-codewords $\mathbf{p}$ of C of monomials $\mathbf{u}^\mathbf{p}$ with positive integral coefficients. See §5 for terminologies and details.

Now we turn to the second topic of this paper. As explained in §2, in MPID half of the girth of the Tanner graph plays the role of minimum distance in the traditional theory of error correcting codes. Thus it is desirable to construct codes from graphs with large girth. Ramanujan graphs introduced by Lubotzky-Phillips-Sarnak [11] are known to have this property. These are connected k-regular graphs whose nontrivial eigenvalues, namely eigenvalues other than $\pm k$, have absolute value at most $2\sqrt{k-1}$. The extremal spectral property makes them optimal expanders, leading to wide applications. In addition, they are sparse and have large girth. Hence they appear to be good candidates for Tanner graphs. Rosenthal and Vontobel [12] were the first to construct LDPC codes based on Ramanujan graphs.

For a fixed prime power q, Lazebnik and Ustimenko [8] constructed an infinite q-regular bipartite graph $D(q)$ with vertex sets

$$X = \{\mathbf{x} = [x, x_1, x_2, \ldots] : x, x_i \in \mathbb{F}_q \text{ for } i \geq 1\}$$

and

$$Y = \{\mathbf{y} = [y, y_1, y_2, \ldots] : y, y_i \in \mathbb{F}_q \text{ for } i \geq 1\};$$

a vertex $\mathbf{x}$ in X and a vertex $\mathbf{y}$ in Y are adjacent if their coordinates satisfy the following relations:

$$y_1 = xy + x_1, \qquad y_2 = xy_1 + x_2, \qquad y_3 = yx_1 + x_3, \qquad y_4 = yx_2 + x_4, \quad (1)$$

and for $i \geq 1$,

$$y_{4i+1} = xy_{4i-1} + x_{4i+1}, \qquad\qquad y_{4i+2} = xy_{4i} + x_{4i+2},$$
$$y_{4i+3} = yx_{4i+1} + x_{4i+3}, \qquad\qquad y_{4i+4} = yx_{4i+2} + x_{4i+4}.$$

By deleting all except the first m coordinates of the vertices, one obtains a truncation graph $D(m, q)$, which is q-regular bipartite of size $2q^m$. Each $D(m + 1, q)$ is a q-fold unramified cover of $D(m, q)$. So they form an infinite tower of covering graphs. It was shown in [8] that for $m \geq 3$, the girth of $D(m, q)$ is large, at least $m + 5$. Recently, using $D(m, q)$ as the Tanner graph, Kim et al [4] constructed and studied the associated LDPC code $LU(m, q)$. Sin and Xiang [13] determined the dimension of the code $LU(3, q)$ for odd q, and Arslan [1] very recently settled the case of even q.

In view of the nice properties that $D(m, q)$ possess, we wonder if they happen to be Ramanujan graphs. It turns out that the first few are.

Theorem 2. *The nontrivial eigenvalues of $D(2, q)$ have absolute value at most $\sqrt{q}$ and those of $D(3, q)$ at most $\sqrt{2q}$. Consequently, they are Ramanujan graphs.*

The proof given in §5 is a new version of the joint work by two of the authors (Li and Wang). More information on eigenvalues and eigenvectors can be found in the thesis of Wang [15].

This paper is organized as follows. In §2 we summarize the behavior of MPID. The conclusion is that, if not too many errors occurred, then after a definite number of iterations, the outputs remain the same and the errors are corrected. The two criteria on pseudo-codewords obtained in [6] are reviewed in §3. The rationality of the zeta function of the code (Theorem 1) is proved in §4, where examples are also exhibited. Finally in §5 we prove Theorem 2.

2   Stability of MPID and Comparison with ML Decoding

Wiberg proves in [16] that if the Tanner graph is a tree, then MPID algorithm stabilizes, that is, gives the same output after finitely many iterations, and it agrees with the ML decoding. In this case, to guarantee the agreement with ML decoding, in an unpublished paper, M-H. Kang shows that the number of iterations should be at least half of the diameter of the tree. Furthermore, he constructs an example to show that, while a codeword is obtained after fewer number of iterations, it is not the one sent. In other words, mis-correction occurs. In real life applications, it is important to avoid mis-corrections.

The case where the Tanner graph is not a tree is considered by Wang in her thesis [15]. Denote by g half of the girth of the Tanner graph, and set $d = \lfloor \frac{g-1}{2} \rfloor$. Wang proves that a similar result holds provided that not too many errors occurred to begin with.

Theorem 3 ((Wang)). *(1) Suppose each bit node has degree 2. If the number of errors in a received word is at most d, then the min-sum decoding will stabilize and can correct the errors after $\frac{2d^2+3d}{2}$ iterations.*

(2) Assume the minimal degree of the bit nodes is $c \geq 3$. If the number of errors in a received word is at most $\lfloor \frac{(c-1)^{d+1}-1}{2c-4} \rfloor$, then the min-sum decoding will stabilize and can correct the errors after $d+1$ iterations.

3   Pseudo-codewords and Graph Edge Zeta Functions

3.1   Edge Zeta Function of a Graph

Let X be an undirected graph with edges $e_1, \ldots, e_n$. Assign the variable u_i to the edge e_i. To each edge path $E = (e_{i_1}, \ldots, e_{i_k})$, associate the monomial

$$g(E) = u_{i_1} \ldots u_{i_k}.$$

The edge zeta function of X is defined as

$$Z_X(u_1, \ldots, u_n) = \prod_{[E]} (1 - g(E))^{-1},$$

where $[E]$ runs through all equivalence classes of backtrackless, tailless and primitive cycles in X.

Endow two directions on each edge of X. Define the out-neighbors of a directed edge $x \to y$ to be the directed edges $y \to z$ with $z \neq x$. The edge adjacency matrix A_e associated to X is a $2n \times 2n$ matrix whose rows and columns are indexed by the directed edges of X such that ee'-entry of A_e records the number of out-edges from e to e'. Let U denote the corresponding column variable matrix such that the same variable u_i is attached to both directions of e_i. In their paper [14], Stark and Terras gave an explicit expression of the edge zeta function as a rational function:

Theorem 4 ((Stark-Terras))

$$Z_X(u_1, ..., u_n) = \frac{1}{\det(I - A_e U)}.$$

3.2   Pseudo-Codewords of a Cycle Code and Graph Edge Zeta Function

Let C be an LDPC code with parity-check matrix H and Tanner graph $T(H)$. If all bit nodes in $T(H)$ have degree two, then we may regard each bit node of $T(H)$ as representing an edge connecting the two check nodes adjacent to this bit node. Thus we obtain a graph X whose vertices are the check nodes of $T(H)$ and whose edges are marked by the bit nodes of $T(H)$. The codewords in C are then in one-to-one correspondence with the simple edge cycles in X, and C is called a cycle code. The relation between pseudo-codewords of a cycle code and the zeta function of the underlying graph was studied in [5].

Theorem 5 ((Koetter-Li-Vontobel-Walker))
Let C be a cycle code of length n on graph X with edge zeta function $Z_X(u_1, \ldots, u_n)$. Then

- *$\mathbf{p} = (p_1, \ldots, p_n)$ is an unscaled pseudo-codeword of C if and only if there is a disjoint union of backtrackless tailless cycles in X which use the edge e_i exactly p_i times for $i = 1, \ldots, n$. In other words,*
- *$\mathbf{p} = (p_1, \ldots, p_n)$ is an unscaled pseudo-codeword of C if and only if the monomial $\mathbf{u^P} = u_1^{p_1} \cdots u_n^{p_n}$ occurs in the power series expansion of $Z_X(u_1, \ldots, u_n)$ with nonzero coefficient.*

Recall that the Newton polygon of a polynomial $f(X, Y)$ is the convex hull of the exponents of the monomials occurring in f. Define the Newton polytope of a power series in several variables in a similar way. In particular, there is a Newton polytope associated to the edge zeta function of a graph. As mentioned in §1, the fundamental cone of an LDPC code is the smallest cone containing all pseudo-codewords of the code. Combined with the above characterization of the pseudo-codewords of a cycle code, we get

Theorem 6. *The fundamental cone of the cycle code C on graph X is the Newton polytope of the edge zeta function Z_X associated to X.*

3.3   Pseudo-codewords of a General LDPC Code and Edge Zeta Function of Its Tanner Graph

Cycle codes are the simplest LDPC codes. For a general LDPC code C with parity-check matrix H, a similar result relating pseudo-codewords of C and the edge zeta function of the Tanner graph $T(H)$ was obtained in [6]. In this case, only those monomials in the edge zeta function satisfying a special property are relevant.

Theorem 7 ((Koetter-Li-Vontobel-Walker))
Let C be an LDPC code of length n, parity-check matrix H and Tanner graph $T(H)$. Suppose that every bit node has even degree. Then

(1) The codewords of C are in one-to-one correspondence with the simple edge cycles in $T(H)$ such that at each bit node, either all or none of the edges from that node occur.

(2) $\mathbf{p} = (p_1, \ldots, p_n)$ is an unscaled pseudo-codeword of C if and only if in the edge zeta function $Z_{T(H)}$ of $T(H)$ there occurs a monomial in which the degree of the variables representing the edges incident to the ith bit node is p_i for $1 \leq i \leq n$.

4   Zeta Function of an LDPC Code

In this section we fix an LDPC code C of length n with parity-check matrix $H = (h_{ji})$ and fundamental cone $\mathcal{K}(H)$. The edge zeta function of its Tanner graph defined in §3 contains monomials that do not correspond to pseudo-codewords. In contrast, we shall prove in this section that

$$Z_C(u_1, ..., u_n) = \sum_{\mathbf{p}=(p_1,...,p_n) \text{ pseudo-codeword of } C} \mathbf{u}^{\mathbf{p}}$$

is a rational function as described in Theorem 1. Moreover, we shall propose a much simpler rational function which also enumerates the pseudo-codewords of C, but allows a pseudo-codeword to appear multiple times.

4.1 Preliminary Results

We will use the following characterization of pseudo-codewords given in [6].

Theorem 8 ((Koetter-Li-Vontobel-Walker)). *Let* $\mathbf{p} = (p_1, \ldots, p_n)$ *be a vector with integral components. Then the following two statements are equivalent:*
(1) $\mathbf{p}$ *is an unscaled pseudo-codeword of* C;
(2) $\mathbf{p} \in \mathcal{K}(H)$ *and* $H\mathbf{p}^t \equiv \mathbf{0} \pmod 2$.

To facilitate our proof, we review some basic results concerning the structure of cones which will be used later. We follow the terminologies in [2]. A pointed cone $\mathcal{K} \subseteq \mathbb{R}^d$ is a set of the form

$$\mathcal{K} = \{\mathbf{v} + \lambda_1 \mathbf{w}_1 + \lambda_2 \mathbf{w}_2 + \cdots + \lambda_m \mathbf{w}_m : \lambda_i \geq 0, i = 1, \ldots, m\},$$

where $\mathbf{v}, \mathbf{w}_1, \mathbf{w}_2, \ldots, \mathbf{w}_m \in \mathbb{R}^d$ are such that there exists a hyperplane P for which $P \cap \mathcal{K} = \{\mathbf{v}\}$, in other words, $\mathcal{K}$ lies strictly on one side of P. The vector $\mathbf{v}$ is called the apex of $\mathcal{K}$, and the $\mathbf{w}_k$'s are the generators of $\mathcal{K}$. The cone is called rational if $\mathbf{v}, \mathbf{w}_1, \mathbf{w}_2, \ldots, \mathbf{w}_m \in \mathbb{Q}^d$, in which case we may choose $\mathbf{w}_1, \ldots, \mathbf{w}_m \in \mathbb{Z}^d$ by clearing the denominators. The dimension of $\mathcal{K}$ is the dimension of the affine space spanned by $\mathcal{K}$; if $\mathcal{K}$ is of dimension d, we call it a d-cone. The d-cone $\mathcal{K}$ is *simplicial* if $\mathcal{K}$ has precisely d linearly independent generators.

A collection T of simplicial d-cones is a *triangulation* of the d-cone $\mathcal{K}$ if it satisfies

(a) $\mathcal{K} = \cup_{\mathcal{S} \in T} \mathcal{S}$; and
(b) For any $\mathcal{S}_1, \mathcal{S}_2 \in T$, $\mathcal{S}_1 \cap \mathcal{S}_2$ is a face common to both $\mathcal{S}_1$ and $\mathcal{S}_2$.

We say that $\mathcal{K}$ can be triangulated using no new generators if there exists a triangulation T such that the generators of any $\mathcal{S} \in T$ are generators of $\mathcal{K}$. An important conclusion on pointed cone is

Theorem 9 (([2], p. 60)). *Any pointed cone can be triangulated into simplicial cones using no new generators.*

The following result says that cones defined by generators are the same as those defined by linear inequalities.

Theorem 10 ([[18], p. 30)). *A cone $\mathcal{K} \subseteq \mathbb{R}^d$ is a finitely generated combination of vectors in $\mathbb{R}^d$ if and only if it is a finite intersection of closed linear half-spaces.*

Our fundamental cone $\mathcal{K}(H)$ is defined by inequalities derived from the check equations as follows. The ith check equation $\sum_j h_{ij} x_j = 0$ gives rise to n inequalities

$$\sum_{j \neq l} h_{ij} x_j \geq h_{il} x_l, \quad \text{for } 1 \leq l \leq n,$$

and the cone consists of points $(x_1, ..., x_n) \in \mathbb{R}^n$ satisfying all $x_j \geq 0$ and all the inequalities above. By Theorem 10 it is a pointed cone with apex at the origin. Its properties are summarized in

Theorem 11. *The fundamental cone $\mathcal{K}(H)$ is a pointed rational cone with apex at the origin, it is generated by finitely many pseudo-codewords, and it can be triangulated into simplicial cones using no new generators.*

Proof. The generators of the fundamental cone lie on its 1-dimensional boundaries, which are the intersections of closed linear half-spaces defined by linear equations with integral coefficients, hence the cone is rational. Moreover, each 1-dimensional boundary contains pseudo-codewords, for instance, the lattice points with all even components; choose as generators a nonzero pseudo-codeword from each 1-dimensional boundary. The third assertion follows from Theorem 9.

4.2   Generating Functions of Rational Cones

The generating function of a subset $S \subset \mathbb{R}^d$ is the sum of monomials recording the lattice points in S:

$$\sigma_S(\boldsymbol{z}) = \sigma_S(z_1, z_2, \ldots, z_d) := \sum_{\boldsymbol{m} \in S \cap \mathbb{Z}^d} \boldsymbol{z}^{\boldsymbol{m}}.$$

The following theorem describes how to enumerate the lattice points in a simplicial cone.

Theorem 12 ([[2], p. 62)). *Suppose*

$$\mathcal{K} := \{\lambda_1 \boldsymbol{w}_1 + \lambda_2 \boldsymbol{w}_2 + \cdots + \lambda_d \boldsymbol{w}_d : \lambda_i \geq 0, i = 1, \ldots, d\}$$

is a simplicial rational d-cone, where $\boldsymbol{w}_1, \boldsymbol{w}_2, \ldots, \boldsymbol{w}_d \in \mathbb{Z}^d$. Then for $\boldsymbol{v} \in \mathbb{R}^d$, the generating function $\sigma_{\boldsymbol{v}+\mathcal{K}}$ of the shifted cone $\boldsymbol{v} + \mathcal{K}$ is the rational function

$$\sigma_{\boldsymbol{v}+\mathcal{K}}(\boldsymbol{z}) = \frac{\sigma_{\boldsymbol{v}+\Pi}(\boldsymbol{z})}{(1 - \boldsymbol{z}^{\boldsymbol{w}_1})(1 - \boldsymbol{z}^{\boldsymbol{w}_2}) \cdots (1 - \boldsymbol{z}^{\boldsymbol{w}_d})},$$

where Π is the (half-open) fundamental parallelepiped of $\mathcal{K}$:

$$\Pi := \{\lambda_1 \boldsymbol{w}_1 + \lambda_2 \boldsymbol{w}_2 + \cdots + \lambda_d \boldsymbol{w}_d : 0 \leq \lambda_1, \lambda_2, \ldots, \lambda_d < 1\}.$$

By Theorem 9 and the fact that the intersection of simplicial cones in a triangulation is again a simplicial cone, the following consequence is evident.

Corollary 1 (([2], p. 63)). *The generating function of a finitely generated pointed rational cone is a rational function.*

Here is a simple example from [2], p. 60, which illustrates how to enumerate lattice points in a two-dimensional cone.

Example 1. Consider the two-dimensional cone

$$\mathcal{K} := \{\lambda_1(1,1) + \lambda_2(-2,3) : \lambda_1, \lambda_2 \geq 0\} \subset \mathbb{R}^2.$$

Its fundamental parallelogram Π is

$$\Pi = \{\lambda_1(1,1) + \lambda_2(-2,3) : 0 \leq \lambda_1, \lambda_2 < 1\} \subset \mathbb{R}^2.$$

The cone $\mathcal{K}$ can be exactly covered by the translations of Π by nonnegative linear combinations of $(1,1)$ and $(-2,3)$. The lattice points in Π are $(0,0), (0,1), (0,2),$ $(-1,2)$ and $(-1,3)$. Hence

$$\sigma_{\mathcal{K}}(x,y) = \frac{1 + y + y^2 + x^{-1}y^2 + x^{-1}y^3}{(1 - xy)(1 - x^{-2}y^3)}.$$

4.3   Generating Functions of Pseudo-codewords

As recalled in §4.1, pseudo-codewords of C are the lattice points in $\mathcal{K}(H)$ congruent to codewords modulo 2. We prove the first main result of this paper, Theorem 1, by adapting the proof of Theorem 12.

Theorem 13. *The zeta function*

$$Z_C(u_1, ..., u_n) = \sum_{\mathbf{p}=(p_1,...,p_n) \text{ pseudo-codeword of } C} \mathbf{u}^{\mathbf{p}}$$

is rational.

Proof. By Theorem 11 the fundamental cone is generated by finitely many pseudo-codewords. We only need to show that the generating function of pseudo-codewords in a simplicial rational cone is rational.

Assume that a simplicial cone $\mathcal{K}$ is generated by the pseudo-codewords $\mathbf{w}_1, \mathbf{w}_2,$ $\ldots, \mathbf{w}_d$ and that all generators have even components. (Multiply them by 2 if necessary). We claim that every pseudo-codeword $\mathbf{p}$ in $\mathcal{K}$ can be uniquely written as

$$\mathbf{p} = \mathbf{p_0} + \lambda_1\mathbf{w}_1 + \lambda_2\mathbf{w}_2 + \cdots + \lambda_d\mathbf{w}_d \tag{2}$$

for some pseudo-codeword $\mathbf{p_0}$ in the fundamental parallelepiped Π of $\mathcal{K}$ and some non-negative integers $\lambda_1, \lambda_2, \ldots, \lambda_d$. Indeed, it follows from the definition of a simplicial cone given in §4.1 that we may write

$$\mathbf{p} = \alpha_1\mathbf{w}_1 + \alpha_2\mathbf{w}_2 + \cdots + \alpha_d\mathbf{w}_d$$

uniquely for some nonnegative real numbers $\alpha_1, \alpha_2, \ldots, \alpha_d$. Let $\lfloor x \rfloor$ and $\{x\}$ denote the integral part and fractional part of a non-negative real number x, respectively. Then

$$p = p_0 + \lfloor \alpha_1 \rfloor w_1 + \lfloor \alpha_2 \rfloor w_2 + \cdots + \lfloor \alpha_d \rfloor w_d,$$

where $p_0 = \{\alpha_1\}w_1 + \{\alpha_2\}w_2 + \cdots + \{\alpha_d\}w_d$ is a lattice point since it is the difference of p and a lattice point. Clearly p_0 lies in the fundamental parallelepiped Π since $0 \leq \{\alpha_i\} < 1$. Again, this representation is unique. Because all the generators have even components, p is congruent to p_0 modulo 2. Therefore, by Theorem 8, p_0 is a pseudo-codeword because p is. This proves the claim with $\lambda_i = \lfloor \alpha_i \rfloor$ for $i = 1, 2, \ldots, d$.

The above unique representation of each pseudo-codeword in $\mathcal{K}$ allows us to express the generating function $\sigma_{\mathcal{K}}(z)$ of the pseudo-codewords in a simplicial (rational) cone $\mathcal{K}$ as

$$\sigma_{\mathcal{K}}(u) := \sum_{p \text{ pseudo-codeword in } \mathcal{K}} u^p$$

$$= \left(\sum_{p_0 \text{ pseudo-codeword in } \Pi} u^{p_0} \right) \left(\sum_{\lambda_1 \geq 0} u^{\lambda_1 w_1} \right) \cdots \left(\sum_{\lambda_d \geq 0} u^{\lambda_d w_d} \right)$$

$$= \frac{\sigma_{\Pi}(u)}{(1 - u^{w_1})(1 - u^{w_2}) \cdots (1 - u^{w_d})},$$

where $\sigma_{\Pi}(u) = \sum_{p_0} u^{p_0}$ is a finite polynomial summing over the pseudo-codewords in the fundamental parallelepiped Π. Therefore $\sigma_{\mathcal{K}}(u)$ is a rational function.

By Theorem 11, the fundamental cone can be triangulated into simplicial cones. Since the intersections of simplicial cones in a triangulation are again simplicial, the generating function of the pseudo-codewords in $\mathcal{K}(H)$, which is Z_C, can be obtained using inclusion-exclusion. It is rational because it is the sum and difference of rational functions.

Remark 1. From the proof we can see that if we do not require that each pseudo-codeword appear exactly once in the generating function, that is, the coefficient of each monomial equal 1, we do not have to decompose the fundamental cone into simplicial cones, which really saves a lot of time.

4.4 An Example

Consider the parity-check matrix

$$H = \begin{pmatrix} 1 & 1 & 1 & 0 \\ 1 & 0 & 1 & 1 \end{pmatrix}.$$

It defines the code

$$C(H) = \{(0,0,0,0), (1,0,1,0), (1,1,0,1), (0,1,1,1)\},$$

and the fundamental cone $\mathcal{K}(H)$ is generated by the following 5 vectors with even components:

$$w_1 = (2, 2, 0, 2),$$

$$w_2 = (2, 4, 2, 0),$$

$$w_3 = (2, 0, 2, 0),$$

$$w_4 = (2, 0, 2, 4),$$

$$w_5 = (0, 2, 2, 2).$$

(These generators are found by the software package POLYMAKE.) A triangulation of $\mathcal{K}(H)$ is

$$T(\mathcal{K}(H)) = \{S_1, S_2\},$$

where

$$S_1 = \text{cone generated by} \{w_1, w_2, w_3, w_5\}$$

and

$$S_2 = \text{cone generated by} \{w_1, w_3, w_4, w_5\}.$$

Let $\sigma_S(u)$ be the generating function of pseudo-codewords in the region S and Π_C be the fundamental parallelepiped of the cone C. then

$$\sigma_{\mathcal{K}(H)}(u) = \frac{\sigma_{\Pi_{S_1}}(u)}{(1 - u^{w_1})(1 - u^{w_2})(1 - u^{w_3})(1 - u^{w_5})} + \frac{\sigma_{\Pi_{S_2}}(u)}{(1 - u^{w_1})(1 - u^{w_3})(1 - u^{w_4})(1 - u^{w_5})}$$
$$- \frac{\sigma_{\Pi_{S_1 \cap S_2}}(u)}{(1 - u^{w_1})(1 - u^{w_3})(1 - u^{w_5})},$$

where

$$\sigma_{\Pi_{S_1}}(u) = 1 + u_1 u_3 + u_2 u_3 u_4 + u_1 u_2 u_4 + u_1 u_2^2 u_3 + u_1 u_2 u_3^2 u_4 + u_1^2 u_2 u_3 u_4$$
$$+ u_1 u_2^2 u_3 u_4^2 + u_1^2 u_2^2 u_3^2 + u_1 u_2^3 u_3^2 u_4 + u_1^2 u_2^3 u_3 u_4 + u_1^2 u_2^2 u_3^2 u_4^2$$
$$+ u_1^2 u_2^3 u_3^3 u_4 + u_1^3 u_2^3 u_3^2 u_4 + u_1^2 u_2^4 u_3^2 u_4^2 + u_1^3 u_2^4 u_3^3 u_4^2,$$

$$\sigma_{\Pi_{S_2}}(u) = 1 + u_1 u_3 + u_1 u_2 u_4 + u_2 u_3 u_4 + u_1 u_3 u_4^2$$
$$+ u_1 u_2 u_3^2 u_4 + u_1^2 u_2 u_3 u_4 + u_1 u_2^2 u_3 u_4^2 + u_1^2 u_3^2 u_4^2$$
$$+ u_1^2 u_1 u_3 u_4^3 + u_1 u_2 u_3^2 u_4^3 + u_1^2 u_2^2 u_3^2 u_4^2$$
$$+ u_1^2 u_2 u_3^3 u_4^3 + u_1^3 u_2 u_3^2 u_4^3 + u_1^2 u_2^2 u_3^2 u_4^4 + u_1^3 u_2^2 u_3^3 u_4^4,$$

and

$$\sigma_{\Pi_{S_1 \cap S_2}}(u) = 1 + u_1 u_3 + u_1 u_2 u_4 + u_2 u_3 u_4$$
$$+ u_1 u_2 u_3^2 u_4 + u_1^2 u_2 u_3 u_4 + u_1 u_2^2 u_3 u_4^2 + u_1^2 u_2^2 u_3^2 u_4^2.$$

The initial terms in the Taylor series of $Z_{C(H)}(\boldsymbol{u})$ are

$$
\begin{aligned}
Z_{C(H)}(\boldsymbol{u}) = {} & 1 + u_1 u_3 + u_1 u_2 u_4 + u_2 u_3 u_4 + u_1 u_2^2 u_3 + u_1 u_3 u_4^2 + u_1^2 u_3^2 \\
& + u_1^2 u_2 u_3 u_4 + u_1 u_2 u_3^2 u_4 + u_1 u_2^2 u_3 u_4^2 + u_1^2 u_2^2 u_3^2 + u_1^2 u_2^2 u_4^2 + u_1^2 u_3^2 u_4^2 + u_2^2 u_3^2 u_4^2 + u_1^3 u_3^3 \\
& + u_1^2 u_2^3 u_3 u_4 + u_1^3 u_2 u_3^2 u_4 + u_1 u_2^3 u_3^2 u_4 + u_1^2 u_2 u_3^3 u_4 + u_1^2 u_2 u_3 u_4^3 + u_1 u_2 u_3^2 u_4^3 \\
& + u_1^2 u_2^4 u_3^2 + u_1^2 u_2^2 u_4^4 + u_1^3 u_2^2 u_3^2 + u_1^3 u_3^2 u_4^2 + u_1^3 u_2^2 u_3 u_4^2 + u_1 u_2^2 u_3^2 u_4^3 + u_1^2 u_2^2 u_3^2 u_4^2 + u_1^4 u_3^4 \\
& + \cdots
\end{aligned}
$$

4.5   Another Zeta Function Enumerating Pseudo-codewords

For a general LDPC code, it is usually complicated to triangulate the fundamental cone, especially when the number of generators of the cone is much greater than the dimension of the cone. So it is desirable to find another rational function that not only records the pseudo-codewords but also has a simpler form. The proof of Theorem 13 shows that if we only want to enumerate pseudo-codewords and have no restrictions on the coefficients of the monomials, then the rational function in the following theorem will serve our purpose.

Theorem 14. *Suppose that the fundamental cone* $\mathcal{K}(H)$ *of* C *is generated by the pseudo-codewords* $\boldsymbol{w}_1, \ldots, \boldsymbol{w}_m$ *with even components. Then the rational function*

$$
Z'_C(u_1, \ldots, u_n) = \frac{\sigma_\Pi(u_1, \ldots, u_n)}{(1 - \mathbf{u}^{\boldsymbol{w}_1}) \cdots (1 - \mathbf{u}^{\boldsymbol{w}_m})}
$$

enumerates all the pseudo-codewords of C, *where* $\sigma_\Pi(u_1, \ldots, u_n)$ *enumerates the pseudo-codewords in*

$$
\Pi = \{\lambda_1 \boldsymbol{w}_1 + \cdots + \lambda_m \boldsymbol{w}_m : 0 \le \lambda_i < 1 \text{ for } 1 \le i \le m\}.
$$

This is because (2) still holds, although the expression may not be unique. In fact, the number of different representations of $\mathbf{p}$ is the coefficient of $\mathbf{u}^{\mathbf{P}}$ in Z'_C.

Example 2. Consider again the code studied in §4.4. We compute $Z'_{C(H)}$. There are 52 pseudo-codewords in $\Pi = \{\lambda_1 \boldsymbol{w}_1 + \cdots + \lambda_5 \boldsymbol{w}_5 : 0 \le \lambda_i < 1 \text{ for } 1 \le i \le 5\}$, and

$$
\begin{aligned}
\sigma_\Pi(\boldsymbol{u}) = {} & 1 + u_1 u_3 + u_2 u_3 u_4 + u_1 u_2 u_4 + u_1 u_3 u_4^2 + \cdots \\
& + u_1^5 u_2^6 u_3^5 u_4^4 + u_1^5 u_2^6 u_3^5 u_4^6 + u_1^6 u_2^5 u_3^5 u_4^5 + u_1^6 u_2^6 u_3^6 u_4^6.
\end{aligned}
$$

Thus

$$
\begin{aligned}
Z'_{C(H)}(\boldsymbol{u}) = {} & \frac{\sigma_\Pi(\mathbf{u})}{(1 - \mathbf{u}^{\boldsymbol{w}_1}) \cdots (1 - \mathbf{u}^{\boldsymbol{w}_5})} \\
= {} & 1 + u_1 u_3 + u_1 u_2 u_4 + u_2 u_3 u_4 + u_1 u_2^2 u_3 + u_1 u_3 u_4^2 + u_1^2 u_3^2 + \cdots \\
& + 2 u_1^2 u_2^3 u_3^3 u_4^3 + 2 u_1^4 u_2^3 u_3^3 u_4^3 + 2 u_1^3 u_2^3 u_3^4 u_4^3 + \cdots \\
& + 3 u_1^4 u_2^5 u_3^3 u_4^3 + 3 u_1^5 u_2^5 u_3^4 u_4^3 + 3 u_1^4 u_2^5 u_3^5 u_4^3 + \cdots.
\end{aligned}
$$

The monomials occurring in $Z_{C(H)}$ agree with those in $Z'_{C(H)}$. Shown above are the lowest degree monomials whose coefficients are more than 1.

5   Eigenvalues of $D(2,q)$ and $D(3,q)$ – A Proof of Theorem 2

In this section we prove Theorem 2. Fix a prime power q. As the eigenvalues of a q-regular graph lie between q and $-q$, and $2\sqrt{q-1} = q$ when $q = 2$, all 2-regular graphs are automatically Ramanujan. We shall assume $q \geq 3$ from now on.

5.1   The Strategy

By definition, $D(m,q)$, $m \geq 1$, is a q-regular bipartite graph with vertex sets the truncated X and Y, namely $X(m) = \{\mathbf{x} = [x, x_1, ..., x_{m-1}]\}$ and $Y(m) = \{\mathbf{y} = [y, y_1, ..., y_{m-1}]\}$. The adjacency matrix of $D(m,q)$ is $A(m) = \begin{pmatrix} 0 & H(m) \\ H(m)^t & 0 \end{pmatrix}$, where the rows and columns of $A(m)$ are first indexed by vertices in $X(m)$ and then those in $Y(m)$, and $H(m)^t$ denotes the transpose of $H(m)$.

Since $D(m,q)$ is bipartite, the eigenvalues of $A(m)$ are symmetric, that is, $\pm\lambda$ occur as eigenvalues simultaneously and with the same multiplicity. Hence it suffices to check the eigenvalues of $A(m)^2 = \begin{pmatrix} H(m)H(m)^t & 0 \\ 0 & H(m)^t H(m) \end{pmatrix}$.

Further, since $H(m)^t H(m)$ and $H(m)H(m)^t$ have the same characteristic polynomial, it suffices to prove that the nontrivial eigenvalues of $H(m)H(m)^t$, which are nonnegative, are no greater than $(2\sqrt{q-1})^2 = 4(q-1)$.

When $m = 1$, $D(1,q)$ is a complete q-regular bipartite graph so that all entries in $H(1)$ are 1 and hence all entries in $H(1)H(1)^t$ are q. This shows that the eigenvalues of $H(1)H(1)^t$ are q^2 (the trivial eigenvalue) of multiplicity one, and 0 of multiplicity $q - 1$. Hence $D(1,q)$ is Ramanujan.

The natural projection $\pi_m : D(m+1,q) \rightarrow D(m,q)$ omitting the last coordinate is a covering map. Thus the lifting of a function f on $D(m,q)$ to a function F on $D(m+1,q)$ given by $F(v) = f(\pi_m(v))$ for all vertices v of $D(m+1,q)$ sends an eigenfunction f of $A(m)$ to an eigenfunction F of $A(m+1)$ with the same eigenvalue. Notice that F takes the same value on each fibre of the projection. The space $V(m+1)$ of the lifted functions F has the same dimension as the space of functions f on $D(m,q)$, which is equal to $2q^m$, the size of $D(m,q)$. Since the space of functions on $D(m+1,q)$ constant on each fibre has dimension equal to $2q^m$, the number of fibres, and it contains $V(m+1)$, we conclude that $V(m+1)$ is generated by functions which are constant on each fibre of the projection π_m. The space $W(m+1)$ of the orthogonal complement of $V(m+1)$ in the space of functions on $D(m+1,q)$ has dimension $2q^{m+1} - 2q^m$, and it is generated by functions which sum to zero on each fibre. Moreover, $W(m+1)$ is invariant under $A(m+1)$ since $A(m+1)$ is diagonalizable by orthogonal functions on $D(m+1,q)$.

Having shown that $D(1,q)$ is Ramanujan, we shall prove that $D(2,q)$ and $D(3,q)$ are Ramanujan inductively. Since the space of functions on $D(m+1,q)$ decomposes into the direct sum $V(m+1) \oplus W(m+1)$ and the eigenvalues of $A(m+1)$ on $V(m+1)$ are the same as those of $A(m)$, it suffices to check the eigenvalues of $A(m+1)$ on $W(m+1)$. As analyzed above, this amounts to showing that the eigenvalues of $H(m+1)H(m+1)^t$ on the subspace $W_X(m+1)$ of functions on $X(m+1)$ which sum to zero on the fibre above each vertex in $X(m)$ are no greater than $4(q-1)$ for $m = 1, 2$.

Observe that, for $m \geq 2$, $H(m)H(m)^t = qI(m) + A_2(m)$, where $I(m)$ is the identity $q^m \times q^m$ matrix and $A_2(m)$ is a $q^m \times q^m$ matrix with rows and columns indexed by the vertices $\mathbf{x}$ in $X(m)$ such that the $\mathbf{xx'}$-entry is 1 if the two vertices have distance 2 in $D(m, q)$, and 0 otherwise. We proceed to compute the eigenvalues of $A_2(m)$ for $m = 2$ and 3.

5.2 The Case $D(2, q)$

Two distinct vertices $\mathbf{x} = [x, x_1]$ and $\mathbf{x'} = [x', x_1']$ in $X(2)$ have distance 2 in $D(2, q)$ if and only if they share a common neighbor $\mathbf{y} = [y, y_1]$ in $Y(2)$. In view of the condition (1), their coordinates satisfy the relations

$$y_1 = xy + x_1 \qquad \text{and} \qquad y_1 = x'y + x_1'.$$

As $\mathbf{x}$ and $\mathbf{x'}$ are two distinct neighbors of $\mathbf{y}$, we have $x \neq x'$ and $x_1' = y(x - x') + x_1$. To find all distance 2 neighbors of $\mathbf{x}$, we let $\mathbf{y}$ run through all neighbors of $\mathbf{x}$, which are uniquely determined by the leading coordinate y of $\mathbf{y}$ running through $\mathbb{F}_q$; and for each $\mathbf{y}$, we collect its neighbors $\mathbf{x'}$ with leading coordinates x' running through all elements in $\mathbb{F}_q$ but x. Hence $\mathbf{x}$ has $q(q-1)$ distance 2 neighbors, which are $[x', x_1']$ with $x' \in \mathbb{F}_q \setminus \{x\}$ and $x_1' \in \mathbb{F}_q$. So as an operator on $W_X(2)$, $A_2(2)$ maps a function $f \in W_X(2)$ to another function $A_2(2)f$ defined by

$$A_2(2)f([x, x_1]) = \sum_{x' \in \mathbb{F}_q \setminus \{x\}} \sum_{x_1' \in \mathbb{F}_q} f([x', x_1']) = 0$$

since the points $[x', x_1']$ with x_1' running through $\mathbb{F}_q$ are the points on the fibre above $x' \in X(1)$ over which f sums to zero. Therefore $A_2(2)$ is the zero operator and the eigenvalues of $H(2)H(2)^t$ on $W_X(2)$ are all equal to q. This proves

Theorem 15. *The eigenvalues of $D(2, q)$ are q of multiplicity one, 0 of multiplicity $2q - 1$, and $\pm\sqrt{q}$ of multiplicity $q^2 - q$. Consequently, it is a Ramanujan graph.*

5.3 The Case $D(3, q)$

The procedure is similar to the previous case, but the situation is more complicated. Two distinct vertices $\mathbf{x} = [x, x_1, x_2]$ and $\mathbf{x'} = [x', x_1', x_2']$ in $X(3)$ have distance 2 in $D(3, q)$ if and only if they share a common neighbor $\mathbf{y} = [y, y_1, y_2]$ in $Y(3)$. By (1), their coordinates satisfy the relations

$$y_1 = xy + x_1, \quad y_2 = xy_1 + x_2, \quad y_1 = x'y + x_1', \quad \text{and} \quad y_2 = x'y_1 + x_2'.$$

This implies that $x_1' = y(x - x') + x_1$ and $x_2' = y_1(x - x') + x_2 = xx_1' - x'x_1 + x_2$. Hence $[x, x_1, x_2]$ has $q(q-1)$ distance 2 neighbors, which are $[x', x_1', xx_1' - x'x_1 + x_2]$

with $x' \in \mathbb{F}_q \setminus \{x\}$ and $x_1' \in \mathbb{F}_q$. The operator $A_2 := A_2(3)$ on $W_X(3)$ sends f to $A_2 f$ defined by

$$A_2 f([x, x_1, x_2]) = \sum_{x' \in \mathbb{F}_q \setminus \{x\}} \sum_{x_1' \in \mathbb{F}_q} f([x', x_1', xx_1' - x'x_1 + x_2]).$$

To proceed, we need another operator C on $W_X(3)$ which sends f to Cf defined by

$$Cf([x, x_1, x_2]) = \sum_{x_1' \in \mathbb{F}_q} f([x, x_1', x(x_1' - x_1) + x_2]).$$

The invariance of $W_X(3)$ under the action of C is easily checked by summing over $x_2 \in \mathbb{F}_q$ in the above formula. The properties of A_2 and C are summarized below.

Lemma 1. *On the space $W_X(3)$ there hold*

(1) $(A_2 + C)^2 = q^2 I$;
(2) $A_2 C = C A_2 = 0$; *and*
(3) $C^2 = qC$.

Proof. (1) By definition, we have

$$(A_2 + C)^2 f([x, x_1, x_2]) = \sum_{x', x_1' \in \mathbb{F}_q} (A_2 + C) f([x', x_1', xx_1' - x'x_1 + x_2])$$
$$= \sum_{x', x_1' \in \mathbb{F}_q} \sum_{z, z_1 \in \mathbb{F}_q} f([z, z_1, x'z_1 - zx_1' + xx_1' - x'x_1 + x_2])$$
$$= \sum_{z, z_1 \in \mathbb{F}_q} \sum_{x', x_1' \in \mathbb{F}_q} f([z, z_1, x_1'(x - z) + x'(z_1 - x_1) + x_2]).$$

With z and z_1 fixed, if either $z \neq x$ or $z_1 \neq x_1$, the sum over x_1' or x' is over the points on the fibre above $[z, z_1]$ and hence is equal to zero. When $z = x$ and $z_1 = x_1$, the inner sum is independent of x' and x_1' and the summand is $f([x, x_1, x_2])$, hence we obtain $(A_2 + C)^2 f([x, x_1, x_2]) = q^2 f([x, x_1, x_2])$, as desired.

(2) We compute

$$A_2 C f([x, x_1, x_2]) = \sum_{x' \in \mathbb{F}_q \setminus \{x\}} \sum_{x_1' \in \mathbb{F}_q} C f([x', x_1', xx_1' - x'x_1 + x_2])$$
$$= \sum_{x' \in \mathbb{F}_q \setminus \{x\}} \sum_{x_1' \in \mathbb{F}_q} \sum_{z \in \mathbb{F}_q} f([x', z, x'(z - x_1') + xx_1' - x'x_1 + x_2])$$
$$= \sum_{x' \in \mathbb{F}_q \setminus \{x\}} \sum_{z \in \mathbb{F}_q} \sum_{x_1' \in \mathbb{F}_q} f([x', z, x_1'(x - x') + x'(z - x_1) + x_2]) = 0.$$

This is because $x - x' \neq 0$, therefore as x_1' runs through $\mathbb{F}_q$, the point $[x', z, x_1' (x - x') + x'(z - x_1) + x_2]$ runs through all points on the fibre above $[x', z] \in X(2)$. Similarly,

$$C A_2 f([x, x_1, x_2]) = \sum_{z \in \mathbb{F}_q} A_2 f([x, z, x(z - x_1) + x_2])$$
$$= \sum_{z \in \mathbb{F}_q} \sum_{x' \in \mathbb{F}_q \setminus \{x\}} \sum_{x_1' \in \mathbb{F}_q} f([x', x_1', xx_1' - x'z + x(z - x_1) + x_2])$$
$$= \sum_{x' \in \mathbb{F}_q \setminus \{x\}} \sum_{x_1' \in \mathbb{F}_q} \sum_{z \in \mathbb{F}_q} f([x', x_1', z(x - x') + x(x_1' - x_1) + x_2]) = 0.$$

(3) This identity follows from the definition of C. Indeed,

$$
\begin{aligned}
C^2 f([x, x_1, x_2]) &= \sum_{x_1' \in \mathbb{F}_q} C f([x, x_1', x(x_1' - x_1) + x_2]) \\
&= \sum_{x_1' \in \mathbb{F}_q} \sum_{z \in \mathbb{F}_q} f([x, z, x(z - x_1') + x(x_1' - x_1) + x_2]) \\
&= \sum_{x_1' \in \mathbb{F}_q} \sum_{z \in \mathbb{F}_q} f([x, z, xz - xx_1 + x_2]) = qCf([x, x_1, x_2]).
\end{aligned}
$$

It follows from the above lemma that $(A_2 + C)^2 = A_2^2 + qC = q^2 I$. Moreover, the minimal polynomial of C on $W_X(3)$ divides $x(x - q)$, which implies that C on $W_X(3)$ is diagonalizable and the eigenvalues are among q and 0. It is easy to verify that both occur as eigenvalues. Since A_2 and C commute with each other and are both diagonalizable, they can be simultaneously diagonalized. This shows that the eigenvalues of A_2^2 on $W_X(3)$ are 0 and q^2, which in turn implies that those of A_2 are 0 and $\pm q$, and consequently, those of $H(3)H(3)^t$ are $2q$, q and 0. We have shown

Theorem 16. *The eigenvalues of $D(3, q)$ are q (of multiplicity one), $\pm\sqrt{2q}$, $\pm\sqrt{q}$, and 0. Consequently $D(3, q)$ is a Ramanujan graph.*

This completes the proof of Theorem 2.

References

1. Arslan, O.: The dimension of LU(3,q) codes (preprint) (2008), http://arxiv.org/abs/0802.0015
2. Beck, M., Robins, S.: Computing the Continuous Discretely: integer-point enumeration in polyhedra. Springer, New York (2007)
3. Gallager, R.G.: Low-density parity-check codes. IRE Trans. Inform. Theory 8, 21–28 (1962)
4. Kim, J.-L., Peled, U.N., Perepelitsa, I., Pless, V., Friedl, S.: Explicit construction of families of LDPC codes with no 4-cycles. IEEE Trans. Inform. Theory 50, 2378–2388 (2004)
5. Koetter, R., Li, W.-C.W., Vontobel, P.O., Walker, J.L.: Pseudo-codewords of cycle codes via zeta functions. In: Proc. IEEE Inform. Theory Workshop, San Antonio, TX, USA, pp. 7–12 (2004)
6. Koetter, R., Li, W.-C.W., Vontobel, P.O., Walker, J.L.: Characterizations of pseudo-codewords of (low-density) parity check codes. Adv. in Math. 213, 205–229 (2007)
7. Koetter, R., Vontobel, P.O.: Graph covers and iterative decoding of finite-length codes. In: Proc. 3rd Intern. Conf. on Turbo Codes and Related Topics, Brest, France, pp. 75–82 (2003)
8. Lazebnik, F., Ustimenko, V.A.: Explicit construction of graphs with arbitrary large girth and of largh size. Discrete Applied Math. 60, 275–284 (1997)
9. Loeliger, H.-A.: An introduction to factor graphs. IEEE Sig. Proc. Mag. 21, 28–44 (2004)
10. Lu, M.: Low-density parity-check codes: asymtotic behavior and zeta functions. Thesis, Penn State University, U.S.A (2009)
11. Lubotzky, A., Phillips, R., Sarnak, P.: Ramanujan graphs. Combinatorica 8, 261–277 (1988)

12. Rosenthal, J., Vontobel, P.O.: Constructions of LDPC codes using Ramanujan graphs and ideas from Margulis. In: Proceedings of the 38th Allerton Conference on Communication, Control, and Computing, pp. 248–257 (2000)
13. Sin, P., Xiang, Q.: On the dimensions of certain LDPC codes based on q-regular bipartite graphs. IEEE Trans. Inform. Theory 52, 3735–3737 (2006)
14. Stark, H.M., Terras, A.A.: Zeta functions of finite graphs and coverings. Adv. in Math. 121, 124–165 (1996)
15. Wang, C.: Analysis of finite-length low-density parity-check codes. Thesis, Penn State University, U.S.A (in preparation)
16. Wiberg, N., Loeliger, H.-A., Kotter, R.: Codes and iterative decoding on general graphs. Europ. Trans. on Telecomm. 6, 513–525 (1995)
17. Wiberg, N.: Codes and Decoding on General Graphs. Thesis, Linköping University, Linköping, Sweden (1996),
http://www.it.isy.liu.se/publikationer/LIU-TEK-THESIS-440.pdf
18. Ziegler, G.M.: Lectures on Polytopes. Springer, New York (1995)

On the Applicability of Combinatorial Designs to Key Predistribution for Wireless Sensor Networks

Keith M. Martin

Information Security Group, Royal Holloway, University of London, Egham, Surrey
TW20 0EX, U.K.

Abstract. The constraints of lightweight distributed computing environments such as wireless sensor networks lend themselves to the use of symmetric cryptography to provide security services. The lack of central infrastructure after deployment of such networks requires the necessary symmetric keys to be predistributed to participating nodes. The rich mathematical structure of combinatorial designs has resulted in the proposal of several key predistribution schemes for wireless sensor networks based on designs. We review and examine the appropriateness of combinatorial designs as a tool for building key predistribution schemes suitable for such environments.

Keywords: Key predistribution, combinatorial designs, sensor networks.

1 Introduction

The management of cryptographic keys in any information system is one of the most challenging aspects of implementing cryptography. One of the most important key management processes is *key establishment*, which governs the placement of cryptographic keys in a network. This is especially relevant in applications of symmetric cryptography, where it is necessary to ensure that all parties who are authorised to access (or verify) a cryptographically protected piece of information have the appropriate key.

Symmetric key establishment almost always involves a trusted third party, which we will term a *key management authority* (KMA), at some stage in the process. In some environments this KMA is online and available at time of use. In such cases the third party is often referred to as a *key distribution centre*. However in many other environments it is not possible for a KMA to form part of a live network and assist in online key establishment. In this case the KMA can only be involved in initialisation processes that take place prior to deployment of the network. At this stage the KMA must equip each node in the network with the necessary cryptographic keys for facilitating security services after the nodes are deployed in the network. Key establishment schemes of this type are usually referred to as *key predistribution schemes* (KPSs) because the keys are distributed in advance and cannot be generated "on the fly".

C. Xing et al. (Eds.): IWCC 2009, LNCS 5557, pp. 124–145, 2009.
© Springer-Verlag Berlin Heidelberg 2009

A major current trend in computing technologies is a shift from centralised, relatively stable, wired networks consisting of powerful devices, to distributed, dynamic (ad hoc), wireless networks consisting of lightweight devices. This is being driven by the development of very small wireless computers, which can either be deployed on their own or embedded into everyday objects. The resulting ubiquitous networks have several important characteristics that typically include the need to conduct basic network services such as routing using the network nodes themselves (rather than via a centralised infrastructure), high unavailability rates of nodes, and the need for highly efficient network protocols due to the power and energy constraints of the nodes. From the perspective of providing security services, these characteristics lend themselves to the use of symmetric cryptography and to key predistribution for key establishment.

Wireless sensor networks are just one class of emerging technologies of this type. While we will frame our discussion around wireless sensor networks, which is the context for almost all of the related research, it is worth noting that many of the schemes we discuss may be equally applicable to other technologies with similar characteristics to wireless sensor networks.

Combinatorial structures are natural objects on which to model many aspects of symmetric key management. For a survey of their contributions to key establishment, see [19]. In this paper we will focus only on key predistribution, and on the application of combinatorial designs in particular.

The paper is organised as follows. In Sect. 2 we discuss wireless sensor networks, outlining aspects which are of relevance to key predistribution. In Sect. 3 we provide a brief background to combinatorial designs. In Sect. 4 we outline a basic model for a KPS and discuss fundamental schemes. In the remaining sections we look at different applications of designs to key predistribution. In Sect. 5 we discuss direct application of designs as KPSs. In Sect. 6 we look at the use of designs as building blocks for KPSs. Finally in Sect. 7 we focus on KPSs for special networking environments. Throughout the discussion we will consider the extent to which combinatorial designs are genuinely useful for building KPSs for wireless sensor networks.

2   Wireless Sensor Networks

A *wireless sensor network* (WSN) is an ad hoc network formed from a collection of low-powered sensor nodes that gather data and use wireless communication to transmit the information that they collect. The number of nodes can vary between dozens to thousands, depending on the application [26]. WSNs are best suited to applications where some form of environmental monitoring is required, but where the scale and hostility of the environment does not lend itself to the deployment of a few expensive monitoring devices (such as humans). Examples include seismic data gathering, remote habitat monitoring, gathering of ecological data, forestry welfare, agriculture, disaster relief operations and military intelligence gathering [22, 1, 2]. The typical characteristics of a WSN are:

- *Highly constrained nodes.* The nodes are very small battery-powered devices and are highly constrained with respect to memory storage and power. They are thus limited in their computational and communication ability.
- *Lack of central control.* Once deployed, most WSNs do not feature any central control node. Thus all network functionality must be achieved through co-operation between the nodes.
- *Requirement to form a network to a sink.* In most WSNs the assumption is that the sensor nodes will take readings and then attempt to communicate this data back to a *sink*, which is a more powerful device that will periodically connect to the WSN and request data. The location of this sink in the network is typically not fixed (it could, for example, be a portable laptop).
- *Hop-based communication.* Most WSNs use radio communication to connect between nodes. The constrained nature of the nodes means that in most cases the communication range of a node will be much smaller than the network diameter. Thus nodes communicate by *hopping*, meaning that a node passes data to a node within range, who then passes it onto a node within its range, etc.
- *Dynamic network structure.* It is generally assumed that WSNs are highly dynamic. Nodes are often assumed to regularly "sleep" to conserve battery power. Nodes expire once their battery is drained. In some WSNs the nodes are mobile, although in most current applications they are static.
- *Nodes vulnerable to compromise.* The constrained nature of sensor nodes mean that strong security protection such as tamper-resistance is usually not viable. Thus it is normally assumed that sensor nodes can be fairly easily captured and that any sensitive information (such as keys) that is stored on them is likely to be exposed.

We will make three restrictions on the type of WSN that we consider for most of this paper:

1. *Homogeneous nodes.* We will assume that all nodes have the same capabilities and constraints.
2. *Communication structure.* We will assume that the main aim of any communication in the WSN is to send data from a node to the sink. We will thus not be attempting to set up fully connected subnetworks or establish group keys.
3. *No mobility.* We will assume (for simplicity) that nodes are not mobile after deployment. In fact many of the solutions discussed here are also appropriate for mobile nodes.

An important issue that affects KPS design is that WSNs vary in the extent to which the location of nodes is known prior to deployment. We will thus follow [20] by classifying WSNs as being either:

1. *Uncontrolled* if the location of sensors cannot be predicted before deployment. This is the default WSN scenario and assumes that the application environment is sufficiently hostile that nodes cannot be positioned in any controlled way. For example, they may be released from the air over a disaster site.

2. *Partially controlled* if some information about the location of sensors is known before deployment. This might be the case when sensors are strategically released from the air in batches.
3. *Fully controlled* if the precise location of sensors is known before deployment. This is likely to be the case, for example, when sensors are deployed in a grid in a vineyard to monitor ground humidity.

We will generally assume that a WSN is uncontrolled, however we will discuss KPSs for other types of WSN in Sect. 7.

There has been some debate about the practicality of using public key cryptography to implement security services in a WSN [18]. While this may indeed become more practical (and where it is, some aspects of key establishment may become easier), the case for designing solutions that only use symmetric cryptography remains strong. Symmetric cryptography is still preferred in many modern applications which are not as resource constrained as WSNs because of the efficiency gains and the unique problems posed by management of public keys. Perhaps more compellingly, it is likely that as soon as public key cryptography is practical on a given sensor node technology, even more constrained sensor technology will be being developed where it is not. In this paper we assume that a fully symmetric solution is required.

3   Combinatorial Designs

In this section we briefly review some definitions and notation that we will employ later. We refer the reader to the combinatorial literature for further details [12].

A *set system* $(\mathcal{I}, \mathcal{B})$ consists of a set $\mathcal{I}$ of v elements (*points*) and a collection $\mathcal{B}$ of subsets (*blocks*) of $\mathcal{I}$. The *degree* of $x \in \mathcal{I}$ is the number of blocks of $\mathcal{B}$ containing x and $(\mathcal{I}, \mathcal{B})$ is *regular* if all points have the same degree r. The *rank* k of $(\mathcal{I}, \mathcal{B})$ is the size of the largest block in $\mathcal{B}$ and we say that $(\mathcal{I}, \mathcal{B})$ is *uniform* if all blocks have size k.

A regular, uniform set system with $|\mathcal{I}| = v$, $|\mathcal{B}| = b$ is known as a (v, b, r, k)-*design*. In such designs it must be the case that $bk = vr$. A (v, b, r, k)-design in which every t points occurs on precisely λ blocks is known as a t-(v, b, r, k, λ)-*design* (we often just refer to a t-(v, k, λ)-*design* since b and r can then be uniquely derived). In a *dual* design, the roles of points and blocks are interchanged. *Symmetric* designs are self-dual and thus have $v = b$, $k = r$ and every t blocks meeting in λ points. A symmetric 2-$(s^2 + s + 1, s^2 + s + 1, s + 1, s + 1, 1)$-design is known as a *projective plane*.

A set system is a *group-divisible design* $\mathrm{GD}(n^u, k)$ if $v = nu$ and there exists a partition $\mathcal{H}$ of $\mathcal{I}$ into u *groups* of size n such that:

1. Every $H \in \mathcal{H}$ intersects a block $B \in \mathcal{B}$ in at most one point;
2. Every pair of points from different groups occur together in precisely one block.

A *transversal design* $\mathrm{TD}(k,n)$ is a $\mathrm{GD}(n^k,k)$ (in this case every $H \in \mathcal{H}$ intersects a block $B \in \mathcal{B}$ in precisely one point). A $\mathrm{TD}(t,k,n)$ is a further generalisation where the second condition is applied to sets of t points, rather than pairs. A $\mathrm{TD}(k,n)$ is *resolvable* if the blocks can be partitioned into sets $\mathcal{B}_1, \mathcal{B}_2, \ldots, \mathcal{B}_s$ such that each point of the design is contained in exactly one block in each set. These sets are known as *parallel classes*.

A *graph* $\mathcal{G} = (\mathcal{I}, \mathcal{E})$ consists of a set of of vertices $\mathcal{I}$ joined by *edges* in $\mathcal{E}$, where $\mathcal{E} \subseteq \mathcal{I} \times \mathcal{I}$. We say that a pair of vertices U and V are *adjacent* if $\{U, V\} \in \mathcal{E}$. The *degree* of a vertex U is the number of vertices adjacent to U. A graph is *regular of degree* r if all vertices have degree r. A *complete t-partite* graph is a graph whose vertices can be partitioned into t disjoint subsets such that two vertices are adjacent if and only if they belong to distinct subsets. An (n, r, λ, μ)-*strongly regular graph* is a regular graph on n vertices with degree r such that any two distinct vertices have λ common neighbours if they are adjacent and μ common neighbours if they are not adjacent.

4   Key Predistribution Schemes for WSNs

In this section we provide an introduction to key predistribution for WSNs.

4.1   Key Predistribution Stages

The lack of any central control nodes in a WSN means that in order to equip sensor nodes with symmetric keys, a KMA will need to load keys onto nodes prior to deployment using a KPS to determine which keys are allocated to which nodes. After deployment, two nodes will be able to use a cryptographic service on a network link (such as encryption or a MAC) if they:

1. are in radio communication range of one another; and
2. share at least one key.

If either of these conditions is not met then the nodes will have to seek a path of network links connecting them such that these conditions are met on each of the intermediate hops. Key establishment in a WSN can thus be regarded as consisting of the following three stages:

1. *Key predistribution.* The KMA chooses a KPS defined on the n nodes $\mathcal{U} = \{U_1, \ldots, U_n\}$ in the network. Following [16], this KPS can de modelled by a set system $(\mathcal{I}, \mathcal{B})$ (sometimes referred to as a *key ring*), where $\mathcal{I} = \{x_i : 1 \leq i \leq v\}$ is a set of v *key identifiers* and $\mathcal{B} = \{B_j : 1 \leq j \leq n\}$ is a set of n *node allocations*. For each key identifier x_i, the KMA randomly selects a key K_i. The KMA then associates each node U_j in the network with a node allocation B_j and issues U_j with the keys $L_j = \{K_i : x_i \in B_j\}$. Note that the association of U_j with B_j need not be a secret, however the instantiation of B_j by L_j must be.

2. *Shared key discovery.* If two nodes within communication range of one another wish to deploy a cryptographic service, they first need to determine if they have any keys in common. The default method is to broadcast their node allocations to one another, but more efficient techniques can sometimes be found. If they have key identifiers in common then a session key can be generated from the common keys associated with these identifiers by means of a suitable key derivation function.

3. *Path-key establishment.* If two nodes fail to identify common keys during shared key discovery then they need to find a secure path between one another that employs intermediate nodes which can. Obviously, the shorter this secure path the better.

4.2   Requirements

The main challenge in designing a KPS that is suitable for this type of environment is that a balance must be sought between competing, and to an extent contradictory, requirements:

- *Storage.* Nodes are memory constrained and thus the number of keys stored on each node should be kept as low as possible.

- *Connectivity.* A WSN is dynamic and communication is expensive, thus each node should store sufficient keys that secure paths through the network can be established when needed. There are various different measures for connectivity that could be applied in the context of WSNs. Measures of *global connectivity* assess the connectivity of the entire network as a whole. If the node allocations for any two nodes have non-empty intersection then we will refer to the network as having *full connectivity*. Measures of *local connectivity*, which assess the ability of nodes to form secure paths with nodes in their close neighbourhood are probably most appropriate. One such, from [14], is the probability that U_i and U_j have at least one key in common (i.e. $B_i \cap B_j \neq \emptyset$). This notion can be generalised to measure local connectivity with respect to secure paths of two hops or more.

- *Resilience.* Nodes are vulnerable to compromise, thus keys should be distributed in such a way that the damage caused by exposure of the keys stored on a node is controlled. It is not clear what the "right" measure of resilience is in a WSN. One suggested measure used in [15] is fail(s), which is the probability that a link between two noncompromised nodes U_i and U_j is affected after s other nodes S are compromised at random, where a link is *affected* if $B_i \cap B_j \neq \emptyset$ and $B_i \cap B_j \subseteq \cup_{U_k \in S} B_k$. Another measure proposed in [27] evaluates the probability that compromise of s nodes exposes all the keys from at least one different (not compromised) node allocation.

- *Efficiency.* There are several processes involved in key establishment for a WSN that it may be desirable to make as efficient as possible since nodes are constrained by limited battery power. These include *computation, shared key discovery* and *path-key establishment*. We will be considering the first two, but note that path-key establishment generally involves consideration of routing algorithms in WSNs, which is out of scope for our discussion.

- *Network size.* Since many applications of WSNs involve large numbers of nodes, it is important that a KPS can support a large number of nodes.

The main challenge in designing KPSs is that several of these requirements tend to compete with one another. For example, increasing the maximum number of nodes that can be supported often involves increasing the storage at each node. Also, many KPSs trade off measures of connectivity against resilience.

4.3   Baseline Schemes

There are several important baseline KPSs. Although these are not all designed for WSNs, they provide benchmark schemes that can also be used to illustrate the requirements tradeoffs.

Single Key KPS. This KPS consists of a single key that is stored by each node in the network. It provides optimal connectivity and storage, but has very poor resilience since all communication links are affected by a single node capture.

Complete Pairwise Key KPS. In this KPS, a unique key is assigned to each pair of nodes. This scheme has optimal connectivity and optimal resilience, since compromise of one node does not affect any pair of non-compromised nodes. However this KPS requires each node to store $n - 1$ keys, which is infeasible if n is large (which will be the case in many WSNs).

Blom's KPS[5, 6]. This scheme uses a symmetric bivariate polynomial over a finite field $GF(q)$, i.e. a polynomial $P(x, y) \in GF(q)[x, y]$ with the property that $P(i, j) = P(j, i)$ for all $i, j \in GF(q)$. Node U_i stores the univariate polynomial $f_i(y) = P(U_i, y)$. In order to establish a common key with U_j, node U_i computes $K_{ij} = f_i(U_j) = f_j(U_i)$. This process enables any two nodes to share a common key. If P has degree w, then each share consists of a degree w univariate polynomial hence each node must store the $w + 1$ coefficients of this polynomial, which requires as much space as storing $w + 1$ keys. Blom's KPS thus has optimal connectivity and reasonable storage. It also has very simple shared key discovery, with two nodes simply needing to broadcast their identities to one another. With respect to resilience, an adversary who captures s nodes, where $s \le t$, does not learn any information about keys established between non-compromised nodes. However an adversary who captures $w + 1$ or more nodes can interpolate the polynomial P and hence learn all the keys.

Note that Blom's KPS does not strictly conform to the model in Sect. 4.1 since each user stores secret information that allows it to generate its node allocation rather than storing the separate key identifiers. Thus it reduces storage at the cost of requiring computation in the form of polynomial evaluations each time a key identifier is established.

Random KPS [13]. This scheme is a probabilistic KPS, with each node drawing keys uniformly without replacement from some finite keypool $\mathcal{K}$. The properties of this scheme depend on the number of keys drawn and the size of $\mathcal{K}$. In the basic scheme any two nodes can communicate securely if they share at least one key. The basic scheme was further parameterised in [11], where an additional threshold parameter was introduced so that two nodes are required to have at least a threshold number of keys in common before they can derive a key.

These baseline KPSs provide suitable motivation for several observations concerning the building of KPSs for WSNs:

1. *Optimal connectivity is not necessary.* Optimal connectivity is a nice feature, but unnecessary in a KPS for a WSN. It is certainly not needed in fully controlled WSNs. However, even in uncontrolled WSNs, since only a minority of sensor nodes will be within communication range of one another, the "costs" of optimal connectivity might not be worth paying.
2. *Deterministic schemes have some advantages.* The obvious advantage of deterministic KPSs is that we can generally make definitive statements about their properties, which aids analysis. The example of Blom's KPS also illustrates that in deterministic schemes it may be possible to have very efficient shared key discovery. Lee and Stinson [14] also point out that deterministic schemes tend to involve fewer expensive pseudorandom computations during the key predistribution stage. In [25] it was argued that in certain cases probabilistic solutions tend to converge to deterministic schemes, thus studying the latter provides valuable insight.
3. *Flexibility is attractive.* An attractive feature of the random KPS is that it is highly configurable with respect to the competing requirements. Blom's KPS, for example, allows only minor tradeoffs to be made between storage and resilience.
4. *Compromise is desirable.* It is unlikely that a WSN application will want the extreme tradeoffs seen in the case of the single and complete pairwise KPS. Even Blom's KPS is probably not enough of a compromise, with low storage coming at the cost of low resilience and computation requirements. Thus even if a KPS cannot offer flexibility, it is desirable that it offers a "reasonable" compromise between the competing requirements of Sect. 4.2.

There have been a large number of proposals for KPSs for WSNs. These tend to either be variants of the random KPS, deterministic KPSs, proposals for combining schemes or KPSs with special properties. There are also several surveys [8, 20, 31], each of which takes a slightly different approach. We will now focus primarily on proposals that utilise combinatorial designs.

5   Direct Application of Designs

Combinatorial designs are very natural objects to consider as candidate key rings for KPSs. They have the advantages of being deterministic and having rich and well understood structure. Indeed, they have been associated with the building of KPSs long before the emergence of WSNs [23]. In this section we consider the direct application of (v, b, r, k)-designs as key rings for a KPS.

5.1 Two Interesting Classes

The basic definition of a (v, b, r, k)-design is too general to guarantee any interesting connectivity or resilience properties of the resulting KPSs. We first identify two potentially interesting classes of designs.

Prioritising local connectivity: Our first class of designs are explicitly constructed for their local connectivity properties. It was shown in [15] that any block in a (v, b, r, k)-design meets (has a non-empty intersection) with at most $k(r-1)$ other blocks. Further, every block meets $k(r-1)$ blocks precisely when the design has the property that any two blocks meet in at most one point, in which case the design is known as a (v, b, r, k)-*configuration*. These (v, b, r, k)-configurations are of interest since if they are used as key rings, the KPSs based on them have optimal local connectivity [15]. Knowing that a design is a (v, b, r, k)-configuration does not, unfortunately, offer any immediate guarantees about its resilience.

Prioritising resilience: A class of designs with built-in resilience properties are *key distribution patterns*. These were first proposed in [23, 24], although we present a slightly more general definition here.

Definition 1. *A w-key distribution pattern (KDP) is a set system $(\mathcal{I}, \mathcal{B})$ with $|\mathcal{B}| = n$ such that for any pair $B_i, B_j \in \mathcal{B}$ with $B_i \cap B_j \neq \emptyset$ and any $\{B_{l_1}, \ldots, B_{l_w}\} \subseteq \mathcal{B} \setminus \{B_i, B_j\}$, we have:*

$$B_i \cap B_j \not\subseteq (B_{l_1} \cup \cdots \cup B_{l_w}).$$

A w-KDP can be used as a key ring for a KPS and offers optimal resilience if no more than w nodes are compromised. A w-KDP is only a design if it is also uniform and regular, which many known examples of KDPs are. However, Definition 1 does not provide any guarantees of connectivity.

5.2 Fully Connected Designs

An obvious class of designs to consider are those that offer full connectivity, which happens if every pair of blocks meet in at least one point.

Fully connected configurations: It is shown in [15] that a (v, b, r, k) configuration is fully connected precisely when it is the dual of a $2 - (b, v, k, r, 1)$-design. It is further shown in [15] that when this happens, $b \leq k(k-1) + 1$. Since b represents the number of nodes in a WSN, and this is likely to be large, it is clear that this latter bound is one that we would like to meet, if at all possible. Fortunately there is an infinite class of configurations with this property, namely the projective planes, which are $2 - (q^2 + q + 1, q^2 + q + 1, q + 1, q + 1, 1)$-designs.

This means that if we wish to directly implement a configuration as a key ring in order to obtain a fully connected KPS with other desirable properties then there is really only one candidate family worth considering, the projective planes. Not only do they have optimal local connectivity, but amongst other advantages they have efficient shared key discovery [25]. They were first proposed as key rings by [7]. However the significant "catch" with using a projective plane is the restriction on the number of nodes relative to the size of the node allocation. This means that facilitating a very large number of nodes comes at the unattractive cost of relatively large key storage for each node (in this case each node allocation contains k identifers, where k is approximately the square root of the maximum number nodes).

Fully connected KDPs: The original concept of a KDP, as proposed in [23, 24], was for fully connected KDPs. In this case a KDP, by definition, has every pair of blocks meeting in at least one point. In [29] structures of this type are known as $(2, w)$-KDPs. Several constructions of uniform and regular $(2, w)$-KDPs are known. In [28, 24] it is shown that a 3-(v, k, λ) -design with $w < (v-2)/(k-2)$ is the dual of a $(2, w)$-KDP. In [24] it is shown that every $t - (v, b, r, k, \lambda)$-design is a $(2, t - 2)$-KDP and every symmetric $2 - (v, k, 2)$-design (*biplane*) is a $(2, 1)$-KDP. We also observe that the complete pairwise KPS is a uniform, regular $(2, n - 2)$-KDP (as well as being a configuration).

Dual designs: By definition, the dual of a $2 - (v, b, r, k, \lambda)$-design is fully connected, since every pair of blocks meet in λ points. A special subclass are the symmetric designs, examples of which are the projective planes and the biplanes. In [27] symmetric *partially balanced* designs were proposed as key rings, however they share the problems of projective planes in being highly constrained in terms of the number of nodes they can support.

Comment: In general, fully connected designs are unsuitable for direct application as KPSs for WSNs. Full connectivity places too many constraints on the parameters. The main resulting problems are:

- *Lack of flexibility*: While there are a number of constructions, they leave little room for flexibility of tradeoff between the important parameters.
- *Restrictions on number of nodes*: The tradeoff between number of nodes and storage tends to be unsatisfactory, with reasonable storage limitations leading to too tight a restriction on the maximum number of nodes.
- *Too much to the extreme*: Full connectivity provides better connectivity than we typically need for a WSN. The cost in terms of storage and resiliency is too high to be worth paying for most WSNs.

Nonetheless, direct application of designs in this way provides more baseline KPSs with special properties for comparison, as well as being potentially useful components in more complex constructions.

5.3   Designs without Full Connectivity

Given that full connectivity is not necessary for a KPS for a WSN, it is worth considering direct application of designs that do not have full connectivity. In such designs there will be blocks that do not intersect. This means that there will be pairs of nodes who do not share a key in the resulting KPS. In the first instance it seems wise to consider configurations, since these at least offer optimal local connectivity.

Generalised Quadrangles: In [7] the use of *generalised quadrangles* as WSN key rings was considered. A $GQ(s,t)$ is a $(v, b, t + 1, s + 1)$ -design, where $v = (s+1)(st+1)$, $b = (t+1)(st+1)$, two points lie on at most one block, two blocks meet in at most one point, and a further property that outlaws the occurrence of "triangles" holds. A $GQ(s,t)$ is thus a configuration and hence offers optimal local connectivity. In [7] several $GQ(s,t)$s were shown to enable KPSs with good resilience compared to the random KPS.

Common Intersection Designs: The idea behind the use of $GQ(s,t)$'s as key rings was generalised in [15]:

Definition 2. *Let* $(\mathcal{I}, \mathcal{B})$ *be a* (v, b, r, k)-*configuration. We say that* $(\mathcal{I}, \mathcal{B})$ *is a* (v, b, r, k, μ)-*common intersection design (CID) if for any distinct pair of blocks* $B_i, B_j \in \mathcal{B}$ *we have:* $|\{B_k \in \mathcal{B} : B_i \cap B_k \neq \emptyset \text{ and } B_j \cap B_k \neq \emptyset\}| \geq \mu$.

Thus any key ring based on a CID provides the guarantee that if two nodes do not share a key, there will be at least μ nodes who could act as intermediaries in a secure two-hop path between the original nodes. From a connectivity perspective it is desirable for μ to be as large as possible since this increases the chance that one of these intermediary nodes is within communication range. Several upper bounds on μ were established in [17] and optimal CIDs were constructed using group-divisible designs, strongly-regular graphs and generalized quadrangles.

Transversal Designs: A useful class of CIDs is provided by the transversal designs, since a $TD(k,n)$ is a $(kn, n^2, n, k, k^2 - k)$-CID. In [15] a particularly useful construction of $TD(k,n)$s that exist for any prime $k \leq n$ was used to construct CIDs. The resulting key rings, termed *linear schemes* in [14] have several interesting properties:

- The values of k and n can be varied to produce key rings with a range of compromises between the storage k, maximum number of nodes n^2, local connectivity $k(n + 1)$ and resilience.
- Local connectivity and resilience can be computed using formulae that were derived in [14].
- They have a very efficient shared-key discovery phase, which involves two nodes exchanging identifiers and making a simple computation.

Generalised Transversal Designs: An example of a class of designs that do not offer full connectivity and are not configurations are the generalised transversal designs $TD(t, k, n)$. For example, a block in a $TD(3, k, n)$ intersects other blocks in either 0, 1 or 2 points, hence is not a configuration. In [14] $TD(3, k, n)$s were used to construct key rings based on a requirement that a pair of nodes shares two keys before they can derive a session key. The performance of these so-called *quadratic schemes* was analysed in [14] and shown to offer some interesting tradeoffs. For example, they offered better resilience than linear schemes for low levels of compromised nodes, while providing similar levels of local connectivity.

Trivial KDPs: Let $\mathcal{G}$ be a connected graph on n vertices with no loops or multiple edges. Associate a vertex with node U_j, assign a unique key identifier x_i to each edge, and define node allocation B_j to be the the set of edges (key identifiers) adjacent to U_j. The result is an $(n - 2)$-KDP, which offers the maximum possible resilience. This is an example of a *trivial inclusion*-KDP [19]. The advantage of designing KPSs in this way is that $\mathcal{G}$ can be analysed for connectivity and path-length properties. Trivial KDPs are $(v, b, 2, k)$ -designs when $\mathcal{G}$ is regular of degree k.

In [16] it was pointed out that one class arise from strongly regular graphs, since these graphs offer a guaranteed number of possible two-hop paths between any disconnected nodes. The cost associated with a trivial KDP is that in order to get good levels of connectivity the graph typically needs to be "dense" with edges, which means that the storage for each user tends to be on the high side. The *IOS* KPSs in [16] employ a trick for reducing this storage which works if $\mathcal{G}$ is a connected regular graph whose vertices have even degree. This comes at a small computational cost, as well as the security cost of relying on a hash function.

Comment: Designs without full connectivity are certainly more promising for designing KPSs for WSNs. The main advantage over fully connected designs is increased room for flexibility. The local connectivity levels can be traded off against other parameters, particularly resilience. This increased relaxation of parameters also tends to facilitate an increase in the maximum number of nodes that can be supported given a particular storage constraint. Nonetheless, a number of problems remain:

- *Lack of flexibility*: While flexibility is generally better than for fully connected designs, it is still severely constrained by the combinatorial requirements.
- *Restrictions on number of nodes*: Despite an improvement, there is still a limit to the number of nodes that can be supported, again due to the combinatorial constraints.

Direct application of designs without full connectivity thus provides another interesting collection of KPSs. It should be noted however that in comparison to fully connected designs, these designs have not been so much studied and so useful constructions may have not yet been discovered. While their direct applicability is limited, they again provide excellent components for building more complex KPSs.

5.4   On Direct Application of Designs

The rich mathematical structure of combinatorial designs makes them suitable for building KPSs with particular properties. However most interesting classes of design probably offer too much structure. Some designs offer "all or nothing" guarantees of properties, when a more gradual curve would be preferable. An example of this is w-KDPs (for small w), whose resilience guarantees are no longer offered when more than w nodes are compromised. Designs are also uniform and regular by definition, although there is no strict need for these properties in a KPS. The main problem however is that straight application of designs tends not to provide enough flexibility to generate a wide range of KPSs suitable for different application requirements.

6   Designs as a Building Block

Although combinatorial designs are not always suitable for direct application as KPSs for WSNs, they are very natural objects to use as components in the construction of a KPS. The resulting KPSs can hopefully be made more flexible, while still inheriting the advantages of designs that were outlined at the start of Sect. 5. Another way of looking at this is to start with a KPS based on direct application of a design and consider in what ways we could transform the original scheme in order to "get more for our money". We now consider a number of different techniques.

6.1   Splitting a KPS

One way of modifying a KPS is to *split* nodes, by associating each node in the original KPS with a set of nodes in a new KPS. The new scheme essentially consists of l versions of the original KPS. The main gain here is that this allows an l-fold increase in the number of possible nodes in the network compared to the original scheme.

The simplest technique is to essentially create l "mirror copies" of the original KPS, where each split node is assigned the same node allocation of keys as its parent node. For general applications, this might seem a strange thing to do since there will now be l nodes with exactly the same keys. However, for many applications of WSNs, particularly those where the main required security service is confidentiality with respect to non-members of the network, this may well be quite acceptable.

The other extreme is to associate each version of the original KPS with a disjoint set of key identifiers. This will result in a significant reduction in the connectivity, since only nodes associated with a particular version will share keys. Partially overlapping sets of key identifiers will allow tradeoff between these two extremes.

The *Multiple IOS* KPSs in [16] used this idea of splitting to increase the maximum network size of the IOS KPS (see Sect. 5.3). Since they are slightly different from our standard notion of a key rings (as defined in Sect. 4.1), the cost of splitting is different (in this case it is a loss of resilience).

6.2   Extending a KPS

As noted in Sect. 5, one of the main problems with straight application of a combinatorial design as a KPS is the restriction on the maximum number of nodes. One technique for overcoming this is to generate a KPS based on a combinatorial design and then extend it by appending additional node allocations that are not part of the original scheme.

This technique was used in [7] to extend KPSs based on projective planes and generalized quadrangles. In order to enforce a degree of separation between the appended node allocations and the originals, the new node allocations were selected as random subsets of blocks of the *complementary design* (whose blocks are the complements of the blocks of the original design). The resulting KPSs were analysed in [7] and shown to have better connectivity than a random KPS, while allowing a greater number of nodes and increased resilience in comparison to the underlying design-based KPSs.

Another possibility is to combine the node allocations of two different KPSs. This approach was taken in [27], where two KPSs arising from direct application of two partially balanced designs were combined. The resulting KPS remained fully connected, while the resiliency of the new scheme was slightly poorer than those of the original KPSs.

6.3   Packing a KPS

Another option is to increase the size of node allocations by adding key identifiers (*packing*). By packing the key identifiers more densely, we can expect better connectivity properties at an expected cost to resilience.

We saw in Sect. 5.3 that some combinatorial designs without full connectivity have attractive properties for adoption as key rings. However, such designs often have low inherent connectivity. In [9] two packing strategies were tested in an attempt to increase the connectivity of the linear KPSs based on transversal designs. Both strategies involved merging KPS node allocations. The first strategy was random, whereas the second was deterministic. The results indeed indicate a small increase in connectivity at a small cost to resilience.

6.4   Breaking Up a KPS

An alternative to making a KPS "bigger" through extending or packing is to break it up in various ways. Two initial suggestions for creating potentially interesting tradeoffs are:

- *Contracting a KPS*: By removing key identifiers, either throughout the KPS or just on certain nodes, storage could be reduced at a cost to connectivity.

– *Block splitting*: By splitting node allocations (for example creating two smaller node allocations from each original node allocation by dividing it in two) the maximum network size could be increased and storage reduced, again at a cost to connectivity.

To our knowledge, the full benefits of these strategies as techniques for building KPSs with interesting properties have not yet been fully explored.

6.5   Modifying a KPS

Designs can be used to make structural modifications to an existing KPS. An interesting example of this is the *Modified Blom KPS* [16]. The Blom KPS, defined in Sect. 4.3, is a fully connected KPS based on a symmetric polynomial of degree w. In the Modified Blom KPS, we first define a complete bipartite graph on the set of nodes, which splits the nodes into two classes $\mathcal{U}_1$ and $\mathcal{U}_2$. We now establish a "Blom KPS" using an asymmetric polynomial (see [16] for details), which results in only pairs of nodes from distinct classes directly being able to establish a key. Nodes from the same class are required to establish a two-hop path via a node in the other class. This loss of connectivity comes at a gain in resilience, since an attacker now needs to compromise w nodes from one of the classes before the KPS is completely broken.

6.6   Joining KPSs

A more sophisticated use of KPSs as building blocks is to join many copies of a KPS together. Of course, we need a "rule" to determine how the integration is done. A natural source of such a rule is another KPS, perhaps with quite different properties. The intention is that the resulting KPS will mix the inherent properties of the component schemes.

Product KPS: In such schemes an *inner* KPS and an *outer* set system $(\mathcal{I}, \mathcal{B}^{\text{out}})$ are integrated in the following way:

1. The outer set system provides the core structure. Each node U_j is associated with the block B_j^{out}.
2. Each key identifier x_i in the set system defines a subset of nodes $N_i = \{U_j : x_i \in B_j^{\text{out}}\}$. An *inner* KPS is then defined on the nodes N_i. In this KPS, each node $U_j \in N_i$ receives the node allocation $B_j^{\text{in}-i}$.
3. Only the node allocations of the inner KPSs are used in the final KPS. Hence each node U_j receives the final node allocation

$$B_j = \cup_{x_i \in B_j^{\text{out}}} B_j^{\text{in}-i}.$$

Hence each node in the product KPS receives a final node allocation that consists of several inner KPS node allocations, one for each key identifier in the block associated with the node in the outer set system.

Note that while we have defined the Product KPS in terms of inner KPSs based on a key ring, there is no reason why other KPSs cannot be used. Indeed the low storage of the Blom KPS, which derives its key ring rather than storing it explicitly is an attractive candidate for the inner KPS, as will see in some of the following instantiations of this generic scheme.

Wei-Wu Product schemes: In [30] a general analysis of the Product KPS was conducted. It was shown that if the block size is fixed then the best resilience can be obtained if the outer set system is a design. Several constructions were proposed that used Blom KPSs as the inner KPS and used designs based on *difference sets* as the outer set system.

Multiple Space Blom scheme: In [14] a product KPS was proposed where the outer set system is a linear KPS (see Sect. 5.3) and the inner KPSs are Blom KPSs. The resulting KPS was shown to have a different resilience curve compared to a linear scheme (better resilience for small numbers of compromised nodes) at the cost of some computation in order to establish keys. The efficient shared-key discovery property of both components is preserved.

Multiple Space Modified Blom scheme: In [16] a product KPS was proposed where the outer set system consists of a trivial KDP based on a strongly regular graph that has been split into l identical copies (as in Sect. 6.1) and the inner KPSs are Modified Blom KPSs. The Modified Blom KPSs were applied using the natural partition defined by the two classes of split nodes adjacent to each edge in the strongly regular graph. The resulting scheme was shown to have a different resilience curve compared to deploying a Modified Blom KPS across all nodes.

Park-Blake schemes: In [25] complete subgraphs of two strongly regular graphs (the *triangular graph* and *lattice graph*) were used to define outer set systems. It was shown that if projective planes are used as the inner KPSs then the new schemes allow a greatly increased network size while still gaining from the efficient shared key discovery of the projective plane. Clearly these constructions also lend themselves to use of Blom schemes as the inner KPSs.

Scope for Joining KPSs: Joining KPSs seems to be an interesting way of generating KPSs with different parameter tradeoffs. There seems to be plenty of scope for further exploring effective ways of combining KPSs, as most of the existing work has focussed on instantiations of the Product KPS and employed Blom's KPS as the inner KPSs.

6.7   The Pros and Cons of Combinatorial Engineering

In Sect. 5 we saw that direct application of combinatorial designs is generally too restrictive to produce KPSs that are suitable for WSNs. In this section we have discussed a number of different techniques for using KPSs based on designs as building blocks. There is no reason why these techniques could not be used to combine deterministic KPSs based on designs with probabilistic random KPSs.

With the exception of layered KPSs, this "combinatorial engineering" is not a normal study area for pure mathematicians and hence little theory on the subject exists. Indeed, for many of the techniques, the underlying combinatorial structure is sufficiently destroyed that the resulting properties can only be determined by simulations.

It might be felt that combinatorial engineering is self-defeating in that many of the advantages of using combinatorial designs may be lost, especially if they are combined with probabilistic KPSs. However it would seem that some combinatorics can be better than no combinatorics, since the properties of the underlying design-based KPS in most cases still provides some structural guarantees. It is also important to keep in mind the observations made in Sect. 4.3 and 5.4, which indicate that KPSs for WSNs are not by definition classical combinatorial objects and thus lend themselves to this type of manipulation.

7   Designs for Special Networking Environments

In this section we examine the application of combinatorial designs to KPSs that do not fully conform to the application environment of uncontrolled homogeneous nodes that we have discussed thus far.

7.1   Partially Controlled KPSs

The KPSs that we have discussed thus far are all uncontrolled (see Sect. 2) with respect to their final location. If we are able to have partial control over the location of nodes then this knowledge can be very useful in building a suitable KPS.

An example of partial control occurs in networks in which nodes are deployed in groups in such a way that nodes from a group are deployed closer together on average than nodes from different groups. This *group deployment* might arise, for example, if nodes are deployed in batches from an aeroplane. It would be reasonable to expect nodes from one group to then be physically located closer to one another than nodes from different groups. As a result, keys can be predistributed more efficiently if this knowledge is taken into account.

A possible paradigm is to assign node allocations to each group using a KPS defined only on that group. This KPS could be more "relaxed" with respect to connectivity than an uncontrolled KPS. We then need to build in some means for nodes from different groups to establish common keys. However it is also important to avoid communication bottlenecks, or the risk that an entire group could become disconnected from the rest of the network, so it is desirable to ensure that the probability of nodes from different groups being able to communicate securely is similar to that of nodes from within a group. This property was referred to as *balanced local connectivity* in [21].

The inherent structure required for group deployment of a KPS that has balanced local connectivity lends itself naturally to use of a combinatorial structure. In [21] such a KPS was proposed that utilises the structure of a resolvable transversal design $TD(k, m)$. The m parallel classes $P_1, P_2, \ldots, P_m$ of blocks of this design are further partitioned into μ sets of parallel classes $S_1, S_2, \ldots, S_\mu$, each containing m/μ blocks. Nodes in group G_i are associated with the blocks in parallel classes contained in S_i. The proposed KPS is based on the Multiple Space Blom scheme of Sect. 6.6, with the outer KPS being a based on the resolvable $TD(k, m)$. However in this KPS there are two inner KPSs:

1. As in the Multiple Space Blom scheme, for each key identifier x_i in the outer KPS, a Blom KPS is defined on the nodes $N_i = \{U_j : x_i \in B_j^{\text{out}}\}$.
2. A further Blom KPS is defined on the set of nodes $M_i = \{U_j : B_j^{\text{out}} \in P_i\}$.

The analysis of the resulting KPS in [21] shows that it offers good balanced connectivity, while providing a flexible set of configurable parameters that allow connectivity and resilience to be traded off against storage costs. The KPS also inherits the efficient shared key discovery of the underlying outer KPS based on the transversal design.

7.2 Fully Controlled KPSs

It is of significant advantage for key predistribution to know the precise deployment location of nodes, since this is even more useful than partial location information, as discussed above. It might seem that in such cases of fully controlled networks it suffices to issue a node with keys for each of its neighbours, since these are known in advance. However in dense networks this is an inefficient technique and there are much better options.

If nodes are deployed in a highly structured physical formation then it again becomes natural to look to combinatorial mathematics for building KPSs. The case of KPSs for WSNs arranged in square and hexagonal grids has been investigated in [4] and [3]. Efficient KPSs were constructed using a special type of combinatorial structure called a *distinct difference configuration*. While these are not combinatorial designs, the node assignments that they generate can be viewed as a type of infinite combinatorial design. Nonetheless, this example serves a warning that there are special networking environments where the "right" combinatorial structure for building efficient deterministic KPSs is not necessarily based on a conventional combinatorial design.

7.3 KPSs for Heterogeneous Networks

Although we restricted our previous discussion to homogeneous networks (see Sect. 2) it is worth making some observations about *heterogeneous networks*, where not all the nodes have the same capabilities. The most interesting class of heterogeneous network is probably *hierarchical networks*, where the nodes are partitioned into an ordered hierarchy, with nodes at a given level being more powerful than nodes at lower levels. The most common scenario is a simple *two-level hierarchy*. We now make a few observations about the applicability of combinatorial designs to building KPSs for heterogeneous networks, and in particular two-level hierarchies.

Simple two-level hierarchies: It is worth observing that many of the manipulations of KPSs discussed in Sect. 6 can result in KPSs that are suitable for simple two-level hierarchies, since the resulting node allocations have different sizes. For example:

– A KPS could be partially packed (see Sect. 6.3) using a strategy that results in node allocations of two different sizes, the original and the packed. Nodes with packed node allocations will require greater storage capability. Further, as they hold more keys it is reasonable to expect them to be more likely to be involved in communication (both directly and as an intermediary). The resulting KPS is suited to applications where there is a two-level hierarchy of sensors where there is a fairly small difference in capability.

– Similarly, if a KPS is extended (see Sect. 6.2) by adding node allocations of a different size to the original, then the resulting KPS will have similar properties to the previous case. For example, the extension to the projective plane discussed in [7] could involve choosing larger subsets of the complementary design, hence creating two classes of node allocations.

Two-level hierarchies with a backbone: A more sophisticated class of two-level hierarchies are formed by networks where the top level of nodes form a fully connected *backbone*. Low level nodes are organised into *subnetworks* (sometime called *clusters*) which "hang off" this backbone and are each associated with a unique high level node. Two low level nodes from the same subnetwork can try to communicate directly. On the other hand, two low level nodes from different subnetworks have to communicate via their high level node representatives. The top level nodes thus need to be significantly more powerful than low level nodes, since they are used as communication intermediaries. It is thus also reasonable to assume that they have significantly increased storage capability.

There are many different possible approaches to designing a two-level hierarchical KPS with a backbone. The fully connected backbone could be realised, for example, by any of the KPSs discussed in Sect. 7.2. The subnetworks could be instantiated by any KPS (based on a design, or otherwise). There has been a significant amount of general research on key management in two-level hierarchical networks but, with the exception of a simple framework proposed in [10], there has been very little analysis of how to build deterministic two-level KPSs. There seems to be plenty of scope for further examining exactly how best to choose both the backbone and subnetwork KPSs in order to achieve two-level hierarchical KPSs with interesting properties. In particular, intelligent application of design-based KPSs would seem quite likely to lead to useful constructions.

7.4   Comment

We have seen in this section that combinatorial designs have had a role to play in building KPSs for WSNs that do not conform to the "classical" model of uncontrolled homogeneous nodes. They have found very natural application to group deployment of nodes, but are apparently less applicable to fully controlled deployment of nodes. What remains largely unexplored is their suitability to the design of deterministic heterogenous WSNs, and this merits further study.

8   Concluding Remarks

We have explored the use of combinatorial designs in building KPS for WSNs. While designs have been widely proposed for use in such schemes, to what extent are these schemes really useful? We argued that for WSNs full connectivity is not really necessary and that a key attribute of any KPS is flexibility to allow parameter tradeoffs. This tends to rule out many straight applications of designs as KPSs, although we have seen several examples of flexible families of designs, such as transversal designs, having several useful applications. However it certainly does not rule out designs either as building blocks or components of KPSs. The unusual combinatorial engineering techniques that have seen the basic structure of a design manipulated in order to provide more flexible KPSs are certainly interesting and merit further study, although formal theoretical analysis of such techniques (as in many engineering processes) is not always possible. Combining KPSs based on designs has proved to be a very successful strategy for obtaining deterministic KPSs that trade off parameters, however again there would seem to be more work to do in fully understanding the best combination rules. Thus we would argue that combinatorial designs most definitely do have an important role to play in building KPSs for WSNs, but that their full potential is not yet fully understood.

References

[1] Integrated smart sensing systems (2007), http://dpi.projectforum.com/isss/11
[2] Institut für Chemie und Dynamik der Geosphäre (ICG), Forschungszentrum Jülich: SoilNet - a Zigbee based soil moisture sensor network (2008), http://www.fz-juelich.de/icg/icg-4/index.php?index=739
[3] Blackburn, S.R., Etzion, T., Martin, K.M., Paterson, M.B.: Distinct-difference configurations: Multihop paths and key predistribution in sensor networks (2008), http://arxiv.org/abs/0811.3896
[4] Blackburn, S.R., Etzion, T., Martin, K.M., Paterson, M.B.: Efficient key predistribution for grid-based wireless sensor networks. In: Safavi-Naini, R. (ed.) ICITS 2008. LNCS, vol. 5155, pp. 54–69. Springer, Heidelberg (2008)
[5] Blom, R.: An optimal class of symmetric key generation systems. In: Beth, T., Cot, N., Ingemarsson, I. (eds.) EUROCRYPT 1984. LNCS, vol. 209, pp. 335–338. Springer, Heidelberg (1985)
[6] Blundo, C., Santis, A.D., Vaccaro, U., Herzberg, A., Kutten, S., Yung, M.: Perfectly secure key distribution for dynamic conferences. In: Brickell, E.F. (ed.) CRYPTO 1992. LNCS, vol. 740, pp. 471–486. Springer, Heidelberg (1993)
[7] Camtepe, S.A., Yener, B.: Combinatorial design of key distribution mechanisms for wireless sensor networks. In: Samarati, P., Ryan, P.Y.A., Gollmann, D., Molva, R. (eds.) ESORICS 2004. LNCS, vol. 3193, pp. 293–308. Springer, Heidelberg (2004)

[8] Camtepe, S.A., Yener, B.: Key distribution mechanisms for wireless sensor networks: a survey. Rensselaer Polytechnic Institute, Computer Science Department, Technical Report TR-05-07 (March 2005)

[9] Chakrabarti, D., Maitra, S., Roy, B.: A key pre-distribution scheme for wireless sensor networks: merging blocks in combinatorial design. International Journal of Information Security 5(2)

[10] Chakrabarti, D., Seberry, J.: Combinatorial structures for design of wireless sensor networks. In: Zhou, J., Yung, M., Bao, F. (eds.) ACNS 2006. LNCS, vol. 3989, pp. 365–374. Springer, Heidelberg (2006)

[11] Chan, H., Perrig, A., Song, D.: Random key predistribution schemes for sensor networks. In: IEEE Symposium on Research in Security and Privacy, pp. 197–213 (May 2003)

[12] Colbourn, C.J., Dinitz, J.H. (eds.): The CRC Handbook of Combinatorial Designs. CRC Press, Boca Raton (2007)

[13] Eschenauer, L., Gligor, V.: A key management scheme for distributed sensor networks. In: Proceedings of 9th ACM Conference on Computer and Communication Security (November 2002)

[14] Lee, J., Stinson, D.R.: On the construction of practical key predistribution schemes for distributed sensor networks using combinatorial designs. ACM Transactions on Information and System Security 11(2)

[15] Lee, J., Stinson, D.R.: A combinatorial approach to key predistribution for distributed sensor networks. In: IEEE Wireless Communications and Networking Conference, pp. 6–11, CD-ROM, paper PHY53-06 (2005),
http://www.cacr.math.uwaterloo.ca/dstinson/pubs.html

[16] Lee, J., Stinson, D.R.: Deterministic key predistribution schemes for distributed sensor networks. In: Handschuh, H., Hasan, M.A. (eds.) SAC 2004. LNCS, vol. 3357, pp. 294–307. Springer, Heidelberg (2004)

[17] Lee, J., Stinson, D.R.: Common intersection designs. Journal of Combinatorial Designs 14(4), 251–269 (2009)

[18] Lopez, J.: Unleashing public-key cryptography in wireless sensor networks. Journal of Computer Security 14(5)

[19] Martin, K.M.: The combinatorics of key establishment. In: Surveys in Combinatorics 2007. London Mathematical Society Lecture Note Series, vol. 346, pp. 223–273. Cambridge University Press, Cambridge (2007)

[20] Martin, K.M., Paterson, M.B.: An application-oriented framework for wireless sensor network key establishment. Electron. Notes Theor. Comput. Sci. 192(2), 31–41 (2008)

[21] Martin, K.M., Paterson, M.B., Stinson, D.R.: Key predistribution scheme for homogeneous wireless sensor networks with group deployment of nodes (2008),
http://www.isg.rhul.ac.uk/~martin/files/gdfinal.pdf

[22] McCulloch, J., McCarthy, P., Guru, S.M., Peng, W., Hugo, D., Terhorst, A.: Wireless sensor network deployment for water use efficiency in irrigation. In: REAL-WSN 2008: Proceedings of the workshop on Real-world wireless sensor networks, pp. 46–50. ACM Press, New York (2008)

[23] Mitchell, C.J., Piper, F.C.: The cost of reducing key storage requirements in secure networks. Computers and Security 6, 339–341 (1987)

[24] Mitchell, C.J., Piper, F.C.: Key storage in secure networks. Discrete Applied Mathematics 21, 215–228 (1988)

[25] Park, E.C., Blake, I.F.: Reducing communication overhead of key distribution schemes for wireless sensor networks. In: Proceedings of ICCCN 2007, pp. 1345–1350. IEEE Press, Los Alamitos (2007)

[26] Römer, K., Mattern, F.: The design space of wireless sensor networks. IEEE Wireless Communications Magazine 11(6), 54–61 (2004)

[27] Ruj, S., Roy, B.: Key predistribution using partially balanced designs in wireless sensor networks. In: Stojmenovic, I., Thulasiram, R.K., Yang, L.T., Jia, W., Guo, M., de Mello, R.F. (eds.) ISPA 2007. LNCS, vol. 4742, pp. 431–445. Springer, Heidelberg (2007)

[28] Stinson, D.R.: On some methods of unconditionally secure key distribution and broadcast encryption. Designs Codes and Cryptography 12, 215–243 (1997)

[29] Stinson, D.R., Wei, R.: Generalized cover free families. Discrete Mathematics 279, 463–477 (2004)

[30] Wei, R., Wu, J.: Product construction of key distribution schemes for sensor networks. In: Handschuh, H., Hasan, M.A. (eds.) SAC 2004. LNCS, vol. 3357, pp. 280–293. Springer, Heidelberg (2005)

[31] Xiao, Y., Rayi, V.K., Sun, B., Du, X., Hu, F., Galloway, M.: A survey of key management schemes in wireless sensor networks. Comput. Commun. 30(11-12), 2314–2341 (2007)

On Weierstrass Semigroups of Some Triples on Norm-Trace Curves

Gretchen L. Matthews

Department of Mathematical Sciences, Clemson University,
Clemson, SC 29634-0975, USA
gmatthe@clemson.edu
www.math.clemson.edu/~gmatthe

Abstract. In this paper, we consider the norm-trace curves which are defined by the equation $y^{q^{r-1}} + y^{q^{r-2}} + \cdots + y = x^{\frac{q^r-1}{q-1}}$ over $\mathbb{F}_{q^r}$ where q is a power of a prime number and $r \geq 2$ is an integer. We determine the Weierstrass semigroup of the triple of points $(P_\infty, P_{00}, P_{0b})$ on this curve.

1 Introduction

Let X be a smooth projective absolutely irreducible curve of genus $g > 1$ over a finite field $\mathbb{F}$, and let $P_1, \ldots, P_m$ be m distinct $\mathbb{F}$-rational points on X. The Weierstrass semigroup $H(P_1, \ldots, P_m)$ of the m-tuple $(P_1, \ldots, P_m)$ is defined by

$$
H(P_1, \ldots, P_m) = \left\{ (\alpha_1, \ldots, \alpha_r) \in \mathbb{N}^m : \exists f \in \mathbb{F}(X) \text{ with } (f)_\infty = \sum_{i=1}^{r} \alpha_i P_i \right\},
$$

where $\mathbb{F}(X)$ denotes the field of rational functions on X, $(f)_\infty$ denotes the divisor of poles of a rational function f, and $\mathbb{N}$ denotes the set of nonnegative integers. The Weierstrass gap set $G(P_1, \ldots, P_m)$ of the m-tuple $(P_1, \ldots, P_m)$ is defined by

$$
G(P_1, \ldots, P_m) = \mathbb{N}^m \setminus H(P_1, \ldots P_m).
$$

If $m = 1$, then $H(P_1)$ is the classically studied Weierstrass semigroup and $G(P_1)$ is the classically studied Weierstrass gap sequence (or gap set). It is well known that $|G(P_1)| = g$, the genus of X, regardless of the choice of point P_1. The gap set $G(P_1, P_2)$ was introduced in [1] where the authors note that the cardinality $|G(P_1, P_2)|$ may depend on the choice of points P_1 and P_2. The study of the Weierstrass gap set of a pair was taken up by Kim [9] and later by Homma and Kim [7]. This was soon followed by the works of Ballico and Kim [2] and Ishii [8].

As suggested by Goppa and verified by Garcia, Kim, and Lax for the $m = 1$ case [5], knowledge of Weierstrass semigroups of m-tuples of points provides insight into the parameters of associated algebraic geometry codes. This theme has been explored by a number of authors, including the present [14], [10] as well as Carvalho and Torres [4]. For a recent survey of such results, see [3].

C. Xing et al. (Eds.): IWCC 2009, LNCS 5557, pp. 146–156, 2009.

In this paper, we determine a minimal generating set for the Weierstrass semigroup of the triple $(P_\infty, P_{00}, P_{0b})$ on the norm-trace curve $y^{q^{r-1}} + y^{q^{r-2}} + \cdots + y = x^{\frac{q^r-1}{q-1}}$ over $\mathbb{F}_{q^r}$, where $r \geq 2$. Notice that when $r = 2$, the Hermitian curve is obtained. Hence, these results may be viewed as a generalization of some of those in [13] where the Weierstrass semigroup of an m-tuple of collinear points on the Hermitian curve was obtained. This paper may also be seen as a sequel to that of Munuera, Tizziotti, and Torres [15] where the semigroup of the pair (P_∞, P_{00}) on the norm-trace curve is found and then applied to two-point algebraic geometry codes. In fact, we rely heavily on the results contained in both [13] and [15].

This paper is organized as follows. Section 2 provides a background on the Weierstrass semigroup of an m-tuple of points. Section 3 consists of necessary background on the norm-trace curve. The main result of this paper is contained in Section 4.

2   Weierstrass Semigroups of m-Tuples

In this section, we describe tools useful in the study of Weierstrass semigroups of m-tuples of points. Several generalize those used to study the gap set of a pair of points [9], [7].

We begin with a brief review of notation. The divisor of a rational function f will be denoted by (f), and $\mathbb{Z}^+$ denotes the set of positive integers. Given $a_1, \ldots, a_k \in \mathbb{Z}^+$, the (numerical) semigroup generated by $a_1, \ldots, a_k$ is

$$\langle a_1, \ldots, a_k \rangle := \left\{ \sum_{i=1}^{k} c_i a_i : c_i \in \mathbb{N} \right\}.$$

As usual, given $v \in \mathbb{Z}^r$ where $r \in \mathbb{Z}^+$, the ith coordinate of v is denoted by v_i.

Define a partial order $\preceq$ on $\mathbb{Z}^r$ by $(n_1, \ldots, n_r) \preceq (p_1, \ldots, p_r)$ if and only if $n_i \leq p_i$ for all i, $1 \leq i \leq r$. When comparing elements of $\mathbb{Z}^r$, we will always do so with respect to the partial order $\preceq$.

In [13] it is shown that if $1 \leq m \leq |\mathbb{F}|$, then there exists a minimal subset $\Gamma(P_1, \ldots, P_m) \subseteq H(P_1, \ldots, P_m)$ such that

$$H(P_1, \ldots, P_m) = \{\mathrm{lub}\, \{\mathbf{u_1}, \ldots, \mathbf{u_r}\} \in \mathbb{N}^m : \mathbf{u_1}, \ldots, \mathbf{u_r} \in \Gamma(P_1, \ldots, P_m)\}$$

where

$$\mathrm{lub}\{\mathbf{u_1}, \ldots, \mathbf{u_m}\} = (\max\{\mathbf{u_{1_1}}, \ldots, \mathbf{u_{m_1}}\}, \ldots, \max\{\mathbf{u_{1_m}}, \ldots, \mathbf{u_{m_m}}\}) \in \mathbb{N}^m$$

is least upper bound of the vectors $\mathbf{u_1}, \ldots, \mathbf{u_m} \in \mathbb{N}^m$. In fact, $\Gamma(P_1, \ldots, P_m)$ may be defined as follows.

Definition 1. *Given m $\mathbb{F}$-rational points $P_1, \ldots, P_m$ on a curve over $\mathbb{F}$ where $2 \leq m \leq |\mathbb{F}|$, set*

$$\Gamma(P_1, \ldots, P_m) := \left\{ \mathbf{n} \in \mathbb{N}^m : \begin{array}{l} \mathbf{n} \text{ is minimal in } \{\mathbf{p} \in H(P_1, \ldots, P_m) : p_i = n_i\} \\ \text{for some } i, 1 \leq i \leq m \end{array} \right\}.$$

The set $\Gamma(P_1,\ldots,P_m)$ is called the minimal generating set of the Weierstrass semigroup $H(P_1,\ldots,P_m)$. Hence, to determine the entire Weierstrass semigroup $H(P_1,\ldots,P_m)$, one only needs to determine the minimal generating set $\Gamma(P_1,\ldots,P_m)$.

When $m = 2$,

$$\Gamma(P_1,P_2) = \{(\alpha,\beta_\alpha) : \alpha \in G(P_1)\}$$

where

$$\beta_\alpha := \min\{\beta \in \mathbb{N} : (\alpha,\beta) \in H(P_1,P_2)\}.$$

This set introduced by Kim [9] where he showed that

$$\{\beta_\alpha : \alpha \in G(P_1)\} \subseteq G(P_2)$$

and in fact

$$\phi : G(P_1) \rightarrow G(P_2)$$
$$\alpha \mapsto \beta_\alpha$$

is a bijection. While the latter fact fails for $m \geq 3$, we do have the following as proven in [13].

Lemma 1. *If $P_1,\ldots,P_m$ are distinct $\mathbb{F}$-rational points on a curve X over a finite field $|\mathbb{F}|$ and $2 \leq m \leq |\mathbb{F}|$, then*

$$\Gamma(P_1,\ldots,P_m) \subseteq G(P_1) \times \cdots \times G(P_m).$$

Another property of the minimal generating set that we will rely on is in the following lemma.

Lemma 2. *If $P_1,\ldots,P_m$ are distinct $\mathbb{F}$-rational points on a curve X over a finite field $|\mathbb{F}|$ and $2 \leq m \leq |\mathbb{F}|$, then*

$$\Gamma(P_1,\ldots,P_m) = \left\{ \mathbf{n} \in \mathbb{N}^m : \begin{array}{l} \mathbf{n} \text{ is minimal in } \{\mathbf{p} \in H(P_1,\ldots,P_m) : p_i = n_i\} \\ \text{for all } i, 1 \leq i \leq m \end{array} \right\}.$$

We will use these properties to compute $\Gamma(P_1,P_2,P_3)$ for the norm-trace curve over $\mathbb{F}_{q^r}$ where $P_1 = P_\infty$, $P_2 = P_{00}$, and $P_3 = P_{0b}$. Before doing so, we discuss relevant properties of the norm-trace curve in the next section.

3 Preliminaries on the Norm-Trace Curve

Let q be a power of a prime number and $r \geq 2$ be an integer. The norm-trace curve X over $\mathbb{F}_{q^r}$ is defined by

$$y^{q^{r-1}} + y^{q^{r-2}} + \cdots + y = x^{a+1}$$

where $a := \frac{q^r-1}{q-1} - 1$. One immediately recognizes that setting $r = 2$ gives the Hermitian curve over $\mathbb{F}_{q^2}$.

In [6], Geil determined that X has q^{2r-1} affine points over $\mathbb{F}_{q^r}$, namely $(\alpha : \beta : 1)$ where the norm of α with respect to the extension $\mathbb{F}_{q^r}/\mathbb{F}_q$ is equal to the trace of β with respect to the extension $\mathbb{F}_{q^r}/\mathbb{F}_q$; that is, the set of affine points of X which are rational over $\mathbb{F}_{q^r}$ is

$$\left\{ (\alpha : \beta : 1) : N_{\mathbb{F}_{q^r}/\mathbb{F}_q}(\alpha) = Tr_{\mathbb{F}_{q^r}/\mathbb{F}_q}(\beta) \right\}.$$

We will denote such points by $P_{\alpha\beta}$. In addition, X has a single point at infinity P_∞. Note that X has q^{r-1} points of the form $P_{0\beta}$ and $a = q^{r-1}+q^{r-2}+\cdots+q^2+q$. Then the genus of X is given by $g = \frac{a(q^{r-1}-1)}{2}$.

By exploiting the facts that

$$(x) = \sum_{\beta} P_{0\beta} - q^{r-1}P_\infty$$

and

$$(y) = (a+1)P_{00} - (a+1)P_\infty,$$

Geil [6] found that the Weierstrass semigroup of the point at infinity is

$$H(P_\infty) = \langle q^{r-1}, a+1 \rangle.$$

Later, using these same principal divisors, Munuera, Tizziotti, and Torres [15] proved that the Weierstrass semigroup of the point P_{00} is

$H(P_{00}) =$

$$\langle a, a+1, qa-1, (2q-1)a-2, (3q-2)a-3, \ldots, ((\lambda+1)q-\lambda)a-(\lambda+1) \rangle$$

where $\lambda := a - q^{r-1} - 1 = q^{r-2} + q^{r-3} + \cdots + q - 1$.

Now, fix $b \in \mathbb{F}_{q^r}$ with $b^{q^{r-1}} + b^{q^{r-2}} + \cdots + b = 0$. A similar argument to that mentioned above, using the fact that

$$(y - b) = (a+1)P_{0b} - (a+1)P_\infty,$$

yields

$$H(P_{0b}) = H(P_{00}).$$

Let us use this information to obtain explicit descriptions for elements of the gap sets of the points P_∞ and P_{00}. Some arguments are provided in [15], but we include these details here for easy reference. We claim that the gap set of the point at infinity is

$G(P_\infty) =$

$$\left\{ (q^{r-1} - i + j - 1)(a+1) - jq^{r-1} : \begin{array}{c} 1 \leq j \leq i \leq a - s \text{ and} \\ (s-1)(q-1) \leq i - j < s(q-1) \\ \text{where } 1 \leq s \leq a+1-q^{r-1} \end{array} \right\}.$$

Suppose there exist $\alpha_1, \alpha_2 \in \mathbb{N}$ with

$$\left(q^{r-1} - i + j - 1\right)(a+1) - jq^{r-1} = \alpha_1(a+1) + \alpha_2 q^{r-1}$$

where $1 \le j \le i \le a - s$, $(s-1)(q-1) \le i - j < s(q-1)$, and $1 \le s \le a + 1 - q^{r-1}$. Then

$$\left(q^{r-1} - i + j - 1 - \alpha_1\right)(a+1) = (\alpha_2 + j)q^{r-1},$$

and, thus, $q^{r-1} - i + j - 1 - \alpha_1 \ge 0$. This leads to a contradiction since $q^{r-1} - i + j - 1 - \alpha_1$ is not a multiple of q^{r-1}. Consequently, each such integer $\left(q^{r-1} - i + j - 1\right)(a+1) - jq^{r-1}$ is an element of the gap set of P_∞. We apply a counting argument to see that each element of $G(P_\infty)$ is of the form $\left(q^{r-1} - i + j - 1\right)(a+1) - jq^{r-1}$ with $1 \le j \le i \le a - s$, $(s-1)(q-1) \le i - j < s(q-1)$, and $1 \le s \le a + 1 - q^{r-1}$; that is, we give a counting argument to show that there are precisely g integers of the form $\left(q^{r-1} - i + j - 1\right)(a+1) - jq^{r-1}$ with $1 \le j \le i \le a - s$, $(s-1)(q-1) \le i - j < s(q-1)$, and $1 \le s \le a + 1 - q^{r-1}$. It is not hard to see that

$$\left(q^{r-1} - i + j - 1\right)(a+1) - jq^{r-1} = \left(q^{r-1} - i' + j' - 1\right)(a+1) - j'q^{r-1}$$

where $1 \le j \le i \le a - 1$ and $1 \le j' \le i' \le a - 1$ implies

$$i = i' \text{ and } j = j'.$$

Hence, the number of such integers $\left(q^{r-1} - i + j - 1\right)(a+1) - jq^{r-1}$ is equal to the number of pairs (i, j) satisfying $1 \le j \le i \le a - 1$ and $i - j < q^{r-1} - 1$. Now, the number of (i, j) pairs with $1 \le j \le i \le a - 1$ and $i - j < q^{r-1} - 1$ is

$$\sum_{i=1}^{a-1}\sum_{j=1}^{i} 1 - \sum_{i=q^{r-1}}^{a-1}\sum_{j=1}^{i-q^{r-1}+1} 1 = \frac{a\left(q^{r-1} - 1\right)}{2},$$

which is the genus of the curve. This completes the proof that $G(P_\infty)$ is as claimed.

Next, we claim that the gap set of the point P_{00} (and of the point P_{0b}) is

$$G(P_{00}) = G(P_{0b}) = \left\{ (i-j)(a+1) + j : \begin{array}{c} 1 \le j \le i \le a - s \text{ and} \\ (s-1)(q-1) \le i - j < s(q-1) \\ \text{where } 1 \le s \le a + 1 - q^{r-1} \end{array} \right\}.$$

To see this, it is helpful to visualize the elements of the semigroup $H(P_{00})$ placed in an array as follows. Arrange the positive elements of $H(P_{00})$ in an array so that each row consists of consecutive integers. Consider $\alpha = (i - j)(a+1) + j$ where $1 \le j \le i \le a - s$, $(s-1)(q-1) \le i - j < s(q-1)$, and $1 \le s \le a + 1 - q^{r-1}$. Write $i - j = (s-1)(q-1) + k$ where $0 \le k \le q - 2$. Then

$$\alpha = ((s-1)q - (s-2))a + (k-1)a + i.$$

Hence, if $\alpha \in H(P_{00})$, then α would be on row $(s-1)(q-1)+k$ of the array. However, the largest number on this row is

$$((s-1)(q-1)+k)a + (s-1)(q-1)+k,$$

and $\alpha > ((s-1)(q-1)+k)a+(s-1)(q-1)+k$ as $i > (s-1)q-(s-2)+k+1$. As a result, $\alpha \in G(P_{00})$. The claim now follows by the same counting argument applied above, because there are g positive integers of the form $(i-j)(a+1)+j$ with $1 \le j \le i \le a-s$, $(s-1)(q-1) \le i-j < s(q-1)$, and $1 \le s \le a+1-q^{r-1}$.

We will use these explicit descriptions of elements of the gap sets in the next section to find the Weierstrass semigroup of the triple $(P_\infty, P_{00}, P_{0b})$.

4   Determination of the Semigroup $H(P_\infty, P_{00}, P_{0b})$

In this section, we find the Weierstrass semigroup of the triple $(P_\infty, P_{00}, P_{0b})$ on the norm-trace curve over $\mathbb{F}_{q^r}$. In fact, we produce the minimal generating set for this Weierstrass semigroup. To do so, we rely heavily on the results of [15]. In particular, we will use that the minimal generating set of the pair (P_∞, P_{00}) of points on the norm-trace curve over $\mathbb{F}_{q^r}$ is

$$\Gamma(P_\infty, P_{00}) = \left\{ v_{ij} : \begin{array}{c} 1 \le j \le i \le a-s, \\ (s-1)(q-1) \le i-j \le s(q-1)-1 \\ \text{for some } 1 \le s \le a+1-q^{r-1} \end{array} \right\}$$

where

$$v_{ij} := \left((a+1)\left(q^{r-1}-i+j-1\right) - jq^{r-1}, (a+1)(i-j)+j \right)$$

as proved in [15]. It is not difficult to see that $\Gamma(P_\infty, P_{00}) = \Gamma(P_\infty, P_{0b})$.

Theorem 1. *The minimal generating set of the Weierstrass semigroup of the triple $(P_\infty, P_{00}, P_{0b})$ of $\mathbb{F}_{q^r}$-rational points on the norm-trace curve over $\mathbb{F}_{q^r}$ is*

$$\Gamma(P_\infty, P_{00}, P_{0b}) = \left\{ \gamma_{i,j,t} : \begin{array}{c} 1 \le t \le i-j, 1 \le j < i \le a-s, \\ (s-1)(q-1) \le i-j \le s(q-1)-1 \\ \text{where } 1 \le s \le a+1-q^{r-1} \end{array} \right\}$$

where
$\gamma_{i,j,t} :=$

$$\left(\left(q^{r-1}-i+j-1\right)(a+1) - jq^{r-1}, (i-j-t)(a+1)+j, (t-1)(a+1)+j \right).$$

Proof. Set

$$S := \left\{ \gamma_{i,j,t} : \begin{array}{c} 1 \le t \le i-j, 1 \le j < i \le a-s, \\ (s-1)(q-1) \le i-j \le s(q-1)-1 \\ \text{where } 1 \le s \le a+1-q^{r-1} \end{array} \right\}$$

and $\Gamma := \Gamma(P_\infty, P_{00}, P_{0b})$. First, we will show that $S \subseteq \Gamma$. Assume

$$s := \gamma_{i,j,t} \in S.$$

Then $s \in H(P_\infty, P_{00}, P_{0b})$ since

$$\left(\frac{x^{a+1-j}}{y^{i-j-t+1}(y-b)^t} \right)_\infty = s_1 P_\infty + s_2 P_{00} + s_3 P_{0b}.$$

Hence, $s \in P := \{p \in H(P_\infty, P_{00}, P_{0b}) : p_1 = s_1\}$ and so $P \neq \emptyset$. To conclude that $s \in \Gamma$, we will prove that s is minimal in P.

Suppose not; that is, suppose there exists $v \in P$ with $v \preceq s$ and $v \neq s$. Let $f \in \mathbb{F}_{q^r}(X)$ be so that

$$(f) = A - v_1 P_\infty - v_2 P_{00} - v_3 P_{0b}$$

where $A \geq 0$.

Suppose $v_2 < s_2$. Then $v_2 = s_2 - k$ with $k \in \mathbb{Z}^+$ and so

$$v_2 = (a+1)(i-j-t) + j - k.$$

If $j \leq k$, then

$$\left(f y^{i-j-t} \right)_\infty = (v_1 + (a+1)(i-j-t)) P_\infty + v_3 P_{0b}.$$

Hence,

$$w := \left((a+1)(q^{r-1} - t - 1) - jq^{r-1}, v_3 \right) \in H(P_\infty, P_{0b}).$$

However,

$$\left((a+1)(q^{r-1} - t - 1) - jq^{r-1}, (a+1)t + j \right) \in \Gamma(P_\infty, P_{0b}),$$

$$w \preceq \left((a+1)(q^{r-1} - t - 1) - jq^{r-1}, (a+1)t + j \right),$$

and

$$w \neq \left((a+1)(q^{r-1} - t - 1) - jq^{r-1}, (a+1)t + j \right).$$

Consequently, it must be that $j > k$. Now,

$$\left(f y^{i-j-t} x^{j-k} \right)_\infty =$$

$$\left(v_1 + (a+1)(i-j-t) + (j-k)q^{r-1} \right) P_\infty + (v_3 - (j-k)) P_{0b}$$

which implies

$$w' := \left(v_1 + (a+1)(i-j-t) + (j-k)q^{r-1}, v_3 - (j-k) \right) \in H(P_\infty, P_{0b}).$$

This yields a contradiction since

$$w' \preceq \left((a+1)(q^{r-1} - t - 1) - kq^{r-1}, (a+1)t + k \right),$$

$$w' \neq \left((a+1)\left(q^{r-1}-t-1\right)-kq^{r-1},(a+1)t+k\right),$$

and

$$\left((a+1)\left(q^{r-1}-t-1\right)-kq^{r-1},(a+1)t+k\right) \in \Gamma\left(P_\infty, P_{0b}\right).$$

As a result, $v_2 = s_2$ and $v_3 < s_3$.

Write $v_3 = s_3 - k$ with $k \in \mathbb{Z}^+$ so that $v_3 = (a+1)(t-1)+j-k$. If $j \leq k$, then considering $\left(f(y-b)^{t-1}\right)$ leads to a contradiction as

$$\left((a+1)\left(q^{r-1}-i+t+j-2\right)-jq^{r-1},(a+1)(i-j-t)+j\right) \in H\left(P_\infty, P_{00}\right),$$

$$\left((a+1)\left(q^{r-1}-i+t+j-2\right)-jq^{r-1},(a+1)(i-j-t)+j\right) \preceq w,$$

$$\left((a+1)\left(q^{r-1}-i+t+j-2\right)-jq^{r-1},(a+1)(i-j-t)+j\right) \neq w,$$

and $w \in \Gamma\left(P_\infty, P_{00}\right)$ where

$$w := \left((a+1)\left(q^{r-1}-(i-t)+j-1\right)-jq^{r-1},(a+1)((i-t)-j)+j\right).$$

Thus, $j > k$. However, considering

$$\left(\frac{f(y-b)^{t-1}x^{j-k}}{y^{j-k+t}}\right)_\infty$$

gives

$$\left((a+1)\left(q^{r-1}-i+k-2\right)-kq^{r-1},(a+1)(i-k)+k\right) \in H\left(P_\infty, P_{00}\right).$$

Once again, this leads to a contradiction since

$$\left((a+1)\left(q^{r-1}-i+k-2\right)-kq^{r-1},(a+1)(i-k)+k\right) \preceq w',$$

$$\left((a+1)\left(q^{r-1}-i+k-2\right)-kq^{r-1},(a+1)(i-k)+k\right) \neq w',$$

and $w' \in \Gamma\left(P_\infty, P_{0b}\right)$ by [15] where

$$w' := \left((a+1)\left(q^{r-1}-i+k-1\right)-kq^{r-1},(a+1)(i-k)+k\right).$$

It follows that s is minimal in P and so $S \subseteq \Gamma$.

Next, we will show that $\Gamma \subseteq S$. Suppose $n \in \Gamma$. According to Lemma 1,

$$n \in G\left(P_\infty\right) \times G\left(P_{00}\right) \times G\left(P_{0b}\right).$$

Hence,

$$n_1 = (a+1)\left(q^{r-1}-i_1+j_1-1\right)-j_1 q^{r-1},$$
$$n_2 = (a+1)(i_2-j_2)+j_2, \text{ and}$$
$$n_3 = (a+1)(i_3-j_3)+j_3$$

where $1 \leq j_k \leq i_k \leq a - s_k$ and $(s_k - 1)(q-1) \leq i_k - j_k \leq s_k(q-1)-1$ for $k = 1, 2, 3$, with $1 \leq s_k \leq a+1-q^{r-1}$. We may assume, without loss of generality, that $j_2 \leq j_3$. Let $f \in \mathbb{F}_{q^r}(X)$ be so that $(f) = A - n_1 P_\infty - n_2 P_{00} - n_3 P_{0b}$ for some $A \geq 0$. Then

$$\left(f(y-b)^{i_3-j_3+1} \right) = A + ((a+1)(i_3 - j_3 + 1) - n_3) P_{0b}$$
$$- (n_1 + (a+1)(i_3 - j_3 + 1)) P_\infty$$
$$- n_2 P_{00}.$$

Thus,

$$(n_1 + (a+1)(i_3 - j_3 + 1), n_2) \in H(P_\infty, P_{00}).$$

Consequently, there exists $u \in \Gamma(P_\infty, P_{00})$ with

$$u \preceq (n_1 + (a+1)(i_3 - j_3 + 1), n_2)$$

and $u_2 = n_2$. According to [15], $u_1 = (a+1)\left(q^{r-1} - i_2 + j_2 - 1 \right) - j_2 q^{r-1}$. Notice that $n_1 < u_1$ since otherwise $(u_1, u_2, 0) \preceq n$, contradicting the minimality of n in $\{p \in H(P_\infty, P_{00}, P_{0b}) : p_2 = n_2\}$. As a result,

$$n_1 < u_1 \leq n_1 + (a+1)(i_3 - j_3 + 1).$$

Set

$$h = \frac{\prod_{\beta \in \mathcal{B}} (y - \beta)}{y^{i_2 - j_2} x^{j_2} (y-b)^{i_3 - j_3}}$$

where $\mathcal{B} = \left\{ \beta \in \mathbb{F}_{q^r} : Tr_{\mathbb{F}_{q^r}/\mathbb{F}_q}(\beta) = 0, \ \beta \neq 0, b \right\}$. Then

$$(h) = \sum_{\beta \neq 0, b} (a + 1 - j_2) P_{0\beta} - (u_1 - (a+1)(i_3 - j_3 + 1)) P_\infty$$
$$- ((a+1)(i_2 - j_2) + j_2) P_{00} - ((a+1)(i_3 - j_3) + j_2) P_{0b}.$$

Thus, $w := (w_1, (a+1)(i_2 - j_2) + j_2, (a+1)(i_3 - j_3) + j_2) \in H(P_\infty, P_{00}, P_{0b})$ where

$$w_1 = \max\{0, u_1 - (a+1)(i_3 - j_3 + 1)\}.$$

However, $w \preceq n$ since $j_2 \leq j_3$. It follows that $w = n$; otherwise n is not minimal in $\{p \in H(P_\infty, P_{00}, P_{0b}) : p_2 = n_2\}$. Since $n_1 > 0$, we must have that

$$u_1 > (a+1)(i_3 - j_3 + 1)$$

and $j_2 = j_3$. In particular,

$$n_1 = (a+1)\left(q^{r-1} - (i_2 + i_3 - j_3 + 1) + j_2 \right) - j_2 q^{r-1}$$
$$n_2 = (a+1)(i_2 - j_2) + j_2$$
$$n_3 = (a+1)(i_3 - j_3) + j_2.$$

It can be checked that $1 \leq i_2 + i_3 - j_3 + 1 \leq a - 1$, from which it follows that $i_2 + i_3 - j_3 + 1 = i_1$ and $j_2 = j_1$. As a result,

$$n = \gamma_{i_2 + i_3 - j_3 + 1, j_2, i_3 - j_3 + 1}$$

and so $n \in S$. Thus, $\Gamma \subseteq S$. This concludes the proof that $\Gamma(P_\infty, P_{00}, P_{0b}) = S$.

Example 1. Consider the norm-trace curve X defined by $y^9 + y^3 + y = x^{13}$ over $\mathbb{F}_{27}$. Notice that X has genus 48, the gap set of the point P_∞ is

$G\left(P_{\infty}\right) = \mathbb{N} \setminus \langle 9, 13 \rangle$

$$= \left\{ \begin{array}{l} 1, 2, 3, 4, 5, 6, 7, 8, 10, 11, 12, 14, 15, 16, 17, 19, 20, 21, 23, 24, 25, 28, \\ 29, 30, 32, 33, 34, 37, 38, 41, 42, 43, 46, 47, 50, 51, 55, 56, 59, 60, 64, \\ 68, 69, 73, 77, 82, 86, 95 \end{array} \right\},$$

and the gap set of the points P_{00} and P_{0b} is

$G\left(P_{00}\right) = G\left(P_{0b}\right) = \mathbb{N} \setminus \langle 12, 13, 35, 58, 81 \rangle$

$$= \left\{ \begin{array}{l} 1, 2, 3, 4, 5, 6, 7, 8, 9, 10, 11, 14, 15, 16, 17, 18, 19, 20, 21, 22, 23, 27, \\ 28, 29, 30, 31, 32, 33, 34, 40, 41, 42, 43, 44, 45, 46, 53, 54, 55, 56, 57, \\ 66, 67, 68, 69, 79, 80, 92 \end{array} \right\}.$$

In [15], it is shown that

$\Gamma\left(P_{\infty}, P_{00}\right) =$

$$\left\{ \begin{array}{l} (1, 23), (2, 46), (3, 69), (4, 92), (5, 11), (6, 34), (7, 57), (8, 80), (10, 22), \\ (11, 45), (12, 68), (14, 10), (15, 33), (16, 56), (17, 79), (19, 21), (20, 44), \\ (21, 67), (23, 9), (24, 32), (25, 55), (28, 20), (29, 43), (30, 66), (32, 8), (33, 31), \\ (34, 54), (37, 19), (38, 42), (41, 7), (42, 30), (43, 53), (46, 18), (47, 41), (50, 6), \\ (51, 29), (55, 17), (56, 40), (59, 5), (60, 28), (64, 16), (68, 4), (69, 27), \\ (73, 15), (77, 3), (82, 14), (86, 2), (95, 1) \end{array} \right\}.$$

According to Theorem 1, the minimal generating set of the Weierstrass semigroup of the triple $\left(P_{\infty}, P_{00}, P_{0b}\right)$ is
$\Gamma\left(P_{\infty}, P_{00}, P_{0b}\right) =$

$$\left\{ \begin{array}{l} (1, 10, 10), (2, 7, 33), (2, 20, 20), (2, 33, 7), (3, 4, 56), \\ (3, 17, 43), (3, 30, 30), (3, 43, 17), (3, 56, 4), (4, 1, 79), \\ (4, 14, 66), (4, 27, 53), (4, 40, 40), (4, 53, 27), (4, 66, 14), \\ (4, 79, 1), (6, 8, 21), (6, 21, 8)(7, 5, 44), (7, 18, 31), \\ (7, 31, 18), (7, 44, 5), (8, 2, 67), (8, 15, 54), (8, 28, 41), \\ (8, 41, 28), (8, 54, 15), (8, 67, 2), (10, 9, 9), (11, 6, 32), \\ (11, 19, 19), (11, 32, 6), (12, 3, 55), (12, 16, 42), (12, 29, 29), \\ (12, 42, 16), (12, 55, 3), (15, 7, 20), (15, 20, 7), (16, 4, 43), \\ (16, 17, 30), (16, 30, 17), (16, 43, 4), (17, 1, 66), (17, 14, 53), \\ (17, 27, 40), (17, 40, 27), (17, 53, 14), (17, 66, 1), (19, 8, 8), \\ (20, 5, 31), (20, 18, 18), (20, 31, 5), (21, 2, 54), (21, 15, 41), \\ (21, 28, 28), (21, 41, 15), (21, 54, 2), (24, 6, 19), (24, 19, 6), \\ (25, 3, 42), (25, 16, 29), (25, 29, 16), (25, 42, 3), (28, 7, 7), \\ (29, 4, 30), (29, 17, 17), (29, 30, 4), (30, 1, 53), (30, 14, 40), \\ (30, 27, 27), (30, 40, 14), (30, 53, 1), (33, 5, 18), (33, 18, 5), \\ (34, 2, 41), (34, 15, 28), (34, 28, 15), (34, 41, 2), (37, 6, 6), \\ (38, 3, 29), (38, 16, 16), (38, 29, 3), (42, 4, 17), (42, 17, 4), \\ (43, 1, 40), (43, 14, 27), (43, 27, 14), (43, 40, 1), (46, 5, 5), \\ (47, 2, 28), (47, 15, 15), (47, 28, 2), (51, 3, 16), (51, 16, 3), \\ (55, 4, 4), (56, 1, 27), (56, 14, 14), (56, 27, 1), (60, 2, 15), \\ (60, 15, 2), (64, 3, 3), (69, 1, 14), (69, 14, 1), (73, 2, 2), \\ (82, 1, 1) \end{array} \right\}.$$

Acknowledgements. The referee's comments and corrections are greatly appreciated.

References

1. Arbarello, E., Cornalba, M., Griffiths, P., Harris, J.: Geometry of Algebraic Curves. Springer, Heidelberg (1985)
2. Ballico, E., Kim, S.J.: Weierstrass multiple loci of n-pointed algebraic curves. J. Algebra 199(2), 455–471 (1998)
3. Carvalho, C., Kato, T.: On Weierstrass semigroups and sets: a review of new results, Geom. Dedicata (to appear)
4. Carvalho, C., Torres, F.: On Goppa codes and Weierstrass gaps at several points. Des. Codes Cryptogr. 35(2), 211–225 (2005)
5. García, A., Kim, S.J., Lax, R.F.: Consecutive Weierstrass gaps and minimum distance of Goppa codes. J. Pure Appl. Algebra 84(2), 199–207 (1993)
6. Geil, O.: On codes from norm-trace curves. Finite Fields Appl. 9(3), 351–371 (2003)
7. Homma, M., Kim, S.J.: Goppa codes with Weierstrass pairs. J. Pure Appl. Algebra 162(2-3), 273–290 (2001)
8. Ishii, N.: A certain graph obtained from a set of several points on a Riemann surface. Tsukuba J. Math. 23(1), 55–89 (1999)
9. Kim, S.J.: On the index of the Weierstrass semigroup of a pair of points on a curve. Arch. Math (Basel) 62(1), 73–82 (1994)
10. Matthews, G.L.: Codes from the Suzuki function field. IEEE Trans. Inform. Theory 50(12), 3298–3302 (2004)
11. Matthews, G.L.: Some computational tools for estimating the parameters of algebraic geometry codes. In: Coding theory and quantum computing, Contemp. Math., vol. 381, pp. 19–26. Amer. Math. Soc., Providence (2005)
12. Matthews, G.L.: Weierstrass semigroups and codes from a quotient of the Hermitian curve. Des. Codes Cryptogr. 37(3), 473–492 (2005)
13. Matthews, G.L.: The Weierstrass semigroup of an m-tuple of collinear points on a Hermitian curve. In: Mullen, G.L., Poli, A., Stichtenoth, H. (eds.) Fq7 2003. LNCS, vol. 2948, pp. 12–24. Springer, Heidelberg (2004)
14. Matthews, G.L.: Weierstrass pairs and minimum distance of Goppa codes. Des. Codes Cryptogr. 22(2), 107–121 (2001)
15. Munuera, C., Tizziotti, G.C., Torres, F.: Two-point codes on Norm-Trace curves. In: Barbero, A. (ed.) ICMCTA 2008. LNCS, vol. 5228, pp. 128–136. Springer, Heidelberg (2008)

ERINDALE: A Polynomial Based Hashing Algorithm

V. Kumar Murty[1] and Nikolajs Volkovs[2]

[1] Department of Mathematics,
University of Toronto,
Toronto, Ontario, CANADA M5S 2E4
murty@math.toronto.edu

[2] GANITA Lab, Department of Mathematical and Computational Sciences,
University of Toronto at Mississauga
3359 Mississauga Road North
Mississauga, Ontario, CANADA L5L 1C6
n.volkovs@utoronto.ca

Abstract. The aim of this article is to describe a new hash algorithm using polynomials over finite fields. In software, it runs at speeds comparable to SHA-384. Hardware implementation of a slightly modified version of the algorithm presented here runs at significantly faster speeds, namely at 2 Gbits/sec on an FPGA Virtex V of frequency 300 MHz. Modelling suggests that this speed can be increased to 3.4 Gbits/sec. Unlike most other existing hash algorithms, our construction does not follow the Damgard-Merkle philosophy. The hash has several attractive features in terms of its flexibility. In particular, the length of the hash is a parameter that can be set at the outset. Moreover, the estimated degree of collision resistance is measured in terms of another parameter whose value can be varied.

1 Introduction

There is much discussion now about how to construct a good hash algorithm. The main difficulty stems from the fact that the design principles of a hash function are not completely understood. However, some desirable features of a future hash function have been enumerated in several NIST workshops. In particular, the algorithm should provide for a changeable length of hash, collision resistance measured in terms of a tunable parameter and perhaps different functions for online and offline usage.

There are some methods that have been studied in the literature to produce new hash functions from old functions. For example, one might consider the concatenation of two existing hash functions. If one of the hash functions is based on the Damgard-Merkle construction, it is known that (Joux [3]) collision resistance is weakened. One might increase the number of rounds in an existing function but it is not clear that this has an essential impact on collision resistance. Several attempts have been made to identify good design principles of hash functions (see Preneel [7]) but this work seems to still be evolving.

C. Xing et al. (Eds.): IWCC 2009, LNCS 5557, pp. 157–170, 2009.

The aim of this article is to describe a new hash algorithm, which incorporates some of the features mentioned above. In particular, the output length of the function is a parameter that can be set at the beginning. Moreover, the degree of collision resistance is expected to depend on another parameter that can also be specified at the outset. An announcement of our work can be found in [6].

The performance of the algorithm is comparable to that of the SHA-family. This comparison was made on an AMD Sempron 2GHz processor 3400+ using 1GB of RAM. In particular, for a 384 bit hash, the speed is comparable to SHA-384. For a 512 bit hash, the speed is 5% faster than SHA-512. The hardware implementation of the algorithm is especially effective. We have shown that the speed of the function reaches 2 Gbits/sec (on an FPGA V running at 300 MHz). Analysis also suggests that with a slightly different choice of parameters and on the same FPGA, we may obtain speeds of approximately 3.4 Gbits/sec. We note that the structure of the algorithm is such that the performance improves for longer files and larger hash sizes. The algorithm has two phases, the second of which does not depend on the length of the message being hashed.

The collision resistance of the function seems to be dependent on the difficulty of solving a family of systems of iterated exponential equations in a finite field. This problem does not seem to have been extensively studied in the literature. However, it does not seem to be tractable by standard methods of analytic number theory.

Summarizing, the hash function that we construct has the following important attributes. Firstly, the length of the output can be changed simply by changing a few steps of the calculation. Secondly, the computation is a bit-stream procedure as opposed to a block procedure. Thirdly, several aspects of the construction, namely the CUrrent Register (CUR) construction (described in Section 3), the compression function and the truncation and exponentiation seem to be novel constructs. Moreover, there seems to be the possibility of some control over collision resistance by the use of "bit strings" as well as the iterated CUR construction, the latter requiring the solution of a system of non-linear iterated exponential equations to invert. As far as we are aware, such equations cannot be solved by standard methods of analytic number theory. Moreover, we are not aware of any other hash function in the literature whose collision resistance involves such iterated exponential equations.

For its performance characteristics,novel design features and dependence on what appears to be an intractible mathematical problem, we believe this hash function is worthy of further attention.

Our construction uses polynomials over finite fields. We note that earlier works have used polynomials over finite fields in the construction of hash algorithms. However, our use of polynomials is very different.

Many well-known hash algorithms that are currently in use are based on the Damgard-Merkle [2], [5] approach. The reader can find a description of such algorithms in the book of Menezes, van Oorschot and Vanstone [4]. While this approach is very elegant, recent work has caused the Damgard-Merkle design methodology to come under close scrutiny. Indeed, the ground-breaking work of Wang [9] has exhibited weaknesses in some of the most popular Damgard-Merkle based hash functions, including MD5 and SHA-1. Moreover, in the case of MD5, multicollisions can be found by exploiting the Damgard-Merkle structure. Moreover, as explained above, multicollisions can be found for any Damgard-Merkle hash function [3]. Our approach, however, is not based on the Damgard-Merkle methodology.

We introduce three constructions which may be of interest in their own right. The first is the CUR construction. This takes as input a sequence of k polynomials over $\mathbb{F}_2$ of degree $< n$ and produces another such sequence. The second is the compression routine. We describe a construction that takes as input a binary sequence of length k, a sequence of k polynomials over $\mathbb{F}_2$ of degree $< n$, an integer r with $2^n < r \leq k$, and a sufficiently large integer λ, and produces r matrices of 2^n rows and $1 + \lambda$ columns. This construction is invertible. In other words, given the matrices, one can reconstruct both the binary sequence as well as the sequence of polynomials. The compression function is obtained by deleting columns 2 to $1 + \lambda$ of selected rows of these matrices. The third construction is truncation followed by exponentiation in a finite group. In essence, our hash function takes a message given as a binary string of length k and performs a preliminary operation on it to transform it into a sequence of k polynomials over $\mathbb{F}_2$ of degree $< n$. It then invokes the CUR construction and the compression routine. The entries of the compressed matrices are then combined in a Cantor enumeration to produce a single integer. This integer is used in the truncation-exponentiation routine to produce a hash value. There are additional steps that provide the transition between the above constructions, but the above description gives the essence.

2   Binary String to Polynomial Sequence

We assume that the message is sufficiently long, something that can be achieved by padding. (We used 4096 bits of a fixed string as padding. A variant of our algorithm uses padding of a length depending on the size of the message.) There are, of course, many ways to apply padding. Some study of this (in terms of "stretching") can be found in the work of Aiello, Haber and Venkatesan [8] in which a randomized function is used to perform stretching. In our algorithm, the purpose of padding is to ensure that we are able to populate certain auxiliary bit strings. Let us denote by k the length in bits of the padded message M.

Choose an integer n with $3 < n < 11$. The message is then split into overlapping segments which are interpreted as polynomials over $\mathbb{F}_2$ of degree $< n$ More precisely, denote by $M(i, j)$ the substring of M beginning with the i-th bit and ending with the j-th bit. Also, denote by $M[i]$ the i-th bit of M. Let us define $S(M, n)$ to be the set $M(1, n), M(2, n + 1), \cdots, M(k - n + 1, k), M(k - n + 2)M[1], M(k - n + 3)M(1, 2), \cdots M[k]M(1, n - 1)$. Each $M(i, i + n - 1)$ may be thought of as a polynomial of degree $< n$ over $\mathbb{F}_2$. Thus, $S(M, n)$ consists of k polynomials of degree $< n$. Note that the construction of the $S(M, n)$ is a stream procedure.

Thus, from the message M which is a binary string of k bits, we produce k polynomials $M_1, \cdots, M_k$ of degree $< n$.

3   Masking: The CUR Construction

Many Damgard-Merkle based hash functions have their compression functions based on block cipher structures. The analogue of that here is the CUR construction which we shall now describe. Given n and a sequence of k polynomials $M_1, \cdots, M_k$ over $\mathbb{F}_2$ of degree $< n$, this construction produces a new sequence of k polynomials, $CUR_1, \cdots, CUR_k$, also of degree $< n$. The CUR construction is one-to-one and length preserving. The construction involves finite field arithmetic. At any given time, we need to store 2^n of these polynomials.

We observe some important aspects of the CUR construction. Firstly, to calculate CUR_i we need to have calculated all CUR_j for $j < i$. Secondly, it is easy to recover $M_1, M_2, ..., M_i$ given the *ordered* sequence $CUR_1, CUR_2, .., CUR_i$, $i < k$. However, during the course of the algorithm and computation of the final hash value, the CUR_i are used to form a weighted sum of integers. From such a sum, it does not seem to be easy to recover the message (see Appendix).

Let $f(x) \in \mathbb{F}_2[x]$ be irreducible of degree n. Thus, there is an isomorphism of fields

$$\mathbb{F}_2[x]/(f(x)) \simeq \mathbb{F}_{2^n}.$$

Denote by ϕ_f the isomorphism of $\mathbb{F}_2$-vector spaces

$$\mathbb{F}_2[x]/(f(x)) \longrightarrow \mathbb{F}_2^n.$$

Let δ (resp. β) be a generator of $(\mathbb{F}_2[x]/(f(x)))^\times$ (resp. $(\mathbb{F}_2[x]/(g(x)))^\times$). We set

$$CUR_1 = M_1 \oplus \phi_f(\delta) \oplus \phi_g(\beta), \tag{1}$$

$$CUR_2 = M_2 \oplus \phi_f(\delta^{int(M_1)}) \oplus \phi_g(\beta^{int(CUR_1)}). \tag{2}$$

For $2 < i \leq 2^n + 1$, we set

$$CUR_i = M_i \oplus$$
$$\phi_f(\delta^{(int(M_{i-1})+int(CUR_{i-2})mod2^n)}) \oplus$$
$$\oplus \phi_g(\beta^{(int(CUR_{i-1})+int(CUR_{i-2})mod2^n)})$$

For $i \geq 2^n + 2$, define two functions d_1 and d_2 as follows. For any bit string B, we define $int(B)$ to be the integer whose base 2 expansion is B. Set

$$d_1 = d_1(i) = i - 2 - int(M_{i-1}), \tag{3}$$

$$d_2 = d_2(i) = i - 2 - int(CUR_{i-1}).$$

Now set

$$CUR_i = M_i \oplus \tag{4}$$

$$\phi_f\left(\delta^{int(M_{i-1})+int(CUR_{d_1})mod2^n}\right)\oplus$$

$$\oplus\phi_g\left(\beta^{(int(CUR_{i-1})+int(CUR_{d_2})mod2^n)}\right)$$

for $i = 2^n + 2, ..., k$ with d_1 and d_2 defined by (3).

Once again, we stress that the procedure just described for calculating the values CUR_i is a stream procedure. Moreover, as the result below indicates, the values of the CUR_i uniquely determine the original message M.

Table 1. Sample calculations of CUR

	1	2	3	4	5	6	7	8
$int(M)$	0	0	0	0	0	0	0	0
$int(CUR)$	4	7	6	15	1	9	9	0
$int(CUR^{(2)})$	0	4	2	12	12	15	11	14

	1	2	3	4	5	6	7	8
$int(M)$	0	0	0	0	0	0	0	1
$int(CUR)$	5	8	4	1	9	0	0	4
$int(CUR^{(2)})$	1	9	0	4	12	3	7	1

Proposition 1. *Let M and M' be messages of length k with $CUR_i(M) = CUR_i(M')$ for $i = 1, \cdots, k$. Then $M = M'$.*

The CUR construction can be iterated. We ran implementations in which we performed seven iterations. We remark that the iterated CUR contruction seems to have good diffusion properties. Below we give a "baby" example with a message of length 8 bits and two iterations without padding.

Example 1. We consider the case $n = 4$ and choose polynomials $f(x) = x^4 + x + 1$ and $g(x) = x^4 + x^3 + 1$. We choose generators $\delta = x + 1$ and $\beta = x^2 + x + 1$. For illustrative purposes, we work with a message M of length 11 bits.

4   The Compression Function

In the Damgard-Merkle methodology, a hash function consists of a compression function and a domain extender. The compression function takes as input a string of fixed length and produces a shorter string of fixed length. The domain extender provides the means of dividing a string of arbitrary length into substrings that can be fed into the compression function. Our method is not based on this approach. However, at the heart of our construction is a compression function, which we describe in this section.

4.1   The Compression Function

We construct a compression function by extracting "features" of the input message. The final hash value will be computed by "packing" this collection of features.

Let k, n be positive integers, r an integer with $2^n < r \le k$, and let λ be a sufficiently large integer. Suppose we are given

- a binary sequence (a message) M of length k, the i-th element of which will be denoted $M[i]$
- a sequence C of length k consisting of integers in the range $\{0, 1, \cdots, 2^n - 1\}$
- a one-to-one "random walk" function (depending on M)

$$h = h_M : \{0, 1 \cdots, k\} \longrightarrow \{1, 2, \cdots, r\lambda\} \subset \mathbb{N}$$

with the property that the composite map

$$\{0, 1 \cdots, k\} \longrightarrow \mathbb{N} \hookrightarrow \mathbb{Z} \longrightarrow \mathbb{Z}/r\mathbb{Z}$$

is surjective.

Then, we define $\mathcal{T}_h(M, C)$. It consists of r matrices $\mathcal{M}_1, \cdots, \mathcal{M}_r$, each having 2^n rows and $1 + \lambda$ columns. Except for the first column, the entries of these matrices are 0 or 1. The first column consists of integers in the range $\{0, 1, \cdots, k\}$. We initialize all of the matrices so that every entry is zero. Set $f(i) = \lceil h(i)/r \rceil$. Note that $0 < f(i) \le \lambda$. For each value of a and b, set

$$\mathcal{M}_a(b, 0) = \sum_{\substack{h(i) \ (\mathrm{mod}\ r)\ =\ a \\ b\ =\ C_i}} M[i].$$

Also, set

$$\mathcal{M}_{h(i) \ (\mathrm{mod}\ r)}(C_i, f(i)) = M[i]. \tag{5}$$

Then we have the following result.

Proposition 2. *Given the entries of* $\mathcal{M}_1, \cdots, \mathcal{M}_r$, *we can reconstruct* M *and* C.

Proof. For each triple (a, b, j) with $1 \le a \le r$, $0 \le b \le 2^n - 1$ and $j > 1$, consider the entry $\mathcal{M}_a(b, j)$. If this is non-zero, it must be equal to 1. In this case, for some i, we must have $h(i) = a + rj$. Such an i is unique as h is one-to-one. For this value of i, we must have $C_i = b$. As we range over a and j, we will get all values of i. Thus, this determines h. Finally, M is determined by (5).

Now we consider a variant of this construction in which we reduce the size of the matrices. This variant may be seen as a compression function. For each $1 \le a \le r$, let Θ_a be a subset of $\{0, 1, \cdots, 2^n - 1\}$. Then, the first variant is to replace (5) with the following rule.

$$\mathcal{M}_{h(i) \ (\mathrm{mod}\ r)}(C_i, f(i)) = \begin{cases} M[i] & \text{if } C_i \in \Theta_{h(i) \ \mathrm{mod}\ r} \\ 0 & \text{otherwise.} \end{cases} \tag{6}$$

The argument given in the proof of Proposition 2 shows that given $\mathcal{T}_h(M, C)$, we can determine $M[i]$, C_i and $h(i)$ for those i for which $h(i) = a + rj$ and $C_i \in \Theta_a$. Note that in order to make this determination, we need to invert h. Assuming that the C_i are equidistributed in $\{0, 1, \cdots, 2^n - 1\}$, this determines $(|\Theta|/2^n r)k$ bits of M. For $b \notin \Theta_a$, we see that the quantity

$$\sum_{\substack{h(i) \pmod{r} = a \\ b = C_i}} M[i]$$

is known (as it is the $(b, 0)$-th entry in $\mathcal{M}_a$). If we assume that the $h(i) \pmod{r}$ and the C_i are independently equidistributed, then the number of values of i in the sum above is $\mathbf{O}(k/2^n r)$. Taking into account the constraint imposed by the above equation, we have $\mathbf{O}(2^{\frac{k}{2^n r} - 1})$ possible values for the relevant $M[i]'s$. Now multiplying this over the possible values of a and $b \notin \Theta_a$ gives

$$\mathbf{O}\left(2^{\left(\frac{k}{2^n r} - 1\right)} 2^n r \left(1 - \frac{|\Theta|}{2^n r}\right)\right)$$

where we have set

$$\Theta = \cup_a \Theta_a.$$

Thus, the number of possible strings M satisfying these constraints is at most $2^{\theta(k - 2^n r)}$ where $\theta = \left(1 - \frac{|\Theta|}{2^n r}\right)$. By our choice of θ, we can thus have some control over collision resistance.

In practice, we will use bit strings that are significantly shorter. This is the second variant of the general construction. In our sample calulations, we chose the length λ to be between 100 and 300 bits. In general, we may choose the length to be a function of the length of the desired output hash. It is therefore possible that $f(i)$ may be larger than λ. If $f(i) \leq \lambda$, we proceed as described above. If $f(i) > \lambda$, then we use the same formulae as above with $f(i)$ replaced by $(f(i) \pmod{\lambda}) + 1$.

In addition to the above table entries, we construct r additional row vectors e_{i0} (for $1 \leq i \leq r$).

Definition 1. *The first column of matrix $\mathcal{M}_i$ will be denoted T_i and called the i-th* table. *Moreover, for each i, and for each $a \in \Theta_i$, the row vectors $(\mathcal{M}_i(a, 2), \cdots, \mathcal{M}_i(a, \lambda + 1))$ will be denoted e_{ia} and called bit strings.*

We conclude this subsection with some general remarks. The length λ and the size of Θ_a are parameters that can be chosen. In our implementation, we used up to $\Theta_a = 5$ for each table, each of length $\lambda = 200$ bits. Changing the number of bit strings sufficiently increases the number of states of the finite state machine that is related to the considered hash function. In this sense, the number of bit strings controls the collision resistance of the function, even when the length of the final hash value is fixed.

At the end of this operation, we produce tables the average entry of which is $k/(2^n * 200)$. In particular, for a message of 1MB and using $n = 4$, the average table entry will be an integer of size 2500. By construction, the sum of the entries in the tables is the same as the message length, namely k.

4.2   The Function h

We used the following function for h. Let q, g be two positive integers chosen so that $q + g < n$. (In our sample calculations, we used $q = g = 1$). Set

$$h_M(i) \;=\; iq \;+\; \alpha_M(i)g$$

where $\alpha_M(i)$ is the number of $j \le i$ for which $M[j] \;=\; 0$. Since α_M is an increasing function, h_M is clearly one-to-one.

5   Further Compression

5.1   Factorization

For every sequence of polynomials CUR_i we define vectors R_i, $i = 1, ..., r$ of length $irr(n)$ where $irr(n)$ is the number of irreducible polynomials with coefficients in $\mathbb{F}_2$ of degree less than n. We will call the vectors R_i *spectrums*. We initialize all of the R_i to be zero. The values of the registers will be determined by means of factoring the polynomials $CUR_1, \cdots, CUR_k$.

5.2   The Spectrum

Recall that T_i is a column vector having 2^n entries. Each integer $0 \le j < 2^n$ may be interpreted as a polynomial P_j over $\mathbb{F}_2$ of degree $< n$. Let us denote by $irr(n)$ the number of irreducible polynomials over $\mathbb{F}_2$ of degree $< n$ and let us list them in some order $Q_1, \cdots, Q_{irr(n)}$ (for example, lexicographically). Factoring the P_j gives us equations

$$P_j \;=\; \prod_k Q_k^{m_{k,j}}.$$

The $\{m_{k,j}\}$ define a $irr(n) \times 2^n$ matrix $\mathcal{R}$ (say). Now we define

$$R_i \;=\; \mathcal{R}T_i.$$

As an example, suppose that $n = 4$ and T_i is the transpose of

$$(13, 10, 11, 11, 10, 9, 12, 4, 5, 15, 12, 8, 4, 13, 14, 13).$$

There are 7 irreducible polynomials which we list in the order

$$Q_1 \;=\; 0, \; Q_2 \;=\; 1, \; Q_3 \;=\; x, \; Q_4 \;=\; x+1,$$

$$Q_5 \;=\; x^2 + x + 1, \; Q_6 \;=\; x^3 + x + 1, \; Q_7 \;=\; x^3 + x^2 + 1.$$

Applying the above construction, we deduce that R_i is the transpose of

$$(13, 10, 92, 123, 33, 8, 13).$$

The maximum value of a spectrum entry is approximately $k/10$ and the average value is $k/(2^{n+1} * 10)$. In particular, for a message of length 1MB, the spectrum from each table contains about 25 bits. It is not necessary to store more than one spectrum at a time.

5.3   Enumeration

Finally, we produce a single integer from the vectors R_i using Cantor enumeration. For each $d \geq 1$, there is a bijective map

$$c_d : \mathbb{N}^d \longrightarrow \mathbb{N}.$$

For $d = 1$, we can take the identity and for $d = 2$, we can take

$$c(x, y) \;=\; c_2(x, y) \;=\; \frac{(x + y)^2 + 3x + y}{2}.$$

Given c_n, the map

$$(x_1, \cdots, x_{n+1}) \;\mapsto\; c_n(c_2(x_1, x_2), x_3, \cdots, x_{n+1})$$

is a candidate for c_{n+1}. However, there are more optimal choices.

We will be applying the enumeration function to the vectors R_i and as such, we need the functions c_m where $m = irr(n)$ for $4 \leq n \leq 10$. We first record the values of the pairs (n, m) in Table 2. Explicit candidates that produce numbers

Table 2. Number of irreducible polynomials

n	4	5	6	7	8	9	10
m	7	10	16	25	43	71	129

of manageable size for $n = 4, 5, 6$ are given in the appendix. For example, for $n = 4$, we used

$$c_7(s_1, s_2, s_3, s_4, s_5, s_6, s_7) \;=\; c(c(c(c(s_1, s_2), s_3), c(s_4, s_5)), c(s_6, s_7)).$$

This number is of the order $(k/10)^{16}$. In particular, for a 1MB file, this is about 40 bytes.

5.4   Knapsack

Based on the bit strings, we compute for each table t, an integer as follows.

$$BS(t) \;=\; \sum_{x \in \Theta_t} (x + 1)int(e_{tx}). \qquad (7)$$

We will use these integers when we calculate the knapsack.

Thus, to each table, we have integers $c(R_t)$ and $BS(t)$. We set $I_t = c(R_t) + BS(t)$. Now we form the sum

$$I \;=\; \sum t I_t$$

which for a 1MB file is an integer of length 320 bits.

6   Truncation and Exponentiation

The third of the "primitives" (the first two being the CUR construction and the compression function) is called truncation followed by exponentiation in a group, and is described as follows. Let

$$H : \{0,1\}^* \longrightarrow \{0,1\}^\kappa$$

be a hash function with output length κ. Let G be a finite abelian group and select an element $g \in G$. Let

$$F : G \longrightarrow \{0,1\}^\tau$$

be a function with $\tau < \kappa$. For a string $M \in \{0,1\}^*$, consider the new function

$$\mathcal{H} : M \mapsto F(g^{int(H(M))}) \in \{0,1\}^\tau.$$

Here, int is the function that associates to a bit string the integer that it represents in base 2.

In our implementation, we used $G = \mathbb{F}_{2^\tau}^\times$ or $G = E(\mathbb{F}_{2^\tau})$ for a cryptographically suitable elliptic curve E over $\mathbb{F}_{2^\tau}$. In the first case F is an identification of $\mathbb{F}_{2^\tau}$ with $\mathbb{F}_2^\tau$. In the second case, F is the function that takes the x-coordinate of a point. Note that F is an isomorphism in the first case, and has a kernel of order 2 (counting multiplicites) in the second case.

Note that in both of these cases, the value of $\mathcal{H}(M)$ only depends on the last τ bits. Thus, the calculation is truncation followed by exponentiation in a group. The idea is that the group operation may help to increase the entropy following truncation. This view can be made somewhat plausible by the following result.

The measure of equidistribution of a sequence of elements of a group is governed by the size of certain exponential sums. If we consider the set of residue classes $\{1 \leq s \leq t\}$ for some parameter $p/2 < t < p$, then the sum

$$\sum_{1 \leq s \leq t} \exp\{2\pi i a s/p\}$$

is

$$e^{2\pi i a/p}(1 - e^{2\pi i a t/p})/(1 - e^{2\pi i a/p})$$

and the denominator shows that for a bounded as a function of p, and t sufficiently large (for example, $t \gg p$), this is not $\ll p^{1-\epsilon}$ for any $\epsilon > 0$. On the other hand, if we let g be a primitive root modulo p, then by a result of Bourgain [1], we have

$$\sum_{1 \leq s \leq t} \exp\{2\pi i a g^s/p\} \ll p^{1-\epsilon}$$

for some $\epsilon > 0$. In fact, this is true under even the weaker hypothess $t > p^\delta$ for some $\delta > 0$. This result suggests that exponentiation may increase the degree of equidistribution.

7   The Final Steps

7.1   Multiple Splitting

We can perform all of the above calculations for several values of n instead of just a single one. Let $c \geq 1$ and let $n_1, \cdots, n_c$ be integers in the interval $[4, 10]$.

7.2   Outline of the Algorithm

Our algorithm, then, can be described in brief as follows.

PARAMETERS: c, $n_1, \cdots, n_c$, $\{r_j, s_j, g_j, q_j\}$, τ
INPUT: Message M of length k
OUTPUT: Hash value H of M of τ bits
1. Compute the stretching and splitting $S(M, n_j)$ $(1 \leq j \leq c)$
2. Compute the masking $CUR_i^{(n_j)}$ for $1 \leq j \leq c$ and $1 \leq i \leq k$.
3. Compute the tables $T_i^{(n_j)}$ for $1 \leq j \leq c$ and $1 \leq i \leq r_j$. Each table has 2^{n_j} entries and each entry has s_j bits.
4. Compute bit strings and their associated integers $BS_i^{(n_j)}$.
5. From the tables, compute the spectra $R_i^{(n_j)}$ and their associated integers $c(R_i^{(n_j)})$.
6. Use both sets of integers to compute an integer I_j (for $1 \leq j \leq c$).
7. Compute H_j in the group.
8. The final hash value H is the sum of the H_j in the group G.

8   Statistical Tests

8.1   Collision Resistance

The collision resistance of the algorithm is difficult to analyze, but seems to depend on the difficulty of solving iterated exponential equations over a finite field. Indeed, if one were to try to begin with the spectra and reconstruct the values of the CUR, one is immediately led to such equations when one goes from $CUR^{(i)}$ to $CUR^{(i-1)}$. There is also the additional complication of reconstructing the bit strings.

While it is difficult to quantify the collision resistance of the algorithm, it is possible to analyze the output statistically.

8.2   Statistical Tests

We tested the algorithm on all of the statistical tests provided by NIST at http://csrc/nist.gov/groups/ST/toolkit/rng/stats tests.html. We ran the tests for output lengths of 160, 224, 256, 384 and 512 bits. In all cases, the algorithm passed the tests with reserve. Further testing (such as avalanche, etc.) is in progress.

9   Implementation

9.1   Software Implementation

We implemented ERINDALE in C on an AMD Sempron 3400+ processor running at 2GHz and using 1 GB of RAM. The most time consuming step of the algorithm is masking which, on average, takes up to 40 percent of the total time of the computation. Another highly time consuming step is the distribution of the elements between tables and updating the bit strings, taking as much as 50 to 55 percent of the total time. The speed of the algorithm is close to one shown by SHA-384.

9.2   Examples and Test Vectors

In all of the examples below, we computed the exponentiation in $GF(2^{160})^*$ using the generator

$$b957ac049c0efdf70b169150cf39a5763f19134f,$$

(presented in hex form) and all the operations in the field were made modulo

$$d1000000000000000000000000000000000000008.$$

Let M be 50000 repetitions of the ten ASCII characters 0123456789. This is a message of length 0.5MB. We computed a 160 bit hash using 10 tables with 5 bit strings per table of length 256 bits each. The result is:

$$\text{ERINDALE}(M) \; = \; 67dfeb5a9ac7c7bd03776dd3abc8963eb3cbf7db$$

Let M be as above except that we changed the last byte (in hexadecimal notation, from 39 to 38). We computed a 160 bit hash with the same parameters as above to get

$$\text{ERINDALE}(M) \; = \; dc48d7cda46cc9dda71424fef93c4201513187ac$$

Let M be $500,000$ repetitions of 0 and the same parameters as above. We get

$$\text{ERINDALE}(M) \; = \; fb879ebd36a72ca27e141129bef05cdab24afed5$$

9.3   Hardware Implementation

We used a FPGA Virtex V (at a frequency 300 MHz). Depending on the choice of parameters, the speed of the function can be parameterized and can vary from approximately 300 Mbit/sec up to 2 Gbits/sec.

10   Summary

ERINDALE offers several attractive features such as tunable output length and a tunable measure of collision resistance. Moreover, in hardware it is running at

2 Gbits/sec on an FPGA (Virtex V) and it is expected to run even faster on an ASIC. The design methodology is not based on the Damgard-Merkle approach and is a bit-stream procedure.

References

1. Bourgain, J.: New bounds on exponential sums related to the Diffie-Hellman distributions. C. R. Acad. Sci. Paris Ser. I 338, 825–830 (2004)
2. Damgard, I.: A design principle for hash functions. In: Brassard, G. (ed.) CRYPTO 1989. LNCS, vol. 435, pp. 416–427. Springer, Heidelberg (1990)
3. Joux, A.: Multicollisions in iterated hash functions. In: Franklin, M. (ed.) CRYPTO 2004. LNCS, vol. 3152, pp. 306–316. Springer, Heidelberg (2004)
4. Menezes, A., van Oorschot, P., Vanstone, S.: Handbook of Applied Cryptography. CRC Press, Boca Raton (1997)
5. Merkle, R.: One way hash functions and DES. In: Brassard, G. (ed.) CRYPTO 1989. LNCS, vol. 435, pp. 428–446. Springer, Heidelberg (1990)
6. Kumar Murty, V., Volkovs, N.: A polynomial based hashing algorithm. In: SECRYPT 2008, pp. 103–106 (2008)
7. Preneel, B.: Analysis and Design of Cryptographic Hash Functions, Ph. D. Thesis (2003)
8. Aiello, W., Haber, S., Venkatesan, R.: New constructions for secure hash functions (extended abstract). In: Vaudenay, S. (ed.) FSE 1998. LNCS, vol. 1372, pp. 150–167. Springer, Heidelberg (1998)
9. Wang, X.: How to break MD5 and other hash function. In: Cramer, R. (ed.) EUROCRYPT 2005. LNCS, vol. 3494, pp. 19–35. Springer, Heidelberg (2005)

A   Formulas for the Enumeration Function

Consider the values $n = 4, \cdots, 10$ with corresponding $m = irr(n)$ given by $m = 7, 10, 16, 25, 43, 71, 129$. We give explicit enumeration functions for the first three values. Variations on these can be used to produce formulas for the remaining values.

CASE 1. $m = 7$. We set

$$c_7 = c(c(c(c(s_1, s_2), s_3),$$
$$c(s_4, s_5)), c(s_6, s_7)).$$

CASE 2. $m = 10$. We set

$$c_{10} = c(c(c(c(s_1, s_2), s_3), c(s_4, s_5)),$$
$$c(c(s_6, s_7), c(c(s_8, s_9), s_{10}))).$$

CASE 3. $m = 16$. We set

$$E'_{16} = c(c(c(s_1, s_2), c(s_3, s_4)),$$
$$c(c(s_5, s_6), c(s_7, s_8))),$$

$$E_{16}'' = c(c(c(s_9, s_{10}), c(s_{11}, s_{12})),$$

$$c(c(s_{13}, s_{14}), c(s_{15}, s_{16}))),$$

and finally

$$c_{16} = c(E_{16}', E_{16}'').$$

A Survey of Algebraic Unitary Codes

Frédérique Oggier[*]

Division of Mathematical Sciences,
School of Physical and Mathematical Sciences,
Nanyang Technological University, Singapore
`frederique@ntu.edu.sg`

Abstract. This survey paper gives an overview of algebraic unitary codes. It describes the coding problem involved, and present several algebraic approaches yielding interesting constructions, as well as the latest bounds.

1 Introduction

The problem addressed in the design of unitary codes can be summarized shortly: find a family $\mathcal{C}$ of $n \times n$ unitary matrices such that

$$\zeta(\mathcal{C}) = \min_{U \neq U' \in \mathcal{C}} \frac{1}{2} |\det(U - U')|^{1/n} \tag{1}$$

is maximized. The above quantity is called the *diversity product*, and it of course depends on both the dimension n and the cardinality of the code $\mathcal{C}$, that we denote by L. When the diversity product is non-zero, the family $\mathcal{C}$ is called *fully diverse*. We call $R = \frac{1}{n} \log_2(L)$ the rate of the code $\mathcal{C}$.

Similarly to classical coding theory, this question originates from an engineering problem, namely the design of so-called non-coherent differential space-time coding, that is coding over a wireless channel with several antennas with no knowledge of the channel at the receiver (the interested reader may refer for example to [5,7] for more details of the engineering problem).

Over the last decade, engineers and mathematicians have tried to provide code constructions using many different approaches. In this survey paper, we focus on algebraic constructions, and present an overview of the algebraic techniques tried so far, as well as some of the best unitary codes and bounds available in the literature.

2 Diagonal Codes

The easiest case one can think of is likely to be families of diagonal matrices. They have been investigated in the first papers [5,7,8], and are usually called *cyclic codes*.

---

[*] The work of Frédérique Oggier is supported in part by the National Research Foundation of Singapore under Research Grant NRF-CRP2-2007-03.

C. Xing et al. (Eds.): IWCC 2009, LNCS 5557, pp. 171–187, 2009.

Let ζ be a primitive Lth root of unity. An $n \times n$ cyclic code $\mathcal{C}$ is given by

$$
\mathcal{C} = \left\{ \begin{pmatrix} \zeta^{u_1} & 0 & \cdots & 0 \\ 0 & \zeta^{u_2} & \cdots & 0 \\ \vdots & \vdots & \ddots & \vdots \\ 0 & 0 & \cdots & \zeta^{u_n} \end{pmatrix}^l \mid l = 0, \ldots, L-1 \right\},
$$

where $u_1, \ldots, u_n$ are integers L between 0 and $L-1$.

Example 1. [Cyclic code with $n = 2$, $L = 8$.] Let ζ_8 be a primitive 8th root of unity. If $(u_1, u_2) = (1, 3)$, a 2×2 cyclic code is given by

$$
\mathcal{C} = \left\{ \begin{pmatrix} \zeta_8 & 0 \\ 0 & \zeta_8^3 \end{pmatrix}^l \mid l = 0, \ldots, 7 \right\}.
$$

We have a diversity product $\zeta(\mathcal{C})$ of 0.5946.

Given the cardinality L, the code design consists of choosing $u_1, \ldots, u_n$ relatively prime to L in order to maximize the diversity product as defined in (1). In particular, it has to be non-zero. In [5,7,8], the authors resort to exhaustive computer search to find optimal $u_1, \ldots, u_n$. Some examples of such codes are reported in Table 1. Note that computer search is restricted to rather small n, and values for $n \geq 6$ have been generated randomly.

Table 1. Some cyclic codes with high diversity product

n	L	R	$\zeta(\mathcal{C})$	$u_1, \ldots, u_n$
2	8	1.5	0.5946	$(1, 3)$
3	63	1.99	0.3301	$(1, 17, 26)$
4	16	1	0.5453	$(1, 3, 5, 7)$
4	256	2	0.2208	$(1, 25, 97, 107)$
5	32	1	0.4095	$(1, 5, 7, 9, 11)$
6	4096	2	0.1428	$(1, 599, 623, 1445, 1527, 1715)$
7	128	1	0.3487	$(1, 13, 17, 27, 29, 45, 49)$
7	16384	2	0.1213	$(1, 1875, 5207, 5551, 7687, 7827, 9013)$

A natural way to give some structure (and to ensure fully diverse codes) is to use cyclotomic fields. As above, $\zeta_L = \zeta$ is a primitive Lth root of unity, and let $F = \mathbb{Q}(\zeta_L)$ be its corresponding cyclotomic field. Assume that F has a subfield K such that F/K is cyclic of degree n, with Galois group generated by σ. We consider cyclic codes $\mathcal{C} = \mathcal{C}(F/K, L, n)$ of the form

$$
\mathcal{C} = \left\{ \begin{pmatrix} \sigma(\zeta) & 0 & \cdots & 0 \\ 0 & \sigma^2(\zeta) & \cdots & 0 \\ \vdots & \vdots & \ddots & \vdots \\ 0 & 0 & \cdots & \sigma^n(\zeta) \end{pmatrix}^l \mid l = 0, \ldots, L-1 \right\}.
$$

It is then easy to give a closed form expression for the diversity product.

Lemma 1. *The diversity product of a cyclic code $\mathcal{C}(F/K, L, n)$ is given by*

$$\frac{1}{2} \min_{l \neq l'} |\det(U^l - U^{l'})|^{1/n} = \frac{1}{2} \min_{l \neq l'} |N_{F/K}(1 - \zeta^{l'-l})|^{1/n}. \tag{2}$$

Proof. We have that

$$\det(U^l - U^{l'}) = \prod_{i=1}^{n} (\sigma^i(\zeta)^l - \sigma^i(\zeta)^{l'}) = \prod_{i=1}^{n} \sigma^i(\zeta^l - \zeta^{l'}) = N_{F/K}(\zeta^l - \zeta^{l'})$$

by definition of the norm $N_{F/K}$. Now

$$N_{F/K}(\zeta^l - \zeta^{l'}) = N_{F/K}(\zeta^l) N_{F/K}(1 - \zeta^{l'-l}),$$

so that

$$|\det(U^l - U^{l'})|^2 = |N_{F/K}(1 - \zeta^{l'-l})|^2$$

since $|N_{F/K}(\zeta^l)|^2 = 1$.   □

If $\mathbb{Q}(\zeta_L)$ contains a quadratic imaginary field $\mathbb{Q}(\sqrt{-d})$ such that $\mathbb{Q}(\zeta_L)/\mathbb{Q}(\sqrt{-d})$ is cyclic, with d some positive square free integer, then we can consider the code $\mathcal{C}(\mathbb{Q}(\zeta_L)/\mathbb{Q}(\sqrt{-d}), L, \varphi(L)/2)$, where φ denotes the Euler totient function.

Lemma 2. *The diversity product of the code $\mathcal{C}(\mathbb{Q}(\zeta_L)/\mathbb{Q}(\sqrt{-d}), L, \varphi(L)/2)$ is given by*

$$\frac{1}{2} \min_{l \neq l'} |\det(U^l - U^{l'})|^{1/n} = \frac{1}{2} |N_{\mathbb{Q}(\zeta)/\mathbb{Q}}(1 - \zeta)|^{1/2\varphi(L)}.$$

Proof. We have that

$$\min_{l \neq l'} |\det(U^l - U^{l'})|^2 = |N_{\mathbb{Q}(\zeta)/\mathbb{Q}(\sqrt{-d})}(1 - \zeta)|^2 = |N_{\mathbb{Q}(\zeta)/\mathbb{Q}}(1 - \zeta)|,$$

by considering the norm of $\mathbb{Q}(\sqrt{-d})/\mathbb{Q}$.   □

Table 2. Some codes based on cyclotomic extensions and their parameters: $n = 2$ and larger L, and $n = 8$

n	L	R	$\zeta(\mathcal{C})$	σ
2	32	2.5	0.19509	$\zeta \mapsto \zeta^{15}$
2	64	3	0.09802	$\zeta \mapsto \zeta^{31}$
2	35	2.56	0.21442	$\zeta \mapsto \zeta^6$
2	63	2.98	0.13914	$\zeta \mapsto \zeta^8$
8	256	1	0.098017	$\zeta \mapsto \zeta^7$
8	255	0.99	0.30688	$\zeta \mapsto \zeta^{15}$

We notice that when $L = 8$ (with $\mathbb{Q}(\zeta_8)/\mathbb{Q}(i)$, $\zeta(\mathcal{C}) = 0.5946$) and $L = 16$ (with $\mathbb{Q}(\zeta_{16})/\mathbb{Q}(i)$, $\zeta(\mathcal{C}) = 0.5453$), we get the same diversity product as for cyclic codes. Actually, we even get the same construction (the code based on a cyclotomic extension with $L = 8$ is indeed the same code as in Example 1).

Apart finding in a different way known constructions, many other codes can be obtained with this technique [15], see Table 2 for some examples.

3   Representation of Finite Fixed Point Free Groups

In order to go beyond diagonal matrices, the seminal paper [19] has looked at unitary representations of finite groups. If the code $\mathcal{C}$ is a group of unitary matrices, we have that

$$|\det(U' - U'')| = |\det(I - U)|$$

for $U = U'^{-1}U'' \in \mathcal{C}$, and in order to have a fully diverse code $\mathcal{C}$, we need to make sure that U has no eigenvalue at 1. Representations with no eigenvalue at 1 have been well studied, and are called *fixed point free representations*. A group is called *fixed point free* if it has a fixed point free representation. A complete classification of finite groups that are fixed point free is given in [19], based on the work of Zassenhaus, who has given an almost complete classification.

For Abelian groups, their irreducible representations are one-dimensional, that is they are characters of the group (characters are multiplicative mappings from the group to complex roots of unity). Furthermore, a character is fixed point free if and only if it is injective (or primitive). Thus for the particular case of a cyclic group $\langle \sigma \rangle$ of order L, its fixed point free characters are exactly those which map a generator of the group to a primitive Lth root of unity. More precisely, primitive characters χ_u are given by $\chi_u(\sigma^l) = e^{2i\pi ul/L}$, $u = 0, \ldots, L-1$, with u and L coprime. An n-dimensional representation Δ of G is given by a direct sum of n characters

$$\Delta(\sigma^l) = \begin{pmatrix} \zeta^{u_1 l} & 0 & 0 \\ 0 & \zeta^{u_2 l} & 0 \\ \vdots & & \ddots & \vdots \\ 0 & 0 & \zeta^{u_n l} \end{pmatrix}, \quad \zeta = \zeta_L = e^{2i\pi/L}.$$

It is interesting to note that we are back to the cyclic code constructions of Section 2. Now a more general Abelian group G is fixed point free if and only if it has a primitive character, that is, if and only if one can embed an isomorphic copy of G in the non-zero complex numbers. This means that G has to be cyclic, which shows that one has to consider non-Abelian groups to get new code constructions.

Let us give one example of representation of a family of fixed point free non-Abelian groups (it is proved in [19] that this is the only family with odd orders, and with irreducible representations of odd dimension n). Let r, m be two positive integers, and define n to be the order of r (mod m), i.e., n is the smallest positive integer such that $r^n \equiv 1$ (mod m). Set $t = m/\gcd(r-1, m)$. Consider the groups given by generators and relations as follows:

$$G_{m,r} = \langle \sigma, \tau \mid \sigma^m, \ \tau^n = \sigma^t, \tau\sigma\tau^{-1} = \sigma^r \rangle.$$

An n-dimensional representation Δ is given by

$$\Delta(G_{m,r}) = \{\Delta(\sigma)^l \Delta(\tau)^k \mid l = 0, \ldots, m-1, \ k = 0, \ldots, n-1\},$$

with

$$\Delta(\tau) = \begin{pmatrix} 0 & 0 & 0 & & \zeta_m^t \\ 1 & 0 & 0 & & 0 \\ 0 & 1 & \ddots & & \vdots \\ 0 & & \ddots & & \\ 0 & & & 1 & 0 \end{pmatrix}^T, \quad \Delta(\sigma) = \begin{pmatrix} \zeta_m & 0 & & 0 \\ 0 & \zeta_m^r & & 0 \\ \vdots & & \ddots & \vdots \\ 0 & 0 & & \zeta_m^{r^{n-1}} \end{pmatrix}.$$

Example 2 (FPF group code with $n = 3$, $L = 63$.). Take $n = 3$, $r = 4$ and $m = 21$. We have the fixed point free group $G_{21,4}$. We get the family of 63 unitary matrices $E^i D^j$, where

$$D = \begin{pmatrix} \zeta_{21} & 0 & 0 \\ 0 & \zeta_{21}^4 & 0 \\ 0 & 0 & \zeta_{21}^{16} \end{pmatrix}, \quad E = \begin{pmatrix} 0 & 0 & \zeta_{21}^7 \\ 1 & 0 & 0 \\ 0 & 1 & 0 \end{pmatrix}.$$

The diversity product is $\zeta(\Delta) = 0.3851$.

Since the matrices obtained via the representation of $G_{m,r}$ are doubly-banded, it is possible to compute explicitly their diversity product.

Lemma 3. *[19, p.18] For any fixed point free representation Δ of $G_{m,r}$, we have*

$$\zeta(\Delta) = \frac{1}{2} \min_{0 \le l \le m-1, 0 \le k \le k-1} \left| \prod_{j=1}^{q} \left(1 - \zeta_m^{\frac{k}{q}t + lr^{j-1}\frac{r^n-1}{r^q-1}} \right) \right|^{1/n}, \quad q = \gcd(n,k).$$

Examples of code constructions based on $G_{m,r}$ and their diversity product are given in Table 3.

Table 3. Some FPF group codes and their parameters

n	L	R	$\zeta(\Delta)$	$G_{m,r}$
3	63	1.99	0.3851	$G_{21,4}$
3	513	3	0.1353	$G_{171,64}$
5	1025	2	0.1679	$G_{205,16}$
5	1021125	3.99	0.0037	$G_{204205,21}$
7	16513	2	0.0955	$G_{2359,8}$
9	513	1	0.3610	$G_{57,4}$

4   The Cayley Transform

As we have already seen through the previous sections, the diversity product is a quantity rather difficult to evaluate. In order to get families of codes with high cardinality, the focus on the diversity product has been relaxed in [4], where Cayley codes have been proposed, based on the Cayley transform that maps the space of (skew)Hermitian matrices to the manifold of unitary matrices.

Let A be a Hermitian matrix, hence iA is automatically skew-Hermitian. The Cayley transform of iA is given by

$$U = (I + iA)^{-1}(I - iA).$$

It is easy to check that U is unitary. The Cayley transform of iA is preferred since all its eigenvalues are strictly imaginary, thus different from 1, so that the inverse of $I + iA$ always exists.

The cardinality of the unitary code $\mathcal{C}$ depends on the choice of the Hermitian matrices. Each matrix A is represented in a basis of Hermitian matrices

$$A = \sum_{q=1}^{Q} \alpha_q A_q,$$

where $\alpha_1, \ldots, \alpha_Q \in \mathbb{R}$ are chosen from a set $\mathcal{S}$ with r possible values, and where A_q are fixed $n \times n$ Hermitian matrices. Let $L = 2^{RM} = r^Q$ be the cardinality of the code. The rate of a Cayley code is given by

$$R = \frac{Q}{M} \log_2 r.$$

At this point, on should choose Q, the Hermitian basis matrices $\{A_q\}$, and the set $\mathcal{S}$ from which each α_q is picked in order to optimize the diversity product (1). Arguing that at high rate this may become quickly intractable, the approach of [4] consists of optimizing a mutual information criterion instead. This leads to a probabilistic characterization of the best parameters. The matrices $A_1, \ldots, A_Q$ are chosen at random such that U is isotropically distributed (an isotropically distributed unitary matrix has a probability density function which is invariant to pre- and post- multiplication by an arbitrary unitary matrix). Consequently, it is shown that the optimal distribution on A is

$$p(A) = \frac{2^{n^2-n}(n-1)! \cdots 1!}{\pi^{n(n+1)/2}} \frac{1}{\det(I + A^2)^n}.$$

The above probability density function is a generalization of the scalar Cauchy distribution, which is why A is said to be Cauchy distributed. Finally, using Cauchy matrices is shown to imply that a good choice for the scalars $\alpha_1, \ldots, \alpha_Q$ is to take them scalar Cauchy distributed.

An alternative approach to Cayley codes has been proposed in [14], where fully diverse Cayley codes have been built. In parallel to the work on unitary codes, progresses have been made in another area of coding theory, where division algebras have been introduced [18] as a mean to design fully diverse codes (which typically are not unitary). Let us first explain how this is done in general (see [18]), before getting back to the particular construction of interest here.

Let L/K be a Galois extension of degree n such that its Galois group $G = Gal(L/K)$ is cyclic, with generator σ.

Definition 1. *Choose a non-zero element $\gamma \in K$. We construct a non-commutative algebra, denoted by $\mathcal{A} = (L/K, \sigma, \gamma)$, as follows:*

$$\mathcal{A} = L \oplus eL \oplus \ldots \oplus e^{n-1}L$$

such that e satisfies

$$e^n = \gamma \quad and \quad \lambda e = e\sigma(\lambda) \text{ for } \lambda \in L.$$

Such an algebra is called a cyclic algebra.

It is a right vector space over L, and as such has dimension $(\mathcal{A} : L) = n$.

Cyclic algebras have been considered for coding applications since they naturally provide families of matrices as follows. Since each $x \in \mathcal{A}$ is expressible as

$$x = x_0 + ex_1 + \ldots + e^{n-1}x_{n-1}, \ x_i \in L \text{ for all } i,$$

there is a correspondence between $x \in \mathcal{A}$ and the matrix of left multiplication by x given by

$$\begin{pmatrix} x_0 & \gamma\sigma(x_{n-1}) & \gamma\sigma^2(x_{n-2}) & \ldots & \gamma\sigma^{n-1}(x_1) \\ x_1 & \sigma(x_0) & \gamma\sigma^2(x_{n-1}) & \ldots & \gamma\sigma^{n-1}(x_2) \\ \vdots & & \vdots & & \vdots \\ x_{n-2} & \sigma(x_{n-3}) & \sigma^2(x_{n-4}) & \ldots & \gamma\sigma^{n-1}(x_{n-1}) \\ x_{n-1} & \sigma(x_{n-2}) & \sigma^2(x_{n-3}) & \ldots & \sigma^{n-1}(x_0) \end{pmatrix}. \tag{3}$$

Example 3 (Cyclic algebra of dimension 2.). For $n = 2$, we have

$$\mathcal{A} = L \oplus eL$$

such that e satisfies

$$e^2 = \gamma \quad and \quad \lambda e = e\sigma(\lambda) \text{ for } \lambda \in L.$$

Now, an element $x \in \mathcal{A}$ can be written $x = x_0 + ex_1$. Let us compute the multiplication by x of any element $y \in \mathcal{A}$.

$$\begin{aligned} xy &= (x_0 + ex_1)(y_0 + ey_1) \\ &= x_0y_0 + x_0ey_1 + ex_1y_0 + ex_1ey_1 \\ &= x_0y_0 + e\sigma(x_0)y_1 + ex_1y_0 + \gamma\sigma(x_1)y_1, \end{aligned}$$

since $e^2 = \gamma$ and using the non-commutative rule $\lambda e = e\sigma(\lambda)$. In matrix form in the basis $\{1, e\}$, this yields

$$xy = \begin{pmatrix} x_0 & \gamma\sigma(x_1) \\ x_1 & \sigma(x_0) \end{pmatrix} \begin{pmatrix} y_0 \\ y_1 \end{pmatrix}.$$

There is thus a correspondence

$$x = x_0 + ex_1 \in \mathcal{A} \leftrightarrow \begin{pmatrix} x_0 & \gamma\sigma(x_1) \\ x_1 & \sigma(x_0) \end{pmatrix}. \tag{4}$$

Cyclic algebras provide us with a family $\mathcal{C}$ of matrices of the form (3), which is clearly linear (since σ is). Thus

$$\det(X' - X'') = \det(X), \ \mathbf{0} \neq X \in \mathcal{C}, \ \text{for all } X' \neq X'' \in \mathcal{C},$$

so that the diversity product (1) simplifies to

$$\zeta(\mathcal{C}) = \frac{1}{2} \min_{X \neq 0} |\det(X)|^{1/n}, \ X \in \mathcal{C}.$$

If now the cyclic algebra we consider is also a division algebra, that is, by definition, that every element in $\mathcal{A}$ is invertible, this guarantees that $\zeta(\mathcal{C}) > 0$.

In order to know whether a cyclic algebra is a division algebra, the following criterion is useful.

Proposition 1. *[16] The algebra $\mathcal{A} = (L/K, \sigma, \gamma)$ of degree n is a division algebra if the smallest positive integer t such that γ^t is the norm of some element in L is n.*

We now show how the general theory exposed above can be used for constructing fully diverse Cayley codes [14]. Let us give an example of constructions of a 2×2 Cayley code built on a cyclic division algebra. Let us first note that for the families of unitary matrices U to be fully diverse, it is equivalent to have the Hermitian matrices A to be fully diverse, which we can find inside any cyclic division algebra.

Let us thus take the cyclic division algebra $\mathcal{A} = (\mathbb{Q}(i, \sqrt{5})/\mathbb{Q}(i), \sigma, i)$, where $\sigma : \sqrt{5} \mapsto -\sqrt{5}$ and $\mathbb{Q}(i, \sqrt{5}) = \{a + b\sqrt{5} \mid a, b \in \mathbb{Q}(i)\}$. Let $x \in \mathcal{A}$,

$$x = x_0 + ex_1, \ x_0, x_1 \in \mathbb{Q}(i, \sqrt{5}).$$

From the matrix representation (4), it is easy to find a subset of Hermitian matrices. Since an arbitrary $n \times n$ Hermitian matrix is parameterized by n^2 real variables, we are free to choose any $Q \leq n^2$ basis matrices. Let us choose $Q = n^2 = 4$. Set $\theta = \frac{1+\sqrt{5}}{2}$, and recall that here $\gamma = i$.

A suitable Hermitian matrix X is given by

$$a_0 I_2 + b_0 \begin{pmatrix} \theta & 0 \\ 0 & 1-\theta \end{pmatrix} + s \begin{pmatrix} 0 & 1-\theta-i\theta \\ i\theta+(1-\theta) & 0 \end{pmatrix} + t \begin{pmatrix} 0 & -\theta+i(1-\theta) \\ -i(1-\theta)-\theta & 0 \end{pmatrix}.$$

The coefficients a_0, b_0, s, t belongs to $\mathbb{Q}$ since $\mathcal{A}$ is defined over a number field. We thus need $\mathcal{S} \subset \mathbb{Q}$.

We finally get a basis of four matrices

$$\left\{ \begin{pmatrix} 1 & 0 \\ 0 & 1 \end{pmatrix}, \begin{pmatrix} 0 & 1 - \theta - i\theta \\ i\theta + (1 - \theta) & 0 \end{pmatrix}, \right.$$
$$\left. \begin{pmatrix} \theta & 0 \\ 0 & 1 - \theta \end{pmatrix}, \begin{pmatrix} 0 & -\theta + i(1 - \theta) \\ -i(1 - \theta) - \theta & 0 \end{pmatrix} \right\},$$

yielding a fully-diverse 2×2 Cayley code.

Values of diversity products are not available for Cayley codes. Note that Cayley codes are the constructions allowing the highest rates. For example, a code for $n = 8$ has been built [4], with a rate of 16, which means that $L = |\mathcal{C}| = 2^{128}$, which is roughly $3.4 \cdot 10^{38}$. By comparison, codes based on cyclotomic fields (see Section 2) are available at rate 1 (256 matrices) but with guaranteed diversity product 0.098017, while the highest rate obtained for codes based on the representation of $G_{m,r}$ (Section 3) is 3.99 in dimension 4, for a diversity product of 0.0037.

5 Representation of Lie Groups

In Section 3, we have briefly discussed how representation of finite groups could help approaching the problem of designing unitary codes. However, in considering finite groups, the cardinality of the code is naturally bounded by the order of the group. This naturally leads to infinite groups and the representation of Lie groups, in order to increase the cardinality of the unitary codes.

This line of research started with a negative result [6]: The only fixed point free Lie groups are $U(1)$, the unitary group of dimension 1, and $SU(2)$, the special unitary group of dimension 2. Furthermore, their only fixed point free irreducible representation are 1- and 2- dimensional, respectively.

To pursue further, one had to relax the condition of being fixed point free, and consider Lie groups of rank 2. The rank of a Lie group equals the maximum number of commuting basis elements of its Lie algebra. The motivation lies in the fact that if a Lie algebra has rank n, then it has at least one element with $n - 1$ eigenvalues at 1. Thus the lower the rank of a Lie group, the more likely to find a subset which is fully-diverse. In particular, fixed point free Lie groups have rank 1. There are three Lie groups of rank 2: the Lie group of unit determinant 3×3 unitary matrices $SU(3)$, the Lie group of 4×4 unitary symplectic matrices $Sp(2)$, and one exceptional Lie group $\mathcal{G}_2$. Both $Sp(2)$ and $SU(3)$ have already been studied in the literature.

In [10], the symplectic group $Sp(2)$ is used for building 4-dimensional unitary codes. Let us first recall the definition of a symplectic group.

Definition 2. *The nth order symplectic group $Sp(n)$ is the set of $2n \times 2n$ complex matrices S obeying*

1. the unitary condition $S^*S = SS^* = I_{2n}$

2. the symplectic condition $S^t J_{2n} S = J_{2n}$, where

$$J_{2n} = \begin{pmatrix} 0 & I_n \\ -I_n & 0 \end{pmatrix}.$$

The symplectic group $Sp(n)$ can be parameterized as follows:

Theorem 1. *[10] A matrix S belongs to $Sp(n)$ if and only if it can be written as*

$$S = \begin{pmatrix} U\Sigma_A V & U\Sigma_B \bar{V} \\ -\bar{U}\Sigma_B V & \bar{U}\Sigma_a \bar{V} \end{pmatrix}$$

where U and V are $n \times n$ unitary matrices, and $\Sigma_A = diag(\cos\theta_1, \ldots, \cos\theta_n)$, $\Sigma_B = diag(\sin\theta_1, \ldots, \sin\theta_n)$, with $\theta_1, \ldots, \theta_n \in [0, \pi/2]$.

Let $\Sigma_A = \Sigma_B = \frac{1}{\sqrt{2}} I_2$ and consider the code given by

$$C_{P,Q,\theta} = \left\{ \frac{1}{\sqrt{2}} \begin{pmatrix} UV & U\bar{V} \\ -\bar{U}V & \bar{U}\bar{V} \end{pmatrix}, \ 0 \le k, l \le P, \ 0 \le m, n \le Q \right\}$$

where

$$U = \frac{1}{\sqrt{2}} \begin{pmatrix} e^{i\frac{2\pi k}{P}} & e^{i\frac{2\pi l}{P}} \\ -e^{-i\frac{2\pi l}{P}} & e^{-i\frac{2\pi k}{P}} \end{pmatrix}, V = \frac{1}{\sqrt{2}} \begin{pmatrix} e^{i(\frac{2\pi m}{Q}+\theta)} & e^{i(\frac{2\pi n}{Q}+\theta)} \\ -e^{-i(\frac{2\pi n}{Q}+\theta)} & e^{-i(\frac{2\pi m}{Q}+\theta)} \end{pmatrix}.$$

Concerning the diversity product (1) of this code, it is proved that there exists a θ such that the code is fully diverse if and only if P and Q are coprime.

In [9], 3-dimensional unitary codes have been obtained based on the special unitary group $SU(3)$, that is the group of 3×3 unitary matrices U such that $\det(U) = 1$. Similarly to $Sp(2)$, we start with a parameterization of $SU(3)$.

Theorem 2. *A matrix U belongs to $SU(3)$ if and only if it can be written*

$$U = \begin{pmatrix} 1 & 0_{12} \\ 0_{21} & \Phi \end{pmatrix} \begin{pmatrix} \sqrt{1-|\alpha|^2} & 0 & \bar{\alpha} \\ 0 & 1 & 0 \\ -\alpha & 0 & \sqrt{1-|\alpha|^2} \end{pmatrix} \begin{pmatrix} \Psi & 0_{21} \\ 0_{12} & 1 \end{pmatrix}$$

where $\Phi, \Psi \in SU(2)$ and α is a complex scalar such that $|\alpha| \le 1$.

In order to get fully diverse codes, the above parametrization is further refined, with U written as

$$\underbrace{\begin{pmatrix} e^{i\theta} & 0 & 0 \\ 0 & \phi_{11} & \phi_{12}e^{-i\theta} \\ 0 & -\bar{\phi}_{12} & \bar{\phi}_{11}e^{-i\theta} \end{pmatrix}}_{A_{\phi,\theta}} \begin{pmatrix} \sqrt{1-|\alpha|^2}e^{-i(\theta+\xi)} & 0 & \bar{\alpha}e^{-i(\theta-\xi)} \\ 0 & 1 & 0 \\ -\alpha e^{i(\theta-\xi)} & 0 & \sqrt{1-|\alpha|^2}e^{i(\theta+\xi)} \end{pmatrix} \underbrace{\begin{pmatrix} \psi_{11}e^{-i\xi} & \psi_{12}e^{-i\xi} & 0 \\ -\psi_{12} & \psi_{11} & 0 \\ 0 & 0 & e^{i\xi} \end{pmatrix}}_{B_{\psi,\xi}}$$

where θ, ξ are any real angles in $[0, 2\pi)$. A first family of codes is obtained by setting the middle matrix to be the identity matrix I_3, and by choosing the matrices Φ and Ψ to be

$$\Phi = \frac{1}{\sqrt{2}} \begin{pmatrix} e^{2i\pi\frac{p}{P}} & e^{2i\pi\frac{q}{Q}} \\ -e^{-2i\pi\frac{q}{Q}} & e^{-2i\pi\frac{p}{P}} \end{pmatrix}, \Psi = \frac{1}{\sqrt{2}} \begin{pmatrix} e^{2i\pi\frac{r}{R}} & e^{2i\pi\frac{s}{S}} \\ -e^{-2i\pi\frac{s}{S}} & e^{-2i\pi\frac{r}{R}} \end{pmatrix},$$

Table 4. Some $SU(3)$ group codes and their parameters

(P,Q,R,S)	Rate	$\zeta(\mathcal{C})$
(5,7,3,1)	2.2381	0.2133
(5,7,9,1)	2.7664	0.1714
(7,9,11,1)	3.1456	0.0961
(3,7,5,11)	3.3912	0.0803
(5,9,7,11)	3.9195	0.0510
(7,11,9,13)	4.3791	0.0316

where P, Q, R, S are positive integers. We thus obtain $U = A_{\phi,\theta} B_{\psi,\xi} = A_{p,q,\theta} B_{r,s,\xi}$, but despite the simplifications made, fully diverse codes still cannot be guaranteed without asking some dependency among the parameters. A linear dependency is proposed:

$$\theta_{p,q} = 2\pi \left(\frac{p}{P} - \frac{q}{Q} \right), \quad \xi_{r,s} = 2\pi \left(\frac{r}{R} - \frac{s}{S} \right).$$

Finally, the code

$$\mathcal{C} = \{ A_{p,q} B_{p,q} \mid p \in (0, P], \ q \in (0, Q], \ r \in (0, R], \ s \in (0, S] \}$$

is proved to be fully diverse if P, Q are coprime, and R, S are coprime.

Examples of code constructions based on $SU(3)$ and their diversity product are given in Table 4.

6 Cyclic Division Algebras

Cyclic division algebras have become a popular algebraic tool to get families of fully diverse codes [18]. However, in order to use cyclic algebras for unitary codes, one needs to understand how to find suitable unitary elements.

As a motivation to consider cyclic division algebras, we first show how the codes obtained from the representation of the groups $G_{m,r}$ (see Section 3) can be found by looking at suitable cyclic algebras.

Let r, m be two positive integers, and define n to be the order of r (mod m), i.e., n is the smallest positive integer such that $r^n \equiv 1 \pmod{m}$. Set $t = m/\gcd(r-1, m)$. Consider the cyclotomic field $L = \mathbb{Q}(\zeta_m)$, where ζ_m is a primitive mth root of unity. It is of degree $\varphi(m)$ over $\mathbb{Q}$ (φ is the Euler totient function).

Proposition 2. *Let* $\sigma : \zeta_m \mapsto \zeta_m^r$. *There exists a field extension K such that $\mathbb{Q}(\zeta_m)$ is a cyclic extension of K of degree n and $\zeta_m^t \in K$.*

Proof. Since $\mathrm{Gal}(\mathbb{Q}(\zeta_m)/\mathbb{Q}) \cong (\mathbb{Z}/m\mathbb{Z})^*$, $r^n \equiv 1 \pmod{m}$ implies that there is a cyclic subgroup of order n in $(\mathbb{Z}/m\mathbb{Z})^*$, which means that $\mathbb{Q}(\zeta_m)/K$ is cyclic of order n. This subgroup of order n is generated by $\sigma : \zeta_m \mapsto \zeta_m^r$.

What is left to prove is that $\zeta_m^t \in K$, i.e., that ζ_m^t is fixed by σ. But

$$\sigma(\zeta_m^t) = \zeta_m^t \iff t(r-1) \equiv 0 \pmod{m},$$

which is satisfied since $t(r-1) = m(r-1)/\gcd(r-1,m)$ and clearly $\gcd(r-1,m) \mid r-1$. □

We therefore consider the cyclic algebra $\mathcal{A} = (\mathbb{Q}(\zeta_m)/K, \sigma, \zeta_m^t)$. The matrix representation of the element e of $\mathcal{A}$ such that $e^n = \zeta_m^t$ is thus (see Section 4)

$$E = \begin{pmatrix} 0 & 0 & 0 & & \zeta_m^t \\ 1 & 0 & 0 & & 0 \\ 0 & 1 & \ddots & & \vdots \\ 0 & & \ddots & & \\ 0 & & & 1 & 0 \end{pmatrix},$$

while the one of $\zeta_m \in L$ is given by

$$D = \begin{pmatrix} \zeta_m & 0 & & 0 \\ 0 & \zeta_m^r & & 0 \\ \vdots & & \ddots & \vdots \\ 0 & 0 & & \zeta_m^{r^{n-1}} \end{pmatrix}.$$

Since $\zeta_m^k \overline{\zeta_m^k} = 1$ for any integer k, the matrices E^i, $i = 0, \ldots, n-1$ and D^j, $j = 0, \ldots, m-1$ are unitary. Thus the set $E^i D^j$ yields a family of $nm-1$ unitary matrices, which is easily recognizable as being similar to the representation of the groups $G_{m,r}$.

In order to identify unitary matrices, one can endow suitable cyclic algebras with a unitary involution. Let us consider the cyclic algebra $\mathcal{A} = (L/K, \sigma, \gamma)$.

Proposition 3. *[12,13] Let $\alpha_L : L \to L$ be a (non-trivial) involution on L such that α_L commutes with all elements of $Gal(L/K)$. Let $\alpha : \mathcal{A} \to \mathcal{A}$ such that*

$$\alpha(x_0 + ex_1 + \ldots + e^{n-1}x_{n-1}) =$$
$$\alpha_L(x_0) + e^{-1}\sigma^{-1}(\alpha_L(x_1)) + \ldots + e^{-(n-1)}\sigma^{-(n-1)}(\alpha_L(x_{n-1})).$$

Then α defines an involution on $\mathcal{A}$ if and only if $\gamma\alpha_L(\gamma) = 1$.

This involution can be extended to the algebra $\mathcal{A} \otimes_K L \cong M_n(L)$, and it can be showed that if α_L is the complex conjugation, then the extension of α to $M_n(L)$ corresponds to the Hermitian conjugation. In other words, unitary matrices in $M_n(L)$ correspond to elements $x \in \mathcal{A}$ such that $x\alpha(x) = x\alpha(x) = 1$.

A family of such elements can be found by restriction to commutative subfields of $\mathcal{A}$ as follows.

Proposition 4. *Let $\mathcal{A} = (L/K, \sigma, \gamma)$ be a cyclic division algebra. Let $x \in \mathcal{A}^*$ such that $x\alpha(x) = 1$, $x \notin K$. Then there exists $u \in \mathcal{A}^*$ such that $u\alpha(u) = \alpha(u)u$ and $x = u\alpha(u)^{-1} = \alpha(u)^{-1}u$.*

Proof. Let $x \in \mathcal{A}^*$ such that $x\alpha(x) = 1$. Let M denote the subfield of $\mathcal{A}$ generated by K and x. It is commutative and satisfies that $\alpha(M) = M$, since $\alpha(K) = K$ and $\alpha(x) = x^{-1}$. Thus

$$M^\alpha = \{y \in M \mid \alpha(y) = y\},$$

the subfield fixed by α is well-defined. Since $M^\alpha \neq M$ ($x \neq \pm 1$), M/M^α is a quadratic extension with Galois group $Gal(M/M^\alpha) = \{Id_M, \alpha|_M\}$. The condition $x\alpha(x) = 1$ is here translated into $N_{M/M^\alpha}(x) = 1$. By the corollary of Hilbert 90 Theorem that exactly characterizes elements of norm 1 in cyclic extension, there exists $u \in M^*$ such that $x = u/\alpha(u)$. $\square$

We thus get the following recipe. Take a commutative subfield $M \neq K$ of $\mathcal{A}$ such that $\alpha(M) = M$ but with $y \in M$ such that $\alpha(y) \neq y$, so that M^α is not M itself. Take $u \in M^*$ and compute $x = u/\alpha(u)$. The element $x \in \mathcal{A}$ will satisfy $x\alpha(x) = 1$.

Example 4. Let $K = \mathbb{Q}(\zeta_3)$ and $L = \mathbb{Q}(\zeta_7 + \zeta_7^{-1}, \zeta_3)$ be the compositum of $\mathbb{Q}(\zeta_3)$ and $\mathbb{Q}(\zeta_7 + \zeta_7^{-1})$, the maximal real subfield of the cyclotomic field $\mathbb{Q}(\zeta_7)$. We have $Gal(L/\mathbb{Q}(\zeta_3)) = \langle \sigma \rangle$, with $\sigma : \zeta_7 + \zeta_7^{-1} \mapsto \zeta_7^2 + \zeta_7^{-2}$. Let $\mathcal{A} = (\mathbb{Q}(\zeta_7 + \zeta_7^{-1}, \zeta_3)/\mathbb{Q}(\zeta_3), \sigma, \zeta_3)$ be the corresponding cyclic division algebra. The involution α on $\mathcal{A}$ is given by

$$\alpha : \mathcal{A} \qquad\qquad \to \mathcal{A}$$
$$x_0 + ex_1 + e^2 x_2 \mapsto \alpha_L(x_0) + e\zeta_3^2 \sigma(\alpha_L(x_2)) + e^2 \zeta_3^2 \sigma^2(\alpha_L(x_1)),$$

where $\alpha_L(b_0 + b_1\zeta_3) = b_0 + b_1\zeta_3^2$. We look for subfields M of $\mathcal{A}$, which possess a quadratic subfield M/M^α fixed by α. For example, consider $K(e)$, with minimal polynomial $\chi_e(X) = X^3 - \zeta_3$. Thus $K(e) = \mathbb{Q}(\zeta_9)$. Since $\alpha(e) = e^{-1}$, we have that $\alpha(\zeta_9) = \zeta_9^{-1}$ and α is the complex conjugation on $K(e) = \mathbb{Q}(\zeta_9)$. Its maximal real subfield $\mathbb{Q}(\zeta_9 + \zeta_9^{-1})$ is fixed by α. We now build unitary matrices in the commutative subfield $\mathbb{Q}(\zeta_9)/\mathbb{Q}(\zeta_3)$ of $\mathcal{A}$.

Take for example the element

$$y = 1 + \zeta_9 + \zeta_9^3 + \zeta_9^5 \in \mathbb{Q}(\zeta_9)$$
$$= (1 + \zeta_3) + e + e^2\zeta_3 \in \mathcal{A}$$

As a matrix, y can be represented as

$$Y = \begin{pmatrix} 1 + \zeta_3 & \zeta_3^2 & \zeta_3 \\ 1 & 1 + \zeta_3 & \zeta_3^2 \\ \zeta_3 & 1 & 1 + \zeta_3 \end{pmatrix}.$$

We have

$$\alpha(y) = -\zeta_9^2 - \zeta_9^3 + \zeta_9^4 - \zeta_9^5 \in \mathbb{Q}(\zeta_9)$$
$$= -\zeta_3 + e\zeta_3 + e^2\zeta_3^2 \in \mathcal{A}$$

Table 5. Some cyclic algebra codes in dimension 3 and their parameters

$|\mathcal{C}|$	R	$\zeta(\mathcal{C})$
63	1.9924	0.38508
126	2.3258	0.29658
441	2.9282	0.18898
819	3.2259	0.06442

Again, as a matrix, $\alpha(y)$ can be represented as

$$\begin{pmatrix} -\zeta_3 & 1 & \zeta_3^2 \\ \zeta_3 & -\zeta_3 & 1 \\ \zeta_3^2 & \zeta_3 & -\zeta_3 \end{pmatrix}$$

which can be checked to be Y^*. We have

$$x = y/\alpha(y) = \frac{1}{19}(-10 + 16\zeta_9 + \zeta_9^2 - 4\zeta_9^3 + 14\zeta_9^4 + 8\zeta_9^5),$$

which has norm 1. Clearly $x\alpha(x) = 1$ and the matrix $Y(Y^*)^{-1}$ is unitary. This can be easily verified, since

$$Y(Y^*)^{-1} = \begin{pmatrix} -0.421 - 0.182i & 0.473 + 0.638i & -0.157 + 0.36i \\ -0.236 - 0.319i & -0.421 - 0.182i & 0.473 + 0.638i \\ -0.789 + 0.09i & -0.236 - 0.319i & -0.421 - 0.182i \end{pmatrix}^t.$$

A few examples of codes obtained using this technique are illustrated in dimension 3 in Table 5.

In a different approach, note that cyclic algebras of degree 3 have been further studied in [1].

7   Some Other Constructions

Apart the Cayley codes, which have been presented because they provide efficient unitary codes despite the lack of understanding of their diversity product, all the constructions presented are part of the same story: cyclic codes are a particular case of fixed point free group codes, which are themselves either a particular case of cyclic algebra codes, or can be generalized towards infinite groups like Lie group codes. Below can be found a few other constructions, the list does not pretend to be exhaustive.

The case of 2×2 *codes.* Many constructions have been proposed for the simple case of 2×2 matrices, and in [11], a systematic parameterization has been done in order to get the highest possible diversity product.

Number fields. Constructions in dimension up to 11 have been proposed based on number fields in [20].

Table 6. Bounds on the diversity product in dimension 2 and 3

n	L	$\zeta(\mathcal{C})$
2	24	0.5730
2	48	0.4989
2	80	0.4504
2	100	0.4301
3	24	0.6431
3	48	0.5942
3	80	0.5628
3	100	0.5482

8 Bounds

We have seen several methods to build unitary codes with good diversity product. A natural question to address at this point is, given a cardinality L of $n \times n$ unitary matrices, what is the best possible diversity product, and vice versa, given a diversity product, what is the best possible cardinality L. This question has been addressed by different authors (for example [3,2]).

The latest (and tightest) bounds have been given in [2], based on linear programming bound techniques:

Theorem 3. *[2] Let $\mathcal{C}$ be a unitary code with diversity product $\zeta(\mathcal{C})$, the following upper bounds hold :*

$$|\mathcal{C}| \quad \leq \frac{2\zeta(\mathcal{C})^2}{2\zeta(\mathcal{C})^2 - 1} \qquad \textit{if } \zeta(\mathcal{C})^2 > \frac{1}{2}$$

$$|\mathcal{C}| \quad \leq \frac{8n\zeta(\mathcal{C})^2}{4n\zeta(\mathcal{C})^2 - (2n-1)} \qquad \textit{if } \zeta(\mathcal{C})^2 > \frac{2n-1}{4n}, \ n \geq 3$$

$$|\mathcal{C}| \leq \frac{8\zeta(\mathcal{C})^6 + 4\zeta(\mathcal{C})^4 + 8\zeta(\mathcal{C})^2}{8\zeta(\mathcal{C})^6 - \frac{1}{4}} \quad \textit{if } \zeta(\mathcal{C})^2 \geq \frac{1}{2}, \ n = 2.$$

In [2], the authors determine upper bounds for $\zeta(\mathcal{C})$ given L, and simultaneously compute these bounds numerically. A few numerical values in dimension 2 and 3 are given in Table 6.

To conclude, let us compare these bounds with some best known codes.

- For $n = 2$ and $L = 24$: the bound predicts $\zeta(\mathcal{C}) \geq 0.57304$, while the best known code is a fixed point free group code, with $\zeta(\mathcal{C}) = 0.5$.
- For $n = 2$ and $L = 48$: we have a bound of $\zeta(\mathcal{C}) \geq 0.4989$, while the best known code is again a fixed point free group code, with $\zeta(\mathcal{C}) = 0.3868$.
- One more value of comparison for $n = 2$ is that for $L = 80$, the bound yields 0.4504, while even for $L = 75$, the best parameterized code [11] has a diversity product of 0.3535.
- For $n = 3$, a construction from number fields [20] gives 0.4845 for $L = 57$. Other code constructions are quite far from the bound. For example, for

$L = 60$, $SU(3)$ codes give a diversity of 0.2977, while we have a fixed point free group yielding 0.3851 for $L = 63$.

9   Open Problems

Despite a fair amount of effort and imagination, the problem of designing good unitary codes is far from being solved. Many techniques have been investigated, yielding nice constructions with good parameters. However, the best codes are still to be found. It can be seen from the known bounds that there is a relatively big gap between the bounds and the known constructions, suggesting that there is still work to be done, to either tighten the bounds, or construct better codes. One of the difficulties of this problem is the fact that the main design criterion, the diversity product, is generally difficult to evaluate.

References

1. Abarbanel, J., Averbuch, A., Rosset, S., Zlotnick, J.: Unitary non-group STBC codes from cyclic algebras. IEEE Trans. Inform. Theory 52, 3903–3912 (2006)
2. Creignou, J., Diete, H.: Linear programming bounds for unitary space time codes. In: Proceedings of International Symposium on Information Theory (ISIT), Toronto (2008)
3. Han, G., Rosenthal, J.: Unitary Space-Time Constellation Analysis: An Upper Bound for the Diversity. IEEE Trans. Inform. Theory 52(10), 4713–4721 (2006)
4. Hassibi, B., Hochwald, B.: Cayley Differential Unitary Space-Time Codes. IEEE Trans. on Information Theory 48 (June 2002)
5. Hochwald, B., Sweldens, W.: Differential unitary space time modulation. IEEE Trans. Commun. 48 (December 2000)
6. Hassibi, B., Khorrami, M.: Fully-diverse multiple-antenna signal constellations and fixed-point free Lee groups. In: Proceedings of International Symposium on Information Theory (ISIT), Washington DC (2001),
   mars.bell-labs.com/cm/ms/what/mars/papers/lie/lie.ps
7. Hughes, B.: Differential space-time modulation. IEEE Trans. Inform. Theory 46 (November 2000)
8. Hughes, B.L.: Optimal space-time constellations from groups. IEEE Trans. Inform. Theory 49, 401–410 (2003)
9. Jing, Y., Hassibi, B.: Three-Transmit-Antenna Space-Time Codes Based on SU(3). IEEE Trans. Signal Processing 53(10) (October 2005)
10. Jing, Y., Hassibi, B.: Design of full-diverse multiple-antenna codes based on Sp(2). IEEE Trans. on Information Theory 50(11) (November 2004)
11. Liang, X.-B., Xia, X.-G.: Unitary Signal Constellations for Differential Space-Time Modulation With Two Transmit Antennas: Parametric Codes, Optimal Designs, and Bounds. IEEE Trans. on Inform. Theory 48(8) (August 2002)
12. Oggier, F., Lequeu, E.: Families of unitary matrices achieving full diversity., Proceedings of the International Symposium on Information Theory (ISIT), Adelaide (2005)
13. Oggier, F.: Cyclic Algebras for Noncoherent Differential Space-Time Coding. IEEE Transactions on Information Theory 53(9) (September 2007)

14. Oggier, F., Hassibi, B.: Algebraic Cayley differential Space-Time Codes. IEEE Transactions on Information Theory 53(5) (May 2007)
15. Oggier, F.: Design of Algebraic Cyclic Codes. In: Proceedings of the Information Theory Workshop (ITW), Porto (2008)
16. Pierce, R.S.: Associative Algebras. Springer, New York (1982)
17. Reiner, I.: Maximal Orders. Academic Press, London (1975)
18. Sethuraman, B.A., Rajan, B.S., Shashidhar, V.: Full-diversity, high-rate space-time block codes from division algebras. IEEE Trans. Inform. Theory 49, 2596–2616 (2003)
19. Shokrollahi, A., Hassibi, B., Hochwald, B.M., Sweldens, W.: Representation Theory for High-Rate Multiple-Antenna Code Design. IEEE Trans. Information Theory 47(6) (September 2001)
20. Xing, C.: Constructions of unitary space-time codes from number fields (preprint)

New Family of Non-Cartesian Perfect Authentication Codes*

Dingyi Pei

Guangzhou University, Guangzhou, China
gztcdpei@scut.edu.cn

Abstract. The authentication codes based on the rational normal curves in projective spaces over finite fields were the first construction of the non-Cartesian t-fold perfect authentication codes for arbitrary positive integer t. In this paper it shows that the subfield rational normal curves provide a new family of such codes, its expected probabilities of successful deception for optimal spoofing attacks are less than those probabilities of former constructed codes in most cases.

Keywords: perfect authentication codes, subfield rational normal curves, partially balanced designs, spoofing attacks.

1 Introduction

Authentication of messages provides protection for messages against impersonating and tampering by an active attacker. Consider the authentication model with three participants [3]: a transmitter, a receiver and an opponent. The transmitter intends to communicate a sequence of source states to the receiver. In order to deceive the receiver, the opponent impersonates the transmitter to send a fraudulent message to the receiver, or tampers with the message sent to the receiver. The transmitter and the receiver should act with the common purpose to deal with the spoofing attack from the opponent. There are two kinds for the security of authentication schemes: computational and unconditional. The formal schemes (for example MAC) usually use hash functions. The authentication codes concerned in this paper have the unconditional security.

Let $\mathscr{S}$ denote the set of all source states which the transmitter may convey to the receiver. In order to protect against attacks from the opponent, source states are encoded using one encoding rule. Let $\mathscr{E}$ denote the set of all encoding rules, and $\mathscr{M}$ denote the set of all possible encoded messages. Usually, the number of encoded messages is much greater than that of source states. Each encoding rule $e \in \mathscr{E}$ is a one-to-one mapping from $\mathscr{S}$ to $\mathscr{M}$. The range of the mapping is called the set of valid messages of e which is denoted by $\mathscr{M}(e) \subset \mathscr{M}$. Prior to transmission the transmitter and receiver agree upon an encoding rule which is kept secret from the opponent. The 3-tuple $(\mathscr{S}, \mathscr{M}, \mathscr{E})$ is used to refer the authentication code.

---

* This work was supported by NSFC (No. 60473017 and 90604034) of China.

C. Xing et al. (Eds.): IWCC 2009, LNCS 5557, pp. 188–201, 2009.
© Springer-Verlag Berlin Heidelberg 2009

After observation of the r $(r > 0)$ messages sent by the transmitter using the same encoding rule, the opponent places a fraudulent message into the channel attempting to make the receiver accept it as authentic. This is called a spoofing attack of order r. Let P_r denote the expected probability of successful deception for an optimum spoofing attack.

It is proved (section 3.3 in [2]) that

$$|\mathcal{E}| \geqslant (P_0 P_1 \cdots P_{r-1})^{-1},$$

for any positive integers r.

Definition 1. *Let t be a positive integer. An authentication code is called t-fold perfect if the number of encoding rules $|\mathcal{E}|$ achieves its lower bound $(P_0 P_1 \cdots P_{t-1})^{-1}$.*

The number of encoding rules is related to the key length used in the authentication schemes. It is also proved that the probabilities P_r $(0 \leqslant r \leqslant t - 1)$ of t-fold perfect authentication codes achieve their information-theoretic lower bounds (Corollary 3.1 in [2]). Hence the family of perfect authentication codes is important.

For each encoding rule $e \in \mathcal{E}$, the set of its valid messages $\mathcal{M}(e)$ is a block (subset) in the set $\mathcal{M}$ of all possible messages. We have a family of blocks

$$\{\mathcal{M}(e) \,|\, e \in \mathcal{E}\}.$$

The combinatorial designs $(\mathcal{M}, \{\mathcal{M}(e) \,|\, e \in \mathcal{E}\})$ corresponding to perfect authentication codes have some special properties.

Definition 2. *Let v, b, k, λ, t be positive integers with $t \leqslant k$. A partially balanced t-design (PBD) $t - (v, b, k; \lambda, 0)$ is a pair $(\mathcal{N}, \mathcal{F})$ where $\mathcal{N}$ is a set of v points and $\mathcal{F}$ is a family of b subsets of $\mathcal{N}$, each of cardinality k (called blocks) such that any t-subset of $\mathcal{N}$ either occurs together in exactly λ blocks or does not occur in any block.*

Definition 3. *If a partially balanced t-design t-$(v, b, k; \lambda_t, 0)$ is a partially balanced r-design r-$(v, b, k; \lambda_r, 0)$ for $1 \leqslant r < t$ as well, then it is called a strong partially balanced t-design (SPBD) and is denoted by t-$(v, b, k; \lambda_1, \lambda_2, \cdots, \lambda_t, 0)$.*

We think of source states, encoded messages and encoding rules as random variables denoted by S, M and E.

Theorem 1 (Theorem 3.1 in [2]). *An authentication code $(\mathcal{S}, \mathcal{M}, \mathcal{E})$ is t-fold perfect if and only if the pair*

$$(\mathcal{M}, \{\mathcal{M}(e) \,|\, e \in \mathcal{E}\})$$

is a SPBD t-$(v, b, k; \lambda_1, \lambda_2, \cdots, \lambda_t, 0)$ with $\lambda_t = 1$, the random variable E has uniform distribution and the random variable S^r $(1 \leqslant r \leqslant t)$ has the message uniform distribution. Where

$$v = |\mathcal{M}|, \quad b = |\mathcal{E}|, \quad k = |\mathcal{S}|,$$
$$\lambda_r = (P_r P_{r+1} \cdots P_{t-1})^{-1}, \quad 1 \leqslant r \leqslant t - 1.$$

The probability distribution of S^r is message uniform if for any sequence of r encoded messages $m^r \in \mathcal{M}^r$ the probability $p(S^r = e^{-1}(m^r))$ is independent of e when $m^r \subset \mathcal{M}(e)$.

An authentication code is called Cartesian if it has the property: once the transmitted message is observed one may know its corresponding conveyed source state.

The construction of authentication codes based on the rational normal curves in projective spaces over finite fields was the first construction of the non-Cartesian t-fold perfect authentication codes for arbitrary positive integers t [1]. A new family of such codes will be constructed based on the subfield rational normal curves in this paper.

2   Preliminary

Let $\mathbb{F}_q$ be the finite field with q elements and n be a positive integer. The projective space of dimension n over $\mathbb{F}_q$ is denoted by $PG(n, \mathbb{F}_q)$. A point of $PG(n, \mathbb{F}_q)$ is denoted by $(x_0, x_1, \cdots, x_n)$ where $x_i \in \mathbb{F}_q$, $0 \leqslant i \leqslant n$ are not all zero. Suppose that T is a nonsingular matrix over $\mathbb{F}_q$ of order $n + 1$, then T generates a one-to-one transformation of points in $PG(n, \mathbb{F}_q)$ defined by

$$PG(n, \mathbb{F}_q) \longrightarrow PG(n, \mathbb{F}_q)$$
$$(x_0, x_1, \cdots, x_n) \longmapsto (x_0, x_1, \cdots, x_n)T.$$

It is called a projective transformation of $PG(n, \mathbb{F}_q)$. The group of all projective transformations of $PG(n, \mathbb{F}_q)$ is denoted by $PGL_{n+1}(\mathbb{F}_q)$.

We define a curve $\mathcal{C}$ in $PG(n, \mathbb{F}_q)$ to be the image set of the map

$$PG(1, \mathbb{F}_q) \longrightarrow PG(n, \mathbb{F}_q)$$
$$(x_0, x_1) \longmapsto (x_0^n, x_0^{n-1}x_1, \cdots, x_1^n).$$

The projective line $PG(1, \mathbb{F}_q)$ consists of the following $q + 1$ points:

$$\{(1, \alpha) \mid \alpha \in \mathbb{F}_q\} \bigcup \{(0, 1)\},$$

Therefore the curve $\mathcal{C}$ consists of the following $q + 1$ points:

$$\{(1, \alpha, \cdots, \alpha^n) \mid \alpha \in \mathbb{F}_q\} \bigcup \{(0, \cdots, 0, 1)\}.$$

We call the image set of the curve $\mathcal{C}$ under any projective transformation in $PGL_{n+1}(\mathbb{F}_q)$ a rational normal curve (RNC).

Lemma 1 (Lemma 5.1 in [2]). *Any $n + 1$ ($n \leqslant q$) points on a RNC in $PG(n, \mathbb{F}_q)$ are linearly independent.*

Lemma 2 (Lemma 5.2 in [2]). *Suppose that $q \geqslant n + 2$. For any $n + 3$ points in $PG(n, \mathbb{F}_q)$, among which any $n + 1$ points are linearly independent, there exists a unique RNC passing through these $n + 3$ points.*

Lemma 3 (Lemma 5.3 in [2]). *Let $n \geqslant 2$ be an integer and $q \geqslant n + 2$ be a prime power. Then the number of RNC in $PG(n, \mathbb{F}_q)$ is*

$$q^{\frac{n(n+1)}{2} - 1} \prod_{i=3}^{n+1} (q^i - 1).$$

Take $\mathscr{M} = PG(n, \mathbb{F}_q)$ as the set of points, and all RNC in $PG(n, \mathbb{F}_q)$ as the family of blocks $\mathscr{E}$. It is proved (Theorem 5.1 in [2]) that the combinatorial design $(\mathscr{M}, \mathscr{E})$ is a strong partially balanced t-design with $t = n + 3$.

Let the set of source states $\mathscr{S}$ be the set of all points on $\mathcal{C}$, which can also be represented by $\mathbb{F}_q \cup \{\infty\}$, the set of messages $\mathscr{M}$ be the set of all points in $PG(n, \mathbb{F}_q)$. For each RNC $\mathcal{C}^* \in \mathscr{E}$, let T be a projective transformation which carries $\mathcal{C}$ to $\mathcal{C}^*$, define an encoding rule σ_T as follows:

$$\sigma_T(\alpha) = (1, \alpha, \cdots, \alpha^n)T, \quad \alpha \in \mathbb{F}_q$$
$$\sigma_T(\infty) = (0, \cdots, 0, 1)T.$$

It is proved (section 5.2 in [2]) that the authentication code $(\mathscr{S}, \mathscr{M}, \mathscr{E})$ is non-Cartesian perfect. Let P_r $(0 \leqslant r \leqslant t)$ be the expected successful probability of the optimal spoofing attack of order r for this authentication code, then we have

$$P_0 = \frac{q^2 - 1}{q^{n+1} - 1},$$
$$P_r = \frac{(q-1)(q-r+1)}{q^{n+1} - q^r}, \quad 2 \leq r \leq n-1.$$
$$P_n = \frac{q - n + 1}{q^n},$$
$$P_{n+1} = \frac{q - n}{(q-1)^n},$$
$$P_{n+2} = \prod_{i=2}^{n} (q - i)^{-1},$$
$$P_{n+3} = 1.$$

$$(1)$$

Let $PO_{n+1}(\mathbb{F}_q)$ denote the subgroup of $PGL_{n+1}(\mathbb{F}_q)$ consisting of the projective transformations which carry the curve $\mathcal{C}$ into itself. We present the following Lemma 4 and its proof, which will be used in the next section.

For $0 \leqslant i, k \leqslant n$, and $t_1, t_2 \in \mathbb{F}_q$, define

$$b_{i,k} = \begin{cases} C_k^i, & i \leqslant k \\ 0, & i > k \end{cases}$$

and

$$c_{i,k} = \sum_{j=\max(0, i-k)}^{\min(i, n-k)} C_{n-k}^j C_k^{i-j} t_1^{k-(i-j)} t_2^{i-j}$$

where C_k^i is the binomial coefficient. For any elements $a \in \mathbb{F}_q^*$ (the set of nonzero element of $\mathbb{F}_q$) and $t \in \mathbb{F}_q$, define two matrices:

$$T_{n+1}(a,t) = (b_{i,k} a^i t^{k-i})_{0 \leq i,k \leq n},$$
$$Q_{n+1}(a,t) = (b_{(n-i),k} a^{n-i} t^{k-n+i})_{0 \leq i,k \leq n}$$

and for any elements $a \in \mathbb{F}_q^*$ and $t_1, t_2 \in \mathbb{F}_q$ with $t_1 \neq t_2$, define the matrix

$$R_{n+1}(a,t_1,t_2) = (c_{i,k} a^i)_{0 \leq i,k \leq n}.$$

Lemma 4 (Theorem 5.2 in [2]). *The group $PO_{n+1}(\mathbb{F}_q)$ consists of the following $q(q^2-1)$ matrices:*

$$T_{n+1}(a,t), \quad Q_{n+1}(a,t), \quad R_{n+1}(a,t_1,t_2)$$

where $a \in \mathbb{F}_q^*$, t, t_1, $t_2 \in \mathbb{F}_q$ with $t_1 \neq t_2$. The number of elements for these three types is $q(q-1)$, $q(q-1)$, $q(q-1)^2$, respectively.

Proof. Since

$$|PO_{n+1}(\mathbb{F}_q)| = \frac{|PGL_{n+1}(\mathbb{F}_q)|}{|\mathscr{E}|}$$

$$= \frac{q^{n(n+1)/2} \prod_{i=2}^{n+1} (q^i - 1)}{q^{n(n+1)/2-1} \prod_{i=3}^{n+1} (q^i - 1)}$$

$$= q(q^2 - 1),$$

it is only necessary to show that the curve $\mathcal{C}$ is carried into itself by any matrix in $T_{n+1}(a,t)$, $Q_{n+1}(a,t)$, $R_{n+1}(a,t_1,t_2)$. The last conclusion of the lemma is obvious.

We have

$$\sum_{i=0}^n b_{i,k} a^i t^{k-i} \alpha^i = \sum_{i=0}^k C_k^i (a\alpha)^i t^{k-i} = (t + a\alpha)^k,$$

hence,

$$(1, \alpha, \cdots, \alpha^n) T_{n+1}(a,t) = (1, (t + a\alpha), \cdots, (t + a\alpha)^n).$$

We also have

$$(0, \cdots, 0, 1) T_{n+1}(a,t) = (0, \cdots, 0, a^n).$$

This shows that $T_{n+1}(a,t) \in PO_{n+1}(\mathbb{F}_q)$.

Similarly, we have

$$(1, \alpha, \cdots, \alpha^n) Q_{n+1}(a,t) = \begin{cases} \alpha^n (1, t + a/\alpha, \cdots, (t + a/\alpha)^n), & \alpha \neq 0 \\ (0, \cdots, 0, a^n), & \alpha = 0 \end{cases}$$

and

$$(0, \cdots, 0, 1) Q_{n+1}(a,t) = (1, t, \cdots, t^n).$$

Therefore, $Q_{n+1}(a,t) \in PO_{n+1}(\mathbb{F}_q)$.

Finally, since

$$\sum_{i=0}^{n} c_{i,k} a^i \alpha^i = \sum_{i=0}^{n} \sum_{j=max(0,i-k)}^{min(i,n-k)} C_{n-k}^{j} C_k^{i-j} t_1^{k-(i-j)} t_2^{i-j} (a\alpha)^i$$

$$= \sum_{j=0}^{n-k} C_{n-k}^{j} (a\alpha)^j \sum_{s=0}^{k} C_k^s t_1^{k-s} (t_2 a\alpha)^s$$

$$= (1+a\alpha)^{n-k} (t_1 + t_2 a\alpha)^k,$$

then

$$(1,\alpha,\cdots,\alpha^n) R_{n+1}(a,t_1,t_2)$$

$$= \begin{cases} (1+a\alpha)^n (1, \frac{t_1+t_2 a\alpha}{1+a\alpha}, \cdots, \left(\frac{t_1+t_2 a\alpha}{1+a\alpha}\right)^n), & 1+a\alpha \neq 0 \\ (0,\cdots,0,(t_1-t_2)^n), & 1+a\alpha = 0 \end{cases}$$

Also

$$(0,\cdots,0,1) R_{n+1}(a,t_1,t_2) = a^n (1, t_2, \cdots, t_2^n).$$

Therefore, $R_{n+1}(a,t_1,t_2) \in PO_{n+1}(\mathbb{F}_q)$. Thus Lemma 4 is proved now.

3   Subfield Rational Normal Curves

Let $PG(n, \mathbb{F}_{q^r})$ be the projective space of dimension $n \geq 1$ over the field $\mathbb{F}_{q^r}$, where r is a positive integer. Obviously, the space $PG(n, \mathbb{F}_q)$ is contained in the space $PG(n, \mathbb{F}_{q^r})$.

Define a curve $\mathcal{C}_q$ in $PG(n, \mathbb{F}_q)$ which consists of the following $q+1$ points:

$$\{(1,\alpha,\cdots,\alpha^n) \,|\, \alpha \in \mathbb{F}_q\} \bigcup \{(0,0,\cdots,0,1)\}.$$

We call the image set of the curve $\mathcal{C}_q$ under any projective transformation of $PG(n, \mathbb{F}_{q^r})$ a subfield rational normal curve (SRNC). There are $q+1$ points on each SRNC.

The curve $\mathcal{C}_q$ is the image set of the map

$$PG(1, \mathbb{F}_q) \longrightarrow PG(n, \mathbb{F}_{q^r})$$
$$(x_0, x_1) \longmapsto (x_0^n, x_0^{n-1} x_1, \cdots, x_1^n)$$

therefore each SRNC is the image set of a map

$$PG(1, \mathbb{F}_q) \longrightarrow PG(n, \mathbb{F}_{q^r})$$
$$(x_0, x_1) \longmapsto \left(A_0(x_0, x_1), A_1(x_0, x_1), \cdots, A_n(x_0, x_1)\right)$$

where $A_0(X_0, X_1), A_1(X_0, X_1), \cdots, A_n(X_0, X_1)$ is an arbitrary basis of the space of homogeneous polynomials of degree n with coefficients in $\mathbb{F}_{q^r}$ over X_0, X_1.

Suppose that $n \geqslant 1$, $q \geqslant n+2$. It is easy to show that for any set with $n+3$ points in $PG(n, \mathbb{F}_{q^r})$, among which any $n+1$ points are linearly independent, there exists a projective transformation of $PG(n, \mathbb{F}_{q^r})$ which maps this set into the following set

$$
\begin{aligned}
p_1 &= (1,0,\cdots,0) \\
p_2 &= (0,1,\cdots,0) \\
&\;\;\vdots \\
p_{n+1} &= (0,\cdots,0,1) \\
p_{n+2} &= (1,1,\cdots,1) \\
p_{n+3} &= (v_0, v_1,\cdots,v_n).
\end{aligned}
\tag{2}
$$

where $v_0, v_1, \cdots, v_n$ are pairwise distinct nonzero elements in $\mathbb{F}_{q^r}$ (Notice that $q \geqslant n+2$).

Let $\mathcal{N}_q$ denote the group of all such point sets, each of which consists of $n+3$ points of $PG(n, \mathbb{F}_{q^r})$ among them any $n+1$ points are linearly independent, and which can be mapped by a projective transformation in $PGL_{n+1}(n, \mathbb{F}_{q^r})$ into a set of the form (2) with $p_{n+3} \in PG(n, \mathbb{F}_q)$.

Lemma 5. *For any two points a and b on the curve $\mathcal{C}_q$, there exists a projective transformation T_0 in $PO_{n+1}(\mathbb{F}_q)$ such that*

$$
aT_0 = (1,0,\cdots,0), \quad bT_0 = (0,\cdots,0,1).
$$

Proof. Using the proof of Lemma 4, we consider the following different cases.

(1) If $a = (1,0,\cdots,0)$ and $b = (0,\cdots,0,1)$, then take T_0 as the identify transformation.

(2) If $a = (1,0,\cdots,0)$ and $b \neq (0,\cdots,0,1)$, then $b = (1, \alpha, \cdots, \alpha^n)$ with $\alpha \neq 0$. We have

$$
\begin{aligned}
aR_{n+1}(-\alpha^{-1},0,-1) &= (1,0,\cdots,0), \\
bR_{n+1}(-\alpha^{-1},0,-1) &= (0,\cdots,0,1).
\end{aligned}
$$

(3) If $a = (0,\cdots,0,1)$ and $b = (1,0,\cdots,0)$, then

$$
\begin{aligned}
aQ_{n+1}(1,0) &= (1,0,\cdots,0), \\
bQ_{n+1}(1,0) &= (0,\cdots,0,1).
\end{aligned}
$$

(4) If $a = (0,\cdots,0,1)$ and $b \neq (1,0,\cdots,0)$, then $b = (1, \alpha, \cdots, \alpha^n)$ with $\alpha \neq 0$. We have

$$
\begin{aligned}
aR_{n+1}(-\alpha^{-1},1,0) &= (-\alpha^{-n})(1,0,\cdots,0), \\
bR_{n+1}(-\alpha^{-1},1,0) &= (0,\cdots,0,1).
\end{aligned}
$$

(5) If $a \neq (1,0,\cdots,0)$ and $b = (0,\cdots,0,1)$, then $a = (1, \alpha, \cdots, \alpha^n)$ with $\alpha \neq 0$. We have

$$
\begin{aligned}
aT_{n+1}(1,-\alpha) &= (1,0,\cdots,0), \\
bT_{n+1}(1,-\alpha) &= (0,0,\cdots,0,1).
\end{aligned}
$$

(6) If $a \neq (0, \cdots, 0, 1)$ and $b = (1, 0, \cdots, 0)$, then $a = (1, \alpha, \cdots, \alpha^n)$ with $\alpha \neq 0$. We have

$$aQ_{n+1}(1, -\alpha^{-1}) = \alpha^n(1, 0, \cdots, 0),$$
$$bQ_{n+1}(1, -\alpha^{-1}) = (0, 0, \cdots, 0, 1).$$

(7) If $a \neq (1, 0, \cdots, 0)$, $a \neq (0, \cdots, 0, 1)$ and $b \neq (1, 0, \cdots, 0)$, $b \neq (0, \cdots, 0, 1)$, then $a = (1, \alpha, \cdots, \alpha^n)$ and $b = (1, \beta, \cdots, \beta^n)$ with $\alpha \neq 0$, $\beta \neq 0$, $\alpha \neq \beta$. We have

$$aR_{n+1}(-\beta^{-1}, \alpha\beta^{-1}, 1) = (1 - \alpha\beta^{-1})^n(1, 0, \cdots, 0),$$
$$bR_{n+1}(-\beta^{-1}, \alpha\beta^{-1}, 1) = (0, \cdots, 0, (\alpha\beta^{-1} - 1)^n).$$

Thus the lemma is proved.

Corollary 1. *For any three points a, b and c on the curve C_q, there exists a projective transformation T_0 in $PO_{n+1}(\mathbb{F}_q)$ such that*

$$aT_0 = (1, 0, \cdots, 0),$$
$$bT_0 = (0, \cdots, 0, 1),$$
$$cT_0 = (1, 1, \cdots, 1).$$

Proof. By Lemma 5, we may assume that

$$a = (1, 0, \cdots, 0), \quad b = (0, \cdots, 0, 1).$$

Then $c = (1, \alpha, \cdots, \alpha^n)$ with $\alpha \neq 0$. Using the proof of Lemma 4 we have

$$aT_{n+1}(\alpha^{-1}, 0) = (1, 0, \cdots, 0),$$
$$bT_{n+1}(\alpha^{-1}, 0) = (0, \cdots, 0, 1),$$
$$cT_{n+1}(\alpha^{-1}, 0) = (1, 1, \cdots, 1).$$

Thus the corollary is proved.

Lemma 6. *Suppose that $q \geqslant n + 2$. For any set in $\mathcal{N}_q$ there exists a unique SRNC passing through the $n + 3$ points of the set. (Obviously each set with $n + 3$ points on any SRBC belongs to $\mathcal{N}_q$.)*

Proof. After taking a projective transformation if necessary, we may assume that the set in $\mathcal{N}_q$ is of the form (2) and $v_0, v_1, \cdots, v_n$ are pairwise distinct nonzero elements in $\mathbb{F}_q$. Define the polynomial in X_0, X_1

$$G(X_0, X_1) = \prod_{i=0}^{n}(X_0 - v_i^{-1}X_1),$$

then the polynomials

$$H_i(X_0, X_1) = \frac{G(X_0, X_1)}{X_0 - v_i^{-1}X_1}, \quad 0 \leqslant i \leqslant n.$$

form a basis for the space of homogeneous polynomials of degree n in X_0 and X_1. Otherwise if there was a linear relation

$$\sum_{i=0}^{n} a_i H_i(X_0, X_1) = 0, \quad a_i \in \mathbb{F}_{q^r},$$

then substituting $(X_0, X_1) = (1, v_i)$ for $0 \leqslant i \leqslant n$ successively we could find that $a_i = 0$ $(0 \leqslant i \leqslant n)$. Thus the SRNC defined by the map

$$\sigma : (x_0, x_1) \longmapsto (H_0(x_0, x_1), \cdots, H_n(x_0, x_1)), \quad (x_0, x_1) \in PG(1, \mathbb{F}_q). \quad (3)$$

passes through the set (2), since $\sigma(1, v_i) = p_i$ $(1 \leqslant i \leqslant n+1)$, $\sigma(1, 0) = p_{n+2}$ and $\sigma(0, 1) = p_{n+3}$.

Now suppose that the SRNC generated by a projective transformation T of $PG(n, \mathbb{F}_{q^r})$ passes through the set (2). We may assume that $v_0 = 1$. Notice that all projective transformations in $PO_{n+1}(\mathbb{F}_q) \cdot T$ carry the curve C_q into the same SRNC. There exist three points a, b and c on the curve C_q such that

$$aT = p_{n+2} = (1, 1, \cdots, 1),$$
$$bT = p_{n+3} = (v_0, v_1, \cdots, v_n),$$
$$cT = p_1 = (1, 0, \cdots, 0).$$

By Corollary 1 we can find a projective transformation T_0 in $PO_{n+1}(\mathbb{F}_q)$ such that

$$aT_0 = (1, 0, \cdots, 0),$$
$$bT_0 = (0, \cdots, 0, 1),$$
$$cT_0 = (1, 1, \cdots, 1).$$

Taking $T_0^{-1}T$ instead of T we may assume that

$$(1, 0, \cdots, 0)T = (1, 1, \cdots, 1),$$
$$(0, \cdots, 0, 1)T = (1, v_1, \cdots, v_n), \qquad (4)$$
$$(1, 1, \cdots, 1)T = (1, 0, \cdots, 0).$$

Let the matrix corresponding to T is

$$(t_{ij})_{0 \leqslant i, j \leqslant n},$$

and

$$(1, \alpha_i, \cdots, \alpha_i^n)T = p_{i+1}, \quad 0 \leqslant i \leqslant n. \qquad (5)$$

where $\alpha_0, \alpha_1, \cdots, \alpha_n$ are pairwise distinct nonzero elements of $\mathbb{F}_q$ and $\alpha_0 = 1$ by the third equality in (4). Put

$$H_i'(X_0, X_1) = \sum_{j=0}^{n} t_{ji} X_0^{n-j} X_1^j, \quad 0 \leqslant i \leqslant n.$$

By (5) we find that

$$H'_i(1, \alpha_s) = 0, \quad 0 \leqslant s \leqslant n, \ s \neq i,$$
$$H'_i(1, \alpha_i) \neq 0, \quad 0 \leqslant i \leqslant n.$$

By the first equality in (4) we may assume that

$$t_{0i} = 1, \quad 0 \leqslant i \leqslant n,$$

it means that

$$H'_i(X_0, X_1) = \frac{\prod\limits_{j=0}^{n}(X_0 - \alpha_j^{-1}X_1)}{X_0 - \alpha_i^{-1}X_1}.$$

Therefore

$$(0, \cdots, 0, 1)T = \left(H'_0(0, 1), H'_1(0, 1), \cdots, H'_n(0, 1)\right)$$
$$= (\alpha_1 \cdots \alpha_n)^{-1}(1, \alpha_1, \cdots, \alpha_n)$$

By the second equality in (4) we have

$$\alpha_i = v_i \quad (0 \leqslant i \leqslant n).$$

Thus Lemma 6 is proved.

Lemma 7. *Let $n \geqslant 1$ be an integer and let $q \geqslant n + 2$ be a prime power. Then the number of SRNC in $PG(n, \mathbb{F}_{q^r})$ is*

$$q^{rn(n+1)/2-1}(q^2 - 1)^{-1} \prod_{i=2}^{n+1}(q^{ri} - 1).$$

Proof. Suppose $\{p_1, p_2, \cdots, p_{n+3}\}$ is an arbitrary set in $\mathcal{N}_q$. There are totally $(q^{r(n+1)} - 1)/(q^r - 1)$ points in $PG(n, \mathbb{F}_{q^r})$. The point p_1 can be any point of $PG(n, \mathbb{F}_{q^r})$. For a fixed p_1, there are

$$\frac{q^{r(n+1)} - 1}{q^r - 1} - 1 = \frac{q^r(q^{rn} - 1)}{q^r - 1}$$

possible choices for p_2, as p_1 and p_2 are linearly independent. Generally, assume that $p_1, p_2, \cdots, p_i, 1 \leqslant i \leqslant n$ have been chosen, the point p_{i+1} may have

$$\frac{q^{r(n+1)} - 1}{q^r - 1} - \frac{q^{ri} - 1}{q^r - 1} = \frac{q^{ri}(q^{r(n+1-i)} - 1)}{q^r - 1}$$

choices, since there are $(q^{ri} - 1)/(q^r - 1)$ points which are linearly dependent on $p_1, p_2, \cdots, p_i$. Now we suppose that the points $p_1, p_2, \cdots, p_{n+1}$ have been chosen. By taking a projective transformation, we can assume that the points

$p_1, p_2, \cdots, p_{n+1}$ are of the form in (2). Hence, the point p_{n+2} is of the form $(a_0, a_1, \cdots, a_n)$ with $a_i \neq 0, 0 \leqslant i \leqslant n$, so it has

$$\frac{(q^r - 1)^{n+1}}{q^r - 1} = (q^r - 1)^n$$

choices. Finally, if the points $p_1, p_2, \cdots, p_{n+2}$ have been chosen, we can also assume that $p_1, p_2, \cdots, p_{n+1}$ are of the form in (2). Then the point p_{n+3} is of the form $(b_0, b_1, \cdots, b_n)$ with $b_i \in \mathbb{F}_q, b_i \neq 0, b_i \neq b_j, 0 \leqslant i < j \leqslant n$; it has

$$(q - 2)(q - 3) \cdots (q - n - 1)$$

choices. Therefore, we have proved that

$$|\mathcal{N}_q| = \prod_{i=0}^{n} \frac{q^{ri}(q^{r(n+1-i)} - 1)}{q^r - 1} \cdot (q^r - 1)^n \prod_{i=2}^{n+1} (q - i) \cdot ((n + 3)!)^{-1}$$

$$= q^{rn(n+1)/2}((n+3)!)^{-1} \prod_{i=2}^{n+1} (q^{ri} - 1)(q - i).$$

There are $q+1$ points on each SRNC, hence, there are C_{q+1}^{n+3} sets from $\mathcal{N}_q$ on each SRNC; here C_{q+1}^{n+3} is the binomial coefficient. The number of SRNC in $PG(n, \mathbb{F}_{q^r})$ must be

$$\frac{|\mathcal{N}_q|}{C_{q+1}^{n+3}} = q^{rn(n+1)/2-1} \prod_{i=2}^{n+1} (q^{ri} - 1) \cdot (q^2 - 1)^{-1}.$$

This completes the proof of Lemma 7.

4   A Family of Non-Cartesian Perfect A-Codes

Let $\mathcal{M}$ denote the set of all points in $PG(n, \mathbb{F}_{q^r})$ and $\mathcal{E}_q$ the set of all SRNC in $PG(n, \mathbb{F}_{q^r})$. The following theorem will show that the pair $(\mathcal{M}, \mathcal{E}_q)$ is a SPBD.

Theorem 2. *Let $n \geq 1$ and $t = n + 3$, let $q \geq t - 1$ be a prime power. There exists a SPBD t-$(v, b, k; \lambda_1, \cdots, \lambda_t, 0)$ where*

$$v = (q^{r(n+1)} - 1)/(q^r - 1),$$
$$k = q + 1,$$
$$b = q^{\frac{rn(n+1)}{2} - 1} \prod_{i=2}^{n+1} (q^{ri} - 1) \cdot (q^2 - 1)^{-1},$$

and

$$\lambda_i = q^{\frac{r(n+1-i)(n+i)}{2}}(q^r-1)^{i-1}\prod_{j=1}^{n+1-i}(q^{rj}-1)\prod_{j=2}^{n+1}(q-j)\prod_{j=i-1}^{n+1}(q-j)^{-1},$$

for $1 \le i \le n,$

$$\lambda_{n+1} = (q^r-1)^n\prod_{j=2}^{n+1}(q-j)\prod_{j=n}^{n+1}(q-j)^{-1},$$

$$\lambda_{n+2} = \prod_{j=2}^{n+1}(q-j)(q-n-1)^{-1},$$

$$\lambda_{n+3} = 1.$$

Proof. Take $\mathcal{M} = PG(n, \mathbb{F}_{q^r})$ as the set of points, and all SRNC in $PG(n, \mathbb{F}_{q^r})$ as the family of blocks $\mathcal{E}_q$. Consider any set with $n+3$ points of $PG(n, \mathbb{F}_{q^r})$, if and only if the set belongs to $\mathcal{N}_q$ there exists a unique SRNC passing through them by Lemma 6. Therefore, $(\mathcal{M}, \mathcal{E}_q)$ is a t-$(v, b, k; 1, 0)$ design.

Now we prove that $(\mathcal{M}, \mathcal{E}_q)$ is a SPBD t-$(v, b, k; \lambda_1, \cdots, \lambda_t, 0)$. Given any i $(1 \le i \le n+1)$ points of $PG(n, \mathbb{F}_{q^r})$, if they are linearly dependent then there is no SRNC passing through them; if they are linearly independent then by the proof of Lemma 7 there are λ_i SRNCs passing through them where

$$\lambda_i = \prod_{j=i}^{n}\frac{q^{r(n+1)}-q^{rj}}{q^r-1}\cdot(q^r-1)^n\prod_{j=2}^{n+1}(q-j)\prod_{j=i-1}^{n=1}(q-j)^{-1}$$

$$= q^{\frac{r(n+1-i)(n+i)}{2}}(q^r-1)^{i-1}\prod_{j=1}^{n+1-k}(q^{rj}-1)\prod_{j=2}^{n+1}(q-j)\prod_{j=k-1}^{n+1}(q-j)^{-1},$$

for $1 \le i \le n,$

$$\lambda_{n+1} = (q^r-1)^n\prod_{j=2}^{n+1}(q-j)\prod_{j=n}^{n+1}(q-j)^{-1}.$$

Finally, for any given $n+2$ points of $PG(n, \mathbb{F}_{q^r})$, if among them there exist $n+1$ points that are linearly dependent, then there is no SRNC passing through the given $n+2$ points; otherwise there are

$$\lambda_{n+2} = \prod_{j=2}^{n+1}(q-j)(q-n-1)^{-1}$$

SRNCs passing through the given $n+2$ points. Thus the theorem is proved.

Remark 1. It was proved that all SPBD in $PG(1, \mathbb{F}_{q^r})$ form a $3-(q^r+1, q+1, 1)$ design called Möbius plane by E. Witt [4]. Theorem 2 shows that $\lambda_3 = 1$ if $n = 1$. It is obvious that any two points in $PG(1, \mathbb{F}_{q^r})$ are linear independent, then the Witt's result can be reduced from Theorem 2.

Based on the SPBD constructed above, one new type of non-Cartesian perfect authentication codes can be constructed.

Let the set of source states $\mathscr{S}$ be the set of all points on the curve C_q, then $k = q+1$. It can also be represented by $\mathscr{S} = \mathbb{F}_q \cup \{\infty\}$. Let the set of messages $\mathscr{M}$ be the set of all points in $PG(n, \mathbb{F}_{q^r})$, then $|\mathscr{M}| = (q^{r(n+1)} - 1)/(q^r - 1)$. For a SRNC C^*, let T^* be a projective transformation which carries C to C^* and define an encoding rule σ_{T^*} as follows:

$$\sigma_{T^*}(\alpha) = (1, \alpha, \cdots, \alpha^n)T^*, \quad \alpha \in \mathbb{F}_q$$
$$\sigma_{T^*}(\infty) = (0, \cdots, 0, 1)T^*.$$

Receiving an encoded message $m \in \mathscr{M}$, the receiver decides to accept m if $m(T^*)^{-1}$ is on C_q (this source state is transmitted) or to reject it otherwise. The number of encoding rules is the number of SRNCs, and, hence, from Lemma 7 it is equal to

$$|\mathscr{E}_q| = q^{\frac{rn(n+1)}{2} - 1} \prod_{i=2}^{n+1} (q^{ri} - 1) \cdot (q^2 - 1)^{-1}.$$

Assume that the random variables E and S^i, $1 \leqslant i \leqslant n + 3$, satisfy the conditions given in Theorem 3.1 in [2]. Then we have

$$P_0 = \frac{\lambda_1}{|\mathscr{E}_q|} = \frac{(q^r - 1)(q + 1)}{q^{r(n+1)} - 1},$$

$$P_i = \frac{\lambda_{i+1}}{\lambda_i} = \frac{(q^r - 1)(q - i + 1)}{q^{r(n+1)} - q^{ir}}, \quad 1 \leq i \leq n$$

$$P_{n+1} = \frac{\lambda_{n+2}}{\lambda_{n+1}} = \frac{q - n}{(q^r - 1)n},$$

$$P_{n+2} = \frac{\lambda_{n+3}}{\lambda_{n+2}} = \prod_{j=2}^{n+1} (q - j)^{-1}(q - n - 1),$$

$$P_{n+3} = 1.$$

Comparing above probabilities with those in (1) (with q^r instead of q) we find that the probabilities P_i $(0 \leqslant i \leqslant n + 1)$ of the codes based on SRNC are less than those probabilities of the codes based on RNC.

5   Conclusions

There were several ways to construct the t-fold perfect authentication codes if $t \leqslant 2$. The construction of authentication codes based on the rational normal curves in projective spaces over finite fields was the first construction of the non-Cartesian t-fold perfect authentication codes for arbitrary positive integers t [1]. Since then we have not seen any new constructions of this kind of codes. The construction based on the subfield rational normal curves presented in this paper provides a new family of such codes. It improves the expected probabilities of successful deception for optimal spoofing attacks.

References

1. Pei, D.: A problem of combinatorial designs related to authentication codes. Journal of Combinatorial Designs 6, 417 (1998)
2. Pei, D.: Authentication codes and combinatorial designs. Chapman & Hall/CRC, Boca Raton (2006)
3. Simmons, G.L.: Authentication theory / coding theory. In: Blakely, G.R., Chaum, D. (eds.) CRYPTO 1984. LNCS, vol. 196, p. 411. Springer, Heidelberg (1985)
4. Witt, E.: Dei fünffan tromsitiven Gruppen von Mathien. Abhandlungen der Mathematik Hamburg 12, 256 (1938)

On the Impossibility of Strong Encryption Over $\aleph_0$

Raphael C.-W. Phan[1,*] and Serge Vaudenay[2]

[1] Loughborough Uni, U.K.
[2] EPFL, Switzerland

Abstract. We give two impossibility results regarding strong encryption over an infinite enumerable domain. The first one relates to statistically secure one-time encryption. The second one relates to computationally secure encryption resisting adaptive chosen ciphertext attacks in streaming mode with bounded resources: memory, time delay or output length. Curiously, both impossibility results can be achieved with either finite or continuous domains. The latter result explains why known CCA-secure cryptosystem constructions require at least two passes to decrypt a message with bounded resources.

1 Introduction

As the cardinality of sets increases, it is a well known fact from set theory that some mathematical problems can suddenly become impossible to solve then become possible again. For instance, for any logical assertion on finite sets we can always decide whether it is true of false. When the set becomes infinite but enumerable (that is, the cardinality of $\aleph_0$ following the Cantor notion) some mathematical statements can become undecidable as shown by Gödel with the Peano arithmetic. That is, statements of the form

$$\forall x \; \exists y \quad f(x, y) = 0$$

may be undecidable even though f has a polynomial form with integral variables and coefficients. When sets become larger, e.g. the cardinality $2^{\aleph_0}$ of continuous sets, predicates based on inequalities can be decided. That is, over logical assertion with elementary formula of form $f(x) = 0$ or $f(x) > 0$ can be decided as shown by Tarski [29].

Assuming the continuum hypothesis we have $\aleph_1 = 2^{\aleph_0}$ but $\aleph_1$ can be smaller otherwise. This hypothesis is undecidable in the standard Zermelo-Fraenkel set theory axiomatic with the axiom of choice.

In cryptography, results on strong encryption are well understood. Since Shannon, we know how to achieve perfect secrecy on finite sets by using the Vernam cipher (aka one-time pad). One-time pad can also be defined on the continuous unit interval $[0, 1]$ by using the modulo 1 addition of a message and a key.

---

* Part of work done while the author was with EPFL, Switzerland.

C. Xing et al. (Eds.): IWCC 2009, LNCS 5557, pp. 202–218, 2009.
© Springer-Verlag Berlin Heidelberg 2009

However, it was shown by Chor and Kushilevitz [9, 10] that it was impossible to achieve over $\aleph_0$ under some ad-hoc generalization of the Shannon secrecy.

Similarly, we can construct computationally secure encryption (in the sense of security against chosen-ciphertext adversaries) using hybrid encryption [12]. However, all proposed constructions require scanning the ciphertext twice for decryption so the decryption algorithm cannot work with finite resources in streaming mode over $\aleph_0$. In practice, this necessarily wastes resources in time and in memory. An open problem [1, 2] is whether strong encryption schemes exist that can be streamlined, and more so the case when the domain is infinite.

In this paper, we first revisit the Chor-Kushilevitz result. We show that their notion of security is unnecessarily strong and show the impossibility over $\aleph_0$ with a weaker notion. We then analyze the practicality of strong encryption, i.e. if it is possible to achieve strong encryption when the scheme's resources are bounded, either in memory or in time. Indeed, when a provably secure scheme especially one for which has infinite domain is implemented in practice, bounded resources are an inevitable artifact. In this setting, one wonders if the provable security results are preserved from theory to practice. If security is preserved, this indicates that the strong encryption scheme even one with infinite domain can be streamlined since bounded resources imply that an infinite input cannot be processed immediately but necessarily requires streamlining.

To be precise, by bounded memory we mean that as the scheme's encryption or decryption process is streamlined, its internal state utilized during the process has a bound which is a polynomial function of the security parameter. By bounded time, we mean that the scheme's process issues the output stream only after some delay, rather than immediately as input streams are received. To this end, we can alternatively model the latter as the process issuing outputs of bounded length.

1.1   Related Work

In a different direction but related to the context of bounded resources, researchers have studied security models in which *adversaries* have bounded memory [19, 26], as a compromise to achieve information theoretic security against computationally unbounded adversaries.

The first known provable security notion for (public-key) encryption is indistinguishability (IND) (or so called polynomial security) [18], which has an equivalent alternative definition called semantic security [18]. These characterizations did not consider adversarial access to the decryption oracle, and thus fall within the chosen-plaintext adversarial model (CPA). Later IND characterizations refined this to the chosen-ciphertext adversarial model (CCA) [8, 27, 28].

Given that the CCA adversarial model allows the adversary access to the decryption oracle, the basic idea in the design of CCA-secure schemes is to make this decryption oracle useless to the adversary in terms of breaking IND. For this, some implicit or explicit form of validity check [25] is typically designed into the decryption algorithms of these schemes. This necessitates having two passes over the text input: for encryption, the first pass over the plaintext to obtain the ciphertext while the second pass over the plaintext is to generate a validity-checking tag for later verification when decrypting ciphertext; for decryption, the first pass over the ciphertext decrypts it to obtain the plaintext and the second pass over the text verifies if it actually corresponds to the received tag before the plaintext is actually output.

While two passes currently seem inevitable for strong security, indeed no known strong encryption schemes exist with a single pass; yet for practical uses (e.g. these days it is common to be downloading hundreds of megabytes of data over the Internet) it is advantageous to achieve a streaming capability, i.e. the second pass can start before the end of the first pass; sort of similar to the concept of streaming video: start watching before the entire movie is downloaded. This has efficiency implications, e.g. there is no need to buffer the entire text, and the encryption/decryption speed increases. Achievement of streamability for strong encryption would indicate that strong encryption schemes over infinite domains exist.

For the symmetric (blockwise) encryption context, the concept of *online* encryption and decryption [3, 4, 13, 14] has been considered. The motivation for this is related to the desire to provide a kind of streaming capability without needing to buffer the entire text or wait until the entire text is received before it starts to be processed. To be precise, online means that the output block can be returned on the fly in one pass, given only the key, the current input block and previous input and output blocks: the rest of the input blocks are not required for returning the output block up to this point. IND notions have been proposed for this particular setting to consider blockwise-adaptive adversaries [15, 21], in both CPA and CCA style adversarial models [4, 14, 15].

It is known [9, 10] that weakly secure (in some statistical sense) symmetric encryption is impossible over infinite sets such as $\{0,1\}^*$ although it is possible over larger sets such as $[0, 1]$. As for public-key encryption, statistical security is of course impossible so we have to consider computational security.

2   Preliminaries

Let $\mathcal{A}^*$ denote the set of finite sequences of elements in a set $\mathcal{A}$; ε denote a sequence of length zero so that $\varepsilon \in \mathcal{A}^*$ for any $\mathcal{A}$; $x \in_U \mathcal{A}$ denote uniformly selecting an element in a set $\mathcal{A}$; and $\|$ denote the concatenation operation in $\mathcal{A}^*$. For $x \in \mathcal{A}^*$, $|x|$ denotes the length of x; $\aleph_0$ denotes the cardinality of the set of natural numbers $\mathbb{N}$, which is the smallest possible infinite set.

In the sequel, we consider encryption over the infinite domain $\{0,1\}^*$. For technical reasons we formalize this domain by a prefix-free language $\mathcal{C} = \{0,1\}^* \| \top$, i.e. the set of words consisting of an arbitrary bitstring terminated by the special $\top$ symbol. When considering messages given as bit streams, this symbol indicates that the message is complete.

2.1   Public Key Encryption (PKE)

A public-key encryption scheme PKE consists of three algorithms, PKE.KeyGen, PKE.Enc, and PKE.Dec. It must be such that there exists some integer κ for which:

1. $\langle pk, sk \rangle \leftarrow$ PKE.KeyGen(1^λ): A probabilistic algorithm that on input the security parameter λ, generates public and private keys $\langle pk, sk \rangle$ by taking time bounded by λ^κ for some integer κ.
2. $c \leftarrow$ PKE.Enc$_{pk}(m; r)$: A probabilistic algorithm that encrypts a message $m \in \mathcal{M}$ into a ciphertext c by using some random coins r and taking time bounded by $(|m| + \lambda)^\kappa$.
3. $m \leftarrow$ PKE.Dec$_{sk}(c)$: An algorithm that decrypts c by taking time bounded by $(|c| + \lambda)^\kappa$. It outputs either $m \in \mathcal{M}$ or a special symbol $\perp \notin \mathcal{M}$. An obvious correctness condition applies.

Let A_E be a polynomial-time oracle machine that plays the following adaptive chosen-ciphertext game:

[IND-ATK PKE] Game
1: $\langle pk, sk \rangle \leftarrow$ PKE.KeyGen(1^λ)
2: $\langle m_0, m_1, \rho \rangle \leftarrow A_E^{\mathcal{O}_1^{\text{ATK}}}(pk)$
3: $b \in_U \{0, 1\}$; $r \in_U \{0, 1\}^{(|m_0|+\lambda)^\kappa}$; $c^* \leftarrow$ PKE.Enc$_{pk}(m_b; r)$
4: $\tilde{b} \leftarrow A_E^{\mathcal{O}_2^{\text{ATK}}}(\rho, c^*)$
Note that it is required for $|m_0| = |m_1|$.

Depending on how the decryption oracles $\mathcal{O}_1^{\text{ATK}}$ and $\mathcal{O}_2^{\text{ATK}}$ are defined, different characterizations of the game can be obtained to capture relevant security notions. For instance, to capture notions related to indistinguishability against adaptive chosen-ciphertext attacks (IND-CCA) [28], $\mathcal{O}_1^{\text{ATK}}$ is defined as:

Oracle $\mathcal{O}_1^{\text{CCA}}(c)$
1: $m \leftarrow$ PKE.Dec$_{sk}(c)$
2: **return** m

Here, the oracle terminates and returns control to the adversary A_E via the **return** statement. Meanwhile it is required that A_E be restricted not to ask c^* to $\mathcal{O}_2^{\text{ATK}}$, i.e. $\mathcal{O}_2^{\text{ATK}}$ only replies to queries that do not equal the challenge ciphertext c^*:

Oracle $\mathcal{O}_2^{\text{CCA}}(c)$
1: **if** $c \neq c^*$ **then**
2:     $m \leftarrow$ PKE.Dec$_{sk}(c)$
3:     **return** m
4: **else**
5:     **return** $\perp$
6: **end if**

This can be relaxed in an IND-rCCA game [8], where the decryption oracle $\mathcal{O}_2^{\text{ATK}}$ behaves as follows:

Oracle $\mathcal{O}_2^{\text{rCCA}}(c)$
1: $m \leftarrow$ PKE.Dec$_{sk}(c)$
2: **if** $m \notin \{m_0, m_1\}$ **then**

3:    **return** m
4: **else**
5:    **return** $\perp$
6: **end if**

In the much weaker IND-CPA game [18], the oracle $\mathcal{O}_2^{\mathsf{ATK}}$ is unavailable by definition:

   Oracle $\mathcal{O}_2^{\mathsf{CPA}}(c)$
   1: **return** $\perp$

Note that in the context of PKEs, the encryption oracle is public by construction. For symmetric encryption, however, access to the encryption oracle characterizes [22, 23] an additional dimension to the adversary's capability and hence corresponding security notion. In that case, an even weaker notion, so-called *one-time encryption* (IND-OTE) game [12, 24] and capturing passive security makes even the encryption oracle unavailable to the adversary.

We define $\mathbf{Adv}_{PKE,A_E}^{\mathsf{IND\text{-}ATK}} = |\Pr[\tilde{b} = b] - \frac{1}{2}|$ and

$$\mathbf{Adv}_{PKE}^{\mathsf{IND\text{-}ATK}} = \max_{A_E}(\mathbf{Adv}_{PKE,A_E}^{\mathsf{IND\text{-}ATK}})$$

where maximum is taken over all ppt machines. We say that a PKE is IND-ATK-secure if $\mathbf{Adv}_{PKE}^{\mathsf{IND\text{-}ATK}}$ is negligible in λ, where $\mathsf{ATK} \in \{\mathsf{CCA}, \mathsf{rCCA}, \mathsf{CPA}\}$.

2.2   Some Streamline Cryptosystems

As already observed, known IND-CCA-secure constructions require decryption to scan the ciphertext at least twice. However, encryption can process by scanning the plaintext once. Examples include the well-known Cramer-Shoup scheme [11] and variants [12], different forms of hybrid encryption [1, 2, 12, 20], and identity-based encryption (IBE) schemes [5–7, 25].

If we now relax the security notion down to IND-CPA security, we can achieve secure stream encryption with bounded resources. Consider a public-key encryption scheme OLDPKE over a finite domain (e.g., RSA). Consider a pseudorandom generator PRG whose input lies in the encryption domain. Namely, for any integer n, PRG_n is a function producing an n-bit string from some random coins. We define an encryption scheme NEWPKE by:

1. $\mathsf{NEWPKE.KeyGen} = \mathsf{OLDPKE.KeyGen}$
2. $\mathsf{NEWPKE.Enc}_{pk}(m; \langle k, r \rangle) = \langle \mathsf{OLDPKE.Enc}_{pk}(k; r), m \oplus \mathsf{PRG}_{|m|}(k) \rangle$
3. $\mathsf{NEWPKE.Dec}_{sk}(\langle c, c' \rangle) = c' \oplus \mathsf{PRG}_{|c'|}(\mathsf{OLDPKE.Dec}_{sk}(c))$

Clearly, both encryption and decryption can process streams with bounded resources.

Theorem 1. *If OLDPKE is* IND-CPA-*secure and PRG is* IND-*secure then NEW-PKE is* IND-CPA-*secure.*

We recall that IND security for PRG is defined by the following game:

[IND PRG] Game
1: $\langle 1^n, \rho \rangle \leftarrow A(1^\lambda)$
2: $b \in_U \{0,1\}$; $k \in_U \mathcal{K}$;
3: **if** $b = 0$ **then**
4:     $c \leftarrow \mathsf{PRG}_n(k)$
5: **else**
6:     $c \in_U \{0,1\}^n$
7: **end if**
8: $\tilde{b} \leftarrow A(\rho, c)$

Proof. Let Γ_1^b be the IND-CPA game conditioned to the value of bit b. Note that the 3rd step of Γ_1^b is

3: $k \in_U \mathcal{K}$; $r \in_U \{0,1\}^{(|m_0|+\lambda)^\kappa}$; $c_1^* \leftarrow \mathsf{OLDPKE.Enc}_{pk}(k; r)$;
$c_2^* \leftarrow m_b \oplus \mathsf{PRG}_{|m_b|}(k)$; $c^* = \langle c_1^*, c_2^* \rangle$

Given a ppt A_E and a game Γ, we denote by $\Gamma(A_E)$ the event that A_E wins. For instance, $\Gamma_1^b(A_E)$ is the event that A_E produces $\tilde{b}$ which equals b. We want to prove that for any A_E, $\Pr[\Gamma_1^0(A_E)] - \Pr[\Gamma_1^1(A_E)]$ is negligible.

Let Γ_2^b be the same game in which the 3rd step of Γ_1^b is replaced by

3: $k \in_U \mathcal{K}$; $k' \in_U \mathcal{K}$; $r \in_U \{0,1\}^{(|m_0|+\lambda)^\kappa}$; $r' \in_U \{0,1\}^{(|m_0|+\lambda)^\kappa}$;
$c_1^* \leftarrow \mathsf{OLDPKE.Enc}_{pk}(k'; r')$; $c_2^* \leftarrow m_b \oplus \mathsf{PRG}_{|m_b|}(k)$; $c^* = \langle c_1^*, c_2^* \rangle$

We construct A_E' an adversary playing the IND-CPA game for OLDPKE by using A_E playing either Γ_1^b or Γ_2^b as follows: the generation of r, r', and the computation of c_1^* are outsourced to the IND-CPA game and A_E' only submit the two plaintexts k and k'. A_E' produces $\tilde{b}$ as a final bit. We let $\Gamma^{b'}$ denote the IND-CPA game for OLDPKE with bit b'. Clearly, we have $\Pr[\Gamma_1^b(A_E)] = \Pr[\Gamma^0(A_E')]$ and $\Pr[\Gamma_2^b(A_E)] = \Pr[\Gamma^1(A_E')]$ since the winning condition is $\tilde{b} = b$ in all cases. As OLDPKE is IND-CPA-secure we obtain that $\Pr[\Gamma_1^b(A_E)] - \Pr[\Gamma_2^b(A_E)]$ is negligible.

Let now Γ_3^b be the Γ_2^b game in which the 3rd step is replaced by

3: $k \in_U \mathcal{K}$; $r \in_U \{0,1\}^{(|m_0|+\lambda)^\kappa}$; random $\in_R \{0,1\}^{|m|}$;
$c_1^* \leftarrow \mathsf{OLDPKE.Enc}_{pk}(k; r)$; $c_2^* \leftarrow m_b \oplus$ random; $c^* = \langle c_1^*, c_2^* \rangle$

We construct A'' an adversary playing the IND game for PRG by using A_E playing either Γ_2^b or Γ_3^b as follows: the generation of k' and the computation of either $\mathsf{PRG}_{|m_b|}(k')$ or random are outsourced to the IND game. We let $\tilde{\Gamma}^{b''}$ denote the IND game for PRG with bit b''. Clearly, we have $\Pr[\Gamma_2^b(A_E)] = \Pr[\tilde{\Gamma}^0(A'')]$ and $\Pr[\Gamma_3^b(A_E)] = \Pr[\tilde{\Gamma}^1(A'')]$. Since PRG is IND-secure we obtain that $\Pr[\Gamma_2^b(A_E)] - \Pr[\Gamma_3^b(A_E)]$ is negligible.

Let now Γ_4^b be the Γ_3^b game in which the 3rd step is replaced by

3: $k \in_U \mathcal{K}$; $r \in_U \{0,1\}^{(|m_0|+\lambda)^\kappa}$; random $\in_R \{0,1\}^{|m|}$;
$c_1^* \leftarrow \mathsf{OLDPKE.Enc}_{pk}(k; r)$; $c_2^* \leftarrow$ random; $c^* = \langle c_1^*, c_2^* \rangle$

Clearly, Γ_3^b and Γ_4^b produce c^* of same distribution so $\Pr[\Gamma_3^b(A_E)] = \Pr[\Gamma_4^b(A_E)]$.

To summarize, we have that $\Pr[\Gamma_1^b(A_E)] - \Pr[\Gamma_4^b(A_E)]$ is negligible. Since we trivially have $\Pr[\Gamma_4^0(A_E)] = \Pr[\Gamma_4^1(A_E)]$ we obtain that $\Pr[\Gamma_1^0(A_E)] - \Pr[\Gamma_1^1(A_E)]$ is negligible. □

3   Statistically-Secure Encryption

Throughout this section, we assume that all sets are enumerable so that we deal with discrete probability theory.

Given two distributions P_0 and P_1 for a random variable X the statistical distance is

$$d(P_0, P_1) = \frac{1}{2} \sum_x \left| \Pr_{P_0}[X = x] - \Pr_{P_1}[X = x] \right|$$

The statistical distance is the exact measure to characterize the advantage of the best distinguisher between the two distributions when using a single sample. When the statistical distance is negligible, the distributions are statistically indistinguishable.

A *cipher* is defined by a distribution for a secret key K and a function Enc mapping (x, k) to $\mathsf{Enc}_k(x) = y$ such that Enc_k is collision-free, so that we can invert it. Given a random plaintext X which is independent from K, we define the random ciphertext $Y = \mathsf{Enc}_K(X)$. We say that a security notion relative to a cipher is *universal* if it does not depend on the distribution of X.

Shannon's notion of perfect secrecy is defined by the statistical independence of X and Y. Although this definition does not look like universal at a first glance, we can easily see that it is equivalent to the property that the function $(x, y) \mapsto \Pr[\mathsf{Enc}_K(x) = y]$ only depends on y. This is a universal notion since it only depends on the cipher design and not on the distribution of X.

Given a possible ciphertext y, the *a posteriori distribution* P_y of X given y is the marginal distribution of X conditioned to $Y = y$. Let P be the a priori distribution of X. We have

$$d(P, P_y) = \frac{1}{2} \sum_x |\Pr[X = x] - \Pr[X = x | Y = y]|$$

$$= \frac{1}{2} \sum_x \Pr[X = x] \left| 1 - \frac{\Pr[\mathsf{Enc}_K(x) = y]}{\Pr[Y = y]} \right|$$

Shannon's secrecy can be defined by saying that $d(P, P_y) = 0$ for all y that can occur. The next question relates to how to relax this security definition so that secrecy is no longer perfect but is still universal and achieves some kind of statistical security. A natural extension would we the following one.

Definition 2. *A cipher provides ε-imperfect secrecy if for any distribution of X and all possible y we have $d(P, P_y) \leq \varepsilon$.*

In [9, 10], Chor and Kushilevitz propose the following definition.

Definition 3. *Given* $\alpha \geq 1$, *a cipher provides* α-*weak secrecy if for all* $x_1, x_2 \in$ Supp(X) *and all* y *we have*

$$\frac{1}{\alpha} \Pr[Y = y | X = x_2] \leq \Pr[Y = y | X = x_1] \leq \alpha \Pr[Y = y | X = x_2]$$

where Supp(X) *is the set of all possible values of* X. *This property is universal.*

Clearly, 1-weak secrecy is equivalent to Shannon's perfect secrecy. If the cipher achieves α-weak secrecy, for any x and any possible y, the ratio between $\Pr[Y = y | X = x]$ and $\Pr[Y = y]$ is between $\frac{1}{\alpha}$ and α, so we have $d(P, P_y) \leq \frac{\alpha - 1}{2}$. We deduce the following theorem.

Theorem 4. α-*weak secrecy implies* $\frac{\alpha - 1}{2}$-*imperfect secrecy.*

One drawback of α-weak secrecy is expressed by the following result.

Theorem 5 (Chor-Kushilevitz [9, 10]). *If a cipher provides* α-*weak secrecy then its domain must be finite.*

Proof. Assuming that a cipher provides α-weak secrecy, we can take a possible plaintext x_2 and a possible key k. We let $y = \text{Enc}_k(x_2)$ so $\Pr[Y = y | X = x_2] \neq 0$. We have that for all x_1 in Supp(X),

$$\Pr[Y = y | X = x_1] \geq \frac{1}{\alpha} \Pr[Y = y | X = x_2] > 0$$

but the left-hand side is just $\Pr[\text{Dec}_K(y) = x_1]$ so summing over many x_1's should be at most 1. We deduce that the plaintext domain must be finite.   □

We can now consider the notion of indistinguishability between the encryption of two arbitrary plaintexts. That is, given a plaintext x we consider the distribution Q_x for $\text{Enc}_K(x)$.

Definition 6. *A cipher is* ε-*statistically indistinguishable under one-time encryption* (IND-OTE) *if for all* x_1 *and* x_2 *we have* $d(Q_{x_1}, Q_{x_2}) \leq \varepsilon$.

Clearly, this notion is universal. We have

$$d(Q_{x_1}, Q_{x_2}) = \frac{1}{2} \sum_y |\Pr[Y = y | X = x_1] - \Pr[Y = y | X = x_2]|$$

Clearly, if we have α-weak secrecy, we have $d(Q_{x_1}, Q_{x_2}) \leq \frac{\alpha - 1}{2}$. We deduce the following theorem.

Theorem 7. α-*weak secrecy implies* $\frac{\alpha - 1}{2}$-*statistically* IND-OTE.

The converse is not true as the following example shows.

Example 8. Let $\{0, 1\}$ be the plaintext domain. Let $k = (\kappa, \beta)$ be composed by an integer κ and a bit β such that $\Pr[K = k] = \frac{1}{2} 2^{-\kappa - 1}$. We define

$$\text{Enc}_k(b) = \begin{cases} \kappa \| (\beta \oplus b) & \text{if } \kappa < n \\ (\kappa \| \beta) + 2b & \text{otherwise} \end{cases}$$

Given a ciphertext $y = z\|t$ with $t \in \{0,1\}$, we have $\Pr[\mathsf{Enc}_K(0) = y] = \frac{1}{2}2^{-z-1}$ and

$$\Pr[\mathsf{Enc}_K(1) = y] = \begin{cases} \Pr[\mathsf{Enc}_K(0) = y] & \text{if } z < n \\ 0 & \text{if } z = n \\ 2\Pr[\mathsf{Enc}_K(0) = y] & \text{if } z > n \end{cases}$$

thus the cipher does not provide α-weak secrecy for any $\alpha < 2$. Assuming an a priori distribution of the plaintext we notice that $\Pr[X = 1|Y = n\|t] = 0$ whereas $\Pr[X = 0|Y = n\|t] = \frac{1}{2}$. So, we have $d(P, P_{n\|t}) = \frac{1}{4}$. The cipher does not provide ε-imperfect secrecy for any $\varepsilon < \frac{1}{4}$. However,

$$d(Q_0, Q_1) = \sum_{t=0}^{1} \left(\sum_{z=n}^{+\infty} \Pr[\mathsf{Enc}_K(0) = n\|t] \right) = \sum_{z=n}^{+\infty} 2^{-z-1} = 2^{-n}$$

so the cipher is 2^{-n}-IND-OTE secure.

The above example shows that α-weak secrecy is sufficient for IND-OTE security but not necessary. Furthermore, the natural extension of Shannon's perfect secrecy by the notion of imperfect secrecy appears insufficient to capture the notion of statistical security. So, the feasibility of IND-OTE secure encryption over an infinite domain is a legitimate question. We answer below by the negative.

Theorem 9. *Let $\varepsilon < 1$. If a given cipher is ε-statistically IND-OTE secure then its plaintext domain is finite.*

Proof. Let x_1 be an arbitrary reference plaintext in the domain. We have that $\sum_y \Pr[\mathsf{Enc}_K(x_1) = y] = 1$ so there must exist a finite set A such that the sum for $y \in A$ is greater than $\frac{1+\varepsilon}{2}$. For any x_2 in the domain we have

$$\sum_{y \in A} \Pr[\mathsf{Enc}_K(x_2) = y] = \sum_{y \in A} \Pr[\mathsf{Enc}_K(x_1) = y] -$$

$$\sum_{y \in A} (\Pr[\mathsf{Enc}_K(x_1) = y] - \Pr[\mathsf{Enc}_K(x_2) = y])$$

$$\geq \frac{1+\varepsilon}{2} - \varepsilon$$

$$= \frac{1-\varepsilon}{2}$$

Since

$$\sum_{x_2 \in \mathsf{Domain}} \sum_{y \in A} \Pr[\mathsf{Enc}_K(x_2) = y] = \sum_{y \in A} \sum_{x_2 \in \mathsf{Domain}} \Pr[\mathsf{Dec}_K(y) = x_2]$$

$$\leq \#A$$

we obtain that $\#A \geq \frac{1-\varepsilon}{2}\#\mathsf{Domain}$ so the domain is finite. $\square$

4   Strong Encryption Over $\aleph_0$ with Bounded Memory

Definition 10. *Let $\mathcal{Z}$ be an alphabet. A **streamline** function with state space $\mathcal{S}$ over $\mathcal{Z}$ is a family $F = (f_c)_{c \in \mathcal{Z}}$ of functions:*

$$f_c : \mathcal{S} \to \mathcal{Z}^* \times \mathcal{S}.$$

By abuse of notation, we define

$$f_c(y, s) = \langle y \| y', s' \rangle$$

where $\langle y', s' \rangle = f_c(s)$. We also define

$$f_x(y, s) = (f_{x_m} \circ \cdots \circ f_{x_1})(y, s)$$

where $x = x_1 \cdots x_m$. Given $x \in \mathcal{Z}^*$ and $s \in \mathcal{S}$, we further define the function $F_s : \mathcal{Z}^* \to \mathcal{Z}^*$ by $f_x(\varepsilon, s) = \langle F_s(x), \cdot \rangle$.

A function over a language is called **monotonic** if for any x and y in the language, the image of x is a prefix of the image of y whenever x is a prefix of y. Note that all functions defined over a prefix code $\mathcal{C}$ are monotonic.

The following fact is pretty trivial:

Lemma 11. *For a streamline function F with state space $\mathcal{S}$ over $\mathcal{Z}$ and $s \in \mathcal{S}$, the function F_s is monotonic.*

Definition 12. *A monotonic function $G : \mathcal{C} \to \mathcal{Z}^*$ over a subset $\mathcal{C}$ of $\mathcal{Z}^*$ is **streamlineable with σ states** if there exists a streamline function F with a state space $\mathcal{S}$ of σ elements and $s \in \mathcal{S}$ such that G equals F_s restricted to $\mathcal{C}$. We call F_s an **implementation** of G with σ states.*

Lemma 13. *All monotonic functions $G : \mathcal{C} \to \mathcal{Z}^*$ on a subset $\mathcal{C}$ of $\mathcal{Z}^*$ are streamlineable with $\aleph_0$ states.*

The idea is to store the input of G in a state and to output something as soon as possible.

Proof. We define $\bar{G} : \mathcal{Z}^* \to \mathcal{Z}^*$ by $\bar{G}(x) = G(y)$ where y is the longest prefix of x in $\mathcal{C}$ if any and $\bar{G}(x) = \varepsilon$ otherwise. Clearly, $\bar{G}$ is monotonic and equal to G when restricted to $\mathcal{C}$. Let $\mathcal{S} = \mathcal{Z}^*$ and define

$$f_c(s) = \langle \mathsf{drop}_{\bar{G}(s)} \bar{G}(s \| c), s \| c \rangle,$$

where $\mathsf{drop}_y(x)$ denotes string x with prefix y dropped (e.g. if $x = y \| z$, then $\mathsf{drop}_y(x) = z$). We easily show that F_ε is an implementation of G for $F = (f_c)_{c \in \mathcal{Z}}$. □

Theorem 14. *Let us consider an $\mathsf{IND\text{-}rCCA}$-secure public-key encryption scheme over $\mathcal{M} = \{0,1\}^* \| \top$. For any key pair, the decryption function is not streamlineable with a finite number of states.*

Proof. Let $\mathcal{Z} = \{0, 1, \top, \bot\}$ and $\mathcal{M} = \{0,1\}^* \| \top$. Here $\top$ is a special character which indicates a word termination so that $\mathcal{M}$ is a prefix code. We consider a PKE over $\mathcal{M}$. Given a key pair, encryption resp. decryption can be defined by a function $\bar{G}_r$ resp. D verifying:

1. $\bar{G}_r : \mathcal{C} \to \mathcal{C}$ is an injective function for any random coins r and
2. $D : \mathcal{C} \to \mathcal{C} \cup \{\bot\}$ is a function such that $D \circ \bar{G}_r(x) = x$ for any $x \in \mathcal{C}$ and any r.

We assume that D is streamlineable with σ states and later show that the PKE is not IND-rCCA-secure.

Let $s \in \mathcal{S}$, and $G = (g_c)_{c \in \mathcal{Z}}$ be such that $|\mathcal{S}| = \sigma$, G_s is an implementation of D with σ states, where g_c corresponding to G_s is as in Definition 10.

Let $\mathsf{bit}_\ell(x)$ denotes the ℓth character of x. Clearly, for $x \in \mathcal{M}$ we have

$$x = \mathsf{trunc}_{|x|-2}(x) \| \mathsf{bit}_{|x|-1}(x) \| \top.$$

Let $\mathcal{C}_k = \{0,1\}^k \| \top$. Let r be fixed. Given x, let ℓ_x be the minimal integer such that $D(\mathsf{trunc}_{\ell_x}(\bar{G}_r(x)) \| \top) = x$. Clearly, $1 \leq \ell_x \leq |\bar{G}_r(x)| - 1$. We define

$$Z(x) = g_{\mathsf{trunc}_{\ell_x - 1}(\bar{G}_r(x))}(\varepsilon, s).$$

For some probabilistic encryption $\bar{G}_r(x)$, we interpret $Z(x)$ to be the pair with a partial decryption of $\bar{G}_r(x)$ together with the internal state of the decryption algorithm D. By definition, we have:

$$\langle x, \cdot \rangle = g_\top \circ g_{\mathsf{bit}_{\ell_x}(\bar{G}_r(x))}(Z(x))$$

and

$$\mathsf{bit}_{\ell_x}(\bar{G}_r(x)) \in \{0, 1\}.$$

Let $\mathcal{X} = \{x \in \mathcal{C}_k : Z(x) \in \{\varepsilon\} \times \mathcal{S}\}$. This is the set of plaintexts x of length k such that the partial decryption of $\bar{G}_r(x)$ is empty. We have

$$\sigma \geq |Z(\mathcal{X})| \geq \frac{|\mathcal{X}|}{2}.$$

Hence

$$\Pr_{x \in U \mathcal{C}_k}[Z(x) \in \{\varepsilon\} \times \mathcal{S}] \leq \frac{2\sigma}{2^k}.$$

If $x \notin \mathcal{X}$ we let $Z(x) = \langle y, s' \rangle$. Since $y \neq \varepsilon$, we have that y is a prefix of $G_s(\mathsf{trunc}_{\ell_x}(\bar{G}_r(x)) \| z)$ for any z. For z such that $\mathsf{trunc}_{\ell_x}(\bar{G}_r(x)) \| z = \bar{G}_r(x)$ we obtain that y is a prefix of x. For $z = \top$, we obtain that y is also a prefix of $D(\mathsf{trunc}_{\ell_x - 1}(\bar{G}_r(x)) \| \top)$. Hence, $D(\mathsf{trunc}_{\ell_x-1}(\bar{G}_r(x)) \| \top) \notin \{\bot, x\}$ and it returns a string whose first bit is $\mathsf{trunc}_1(x)$.

Let T denote the event that

$$\mathsf{trunc}_1(D(\mathsf{trunc}_{\ell_x - 1}(\bar{G}_r(x)) \| \top)) = \mathsf{trunc}_1(x).$$

Therefore

$$\Pr_{x \in_U C_k}[T] \geq 1 - \frac{2\sigma}{2^k}.$$

This holds for any implementation of D with σ states, and for any r so it holds for random choices of r as well. For $k \geq \log_2 \sigma + 3$, we have

$$\Pr[T] \geq \frac{3}{4}. \tag{1}$$

Let A_E be defined as follows:

$A_E^{\mathcal{O}_1^{\mathsf{rCCA}}}(pk)$

  1: pick m_0 and m_1 of length $k = \lceil \log_2 \sigma + 3 \rceil$ with different first bit at random
  2: **return** m_0, m_1, and $\rho = \mathsf{trunc}_1(m_0)$

$A_E^{\mathcal{O}_2^{\mathsf{rCCA}}}(\rho, c^*)$

  1: $\ell \leftarrow |c^*|$
  2: **repeat**
  3:     $\ell \leftarrow \ell - 1$
  4:     $\tilde{m} \leftarrow \mathcal{O}_2^{\mathsf{rCCA}}(\mathsf{trunc}_{\ell-1}(c^*)\|\mathsf{T})$
  5: **until** $\tilde{m} \neq \bot$ or $\ell = 1$
  6: **if** $\mathsf{trunc}_1(\tilde{m}) \in \{0,1\}$ **then**
  7:     **return** $\mathsf{trunc}_1(\tilde{m}) \oplus \rho$
  8: **else**
  9:     **return** a random bit
 10: **end if**

Clearly, while $\ell > \ell_{m_b}$ the rCCA oracle answers $\bot$ because the decryption leads to m_b. When $\ell = \ell_{m_b}$, the decryption is different from m_b so the rCCA oracle is not censored. Then T holds with probability at least $\frac{3}{4}$. In this case, the oracle returns a string whose first bit is the one of m_b Hence, if T occurs as soon as $\ell = \ell_{m_b}$, we must have $b = \tilde{b}$. We obtain the advantage of A_E to win the game as

$$\mathbf{Adv}_{PKE,A_E}^{\mathsf{IND\text{-}rCCA}} \geq \Pr[\tilde{b} = b] - \frac{1}{2}$$

$$\geq \Pr[T] - \frac{1}{2}$$

$$\geq \frac{1}{4}.$$

Thus with non-negligible advantage an adversary A_E will win the IND-rCCA PKE game; and so a PKE with infinite domain and streamlineable decryption with σ states cannot achieve IND-rCCA security when σ is polynomially bounded. $\square$

This result is further supported by the fact that definitions of the decryption algorithm for all known IND-CCA PKE schemes over infinite domains require two passes: one for decrypting the ciphertext and one for validity check of the decryption result.

In essence, this answers the question about the (in)existence of strong encryption schemes with streaming capability. We have shown that IND-rCCA-secure encryption schemes with streaming encryption exist, but the decryption cannot be streamlined.

Clearly, the same result applies to IND-rCCA-secure symmetric encryption.

5   Strong Encryption Over $\aleph_0$ with Bounded Time

We consider here a more general definition of functions on streams with bounded resources. Instead of imposing a finite memory (or equivalently a finite number of states) we require that for each new symbol the number of output symbols is bounded; that is, in other words, the delay for returning something given an input stream is bounded.

Definition 15. *Let $\mathcal{Z}$ be an alphabet, and Δ be some non-negative integer. A streamline function $F = (f_c)_{c \in \mathcal{Z}}$ is called Δ-**delayed** if for any s we have*

$$|(f_c(s))_1| \leq \Delta.$$

Clearly, given a streamline function F over a state space $\mathcal{S}$, if $\mathcal{Z}$ and $\mathcal{S}$ are finite, there exists Δ such that F is Δ-delayed.

Theorem 16. *Let us consider an* IND-rCCA-*secure public-key encryption scheme over* $\mathcal{M} = \{0,1\}^*\top$. *For any key pair and any* Δ, *the decryption function is not Δ-delayed.*

Proof. With the same notations as in the proof of Theorem 14, for $k > 2\Delta$ we have

$$g_{\mathsf{trunc}_{\ell_x - 1}(\bar{G}_r(x))}(\varepsilon, x) \notin \{\varepsilon\} \times \mathcal{S}$$

since $g_{\mathsf{trunc}_{\ell_x - 1}(\bar{G}_r(x))}$ and $g_\top$ have outputs limited to length Δ each and $D(\bar{G}_r(x)) = x$ of length $|x| = k$. Hence, T occurs with probability 1 and our IND-rCCA adversary has advantage at least $\frac{1}{2}$.   □

6   Secure Encryption Over a Continuous Domain

The previous results show the infeasibility of weakly secure encryption over the infinite but enumerable $\aleph_0$ domain. As the domain grows up we have to deal with non-enumerable sets (e.g. the set of real numbers of cardinality $2^{\aleph_0}$) so we have to revisit all definitions for this latter case.

The standard Shannon notion of perfect secrecy adapts well by the notion of statistical independence of X and Y for any distribution of X. We first show that perfect secrecy can be achieved over the continuous set of cardinality $2^{\aleph_0}$ taken from the unit interval $[0, 1]$. We take a uniformly distributed key K in $[0, 1]$ and define $\mathsf{Enc}_k(x) = (x + k) \bmod 1$.

For any density probability function f for X and for any measurable sets A and B of $[0, 1]$ we have that

$$\Pr[X \in A, Y \in B] = \int_{x \in A} f(x) \Pr[x + K \in B] \, dx$$

since $\Pr[x+K \in B] = \mu(B)$ is not dependent on x we obtain $\Pr[X \in A, Y \in B] = \Pr[X \in A]\mu(B)$. Applying this equation for $A = [0, 1]$ we obtain $\Pr[Y \in B] = \mu(B)$ thus for all A and B we have $\Pr[X \in A, Y \in B] = \Pr[X \in A]\Pr[Y \in B]$: X and Y are statistically independent for any distribution of X. Thus, the cipher provides perfect secrecy.

The question of computational security for this case is a little harder since the computational model is not adapted to operations with real numbers.

To define a computational model able to deal with $2^{\aleph_0}$ we should be able to handle algorithms taking infinite sequences of bits as input and returning another infinite sequence of bits. More precisely, an infinite sequence $s = \{s_i\}_{i=0}^n$ of bits is an encoding of a real number from the interval $[0, 1]$. (To avoid confusion we refer to s as a *real number* instead of a sequence.) This implies having memory units able to store such reals and elementary operations on this type of data.

More precisely, given an algorithm $\mathcal{A}$ mapping m bits $a_1, \ldots, a_m$ to n bits $b_1, \ldots, b_n$ in t steps the new Turing machine shall be able to map m reals $\alpha_1, \ldots, \alpha_m$ into n reals $\beta_1, \ldots, \beta_n$ in such a way that the ith bits of the β's are obtained by using $\mathcal{A}$ applied to the ith bits of the α's. We obtain a kind of "free" parallelism this way.

We shall also use arithmetic operations on real numbers in $[0, 1]$ as well as simple bit manipulations in the sequence. Assuming a prefix-free encoding of an arbitrary bitstring, an infinite sequence of bits can be interpreted as a sequence of bitstrings. Therefore we can use operations over sequence of bits to define operations over sequences of bitstrings. We can do this to extend the operations on bitstrings to operations on sequences of bitstrings by using the free parallelism. For instance, we can concatenate two sequence of bitstrings coordinate-wise. We can extract the sequence of the ith bit of a sequence of bitstrings, etc.

The notion of stream of reals which cannot be stored with constant memory becomes irrelevant since a list of reals can be encoded into a single sequence of bits: a list of reals can be compressed into a single real. Clearly, the problem of handling stream of reals with constant time and memory boils down to the problem of encrypting a single real number.

Assuming that IND-CPA secure public key encryption over the domain $\{0, 1\}$ is feasible with these kinds of devices, we can transform it into an IND-CCA secure cryptosystem over the set of reals by adapting the Fujisaki-Okamoto transform [16, 17] in the random oracle model.

In more detail, given an IND-CPA secure public key encryption OLDPKE over the bit domain $\{0, 1\}$, we can define an IND-CPA secure public key encryption TMPPKE over reals where the encryption algorithm $\mathsf{TMPPKE.Enc}_{pk}(m; r)$ takes as input a (potentially infinite) sequence $m = \{m_i\}_{i=0}^n$ of bits m_i and a sequence $r = \{r_i\}_{i=0}^n$ of random strings r_i; more precisely we have:

1. TMPPKE.KeyGen = OLDPKE.KeyGen
2. TMPPKE.$\text{Enc}_{pk}(m;r) = \{\text{OLDPKE.Enc}_{pk}(m_i;r_i)\}_{i=0}^n$
3. TMPPKE.$\text{Dec}_{sk}(c) = \{\text{OLDPKE.Dec}_{sk}(c_i)\}_{i=0}^n$

Next, we construct an IND-CCA secure public key encryption NEWPKE as:

1. NEWPKE.KeyGen = TMPPKE.KeyGen
2. NEWPKE.$\text{Enc}_{pk}(m;r) = \text{TMPPKE.Enc}_{pk}(\langle m,r\rangle;H(\langle m,r\rangle))$
3. NEWPKE.$\text{Dec}_{sk}(c) =$
   (a) $\langle m',h'\rangle = \text{TMPPKE.Dec}_{sk}(c)$
   (b) **if** $c = \text{TMPPKE.Enc}_{pk}(m';h')$, **return** m'

where $\langle a,b\rangle$ denotes the concatenation of the prefix-free encoding of a and b applied in parallel on all coordinates; and $H(x)$ denotes a random oracle call used to compute a sequence of bitstrings of appropriate length.

We could investigate further and consider constructions without random oracles, but this would be beyond our purpose. Our point is that secure and efficient encryption over a domain larger than $\aleph_0$ is feasible (modulo the required adaptations of the computational model).

7   Conclusion

We studied the imperfect notion of secrecy for one-time encryption. We proved that the one by Chor and Kushilevitz is too strong to capture statistical indistinguishability. We extended their impossibility result for encryption over $\aleph_0$ to a weaker notion.

We have shown the decryption (of bitstrings) cannot be implemented in streaming mode with bounded resources without losing the security against adaptive chosen-ciphertext attacks. These results explain the reason why existing CCA-secure encryption schemes are designed with decryption that necessarily performs two passes over the ciphertext before a plaintext is output; and indicate the inexistence of strong encryption schemes over infinite domains. The practical implications of this is that one needs to make a decision tradeoff: between strong encryption (if streamlineability is not required) versus efficiency in practice i.e. streaming capability (if strong encryption is not absolutely mandatory).

We finally observed that those impossibility results are contradicted when the domain is larger, e.g. with $2^{\aleph_0}$. This kind of paradoxical situation reminds some classical results from logic about decidability which can be lost over infinite domains and recovered over yet larger ones.

References

1. Abe, M., Gennaro, R., Kurosawa, K., Shoup, V.: Tag-KEM/DEM: A New Framework for Hybrid Encryption and a New Analysis of Kurosawa-Desmedt KEM. In: Cramer, R. (ed.) EUROCRYPT 2005. LNCS, vol. 3494, pp. 128–146. Springer, Heidelberg (2005)

2. Abe, M., Gennaro, R., Kurosawa, K.: Tag-KEM/DEM: A New Framework for Hybrid Encryption. Journal of Cryptology 21(1), 97–130 (2008); Available at IACR ePrint Archive, http://eprint.iacr.org/2005/027

3. Bellare, M., Boldyreva, A., Knudsen, L., Namprempre, C.: On-line Ciphers and the Hash-CBC Constructios. In: Kilian, J. (ed.) CRYPTO 2001. LNCS, vol. 2139, pp. 292–309. Springer, Heidelberg (2001), http://www-cse.ucsd.edu/users/mihir/papers/olc.html

4. Boldyreva, A., Taesombut, N.: On-line Encryption Schemes: New Security Notions and Constructions. In: Okamoto, T. (ed.) CT-RSA 2004. LNCS, vol. 2964, pp. 1–14. Springer, Heidelberg (2004)

5. Boneh, D., Katz, J.: Improved Efficiency for CCA-Secure Cryptosystems Built using Identity-based Encryption. In: Menezes, A. (ed.) CT-RSA 2005. LNCS, vol. 3376, pp. 87–103. Springer, Heidelberg (2005)

6. Boneh, D., Canetti, R., Halevi, S., Katz, J.: Chosen-Ciphertext Security from Identity-based Encryption. SIAM Journal of Computing 36(5), 1301–1328 (2007)

7. Canetti, R., Halevi, S., Katz, J.: Chosen-ciphertext Security from Identity-based Encryption. In: Cachin, C., Camenisch, J.L. (eds.) EUROCRYPT 2004. LNCS, vol. 3027, pp. 207–222. Springer, Heidelberg (2004)

8. Canetti, R., Krawczyk, H., Nielsen, J.B.: Relaxing Chosen-Ciphertext Security. In: Boneh, D. (ed.) CRYPTO 2003. LNCS, vol. 2729, pp. 565–582. Springer, Heidelberg (2003); Full version available at IACR ePrint Archive, http://eprint.iacr.org/2003/174

9. Chor, B., Kushilevitz, E.: Secret Sharing over Infinite Domains (Extended Abstract). In: Brassard, G. (ed.) CRYPTO 1989. LNCS, vol. 435, pp. 299–306. Springer, Heidelberg (1990)

10. Chor, B., Kushilevitz, E.: Secret Sharing over Infinite Domains. Journal of Cryptology 6(2), 87–95 (1993)

11. Cramer, R., Shoup, V.: Design and Analysis of Practical Public-Key Encryption Schemes Secure against Adaptive Chosen Ciphertext Attack. In: Krawczyk, H. (ed.) CRYPTO 1998. LNCS, vol. 1462, pp. 13–25. Springer, Heidelberg (1998)

12. Cramer, R., Shoup, V.: Design and Analysis of Practical Public-Key Encryption Schemes Secure against Adaptive Chosen Ciphertext Attack. SIAM Journal of Computing 33(1), 167–226 (2004)

13. Fouque, P.-A., Joux, A., Martinet, G., Valette, F.: Authenticated On-line Encryption. In: Matsui, M., Zuccherato, R.J. (eds.) SAC 2003. LNCS, vol. 3006, pp. 145–159. Springer, Heidelberg (2004)

14. Fouque, P.-A., Joux, A., Poupard, G.: Blockwise Adversarial Model for On-line Ciphers and Symmetric Encryption Schemes. In: Handschuh, H., Hasan, M.A. (eds.) SAC 2004. LNCS, vol. 3357, pp. 212–226. Springer, Heidelberg (2004)

15. Fouque, P.-A., Martinet, G., Poupard, G.: Practical Symmetric On-line Encryption. In: Johansson, T. (ed.) FSE 2003. LNCS, vol. 2887, pp. 362–375. Springer, Heidelberg (2003)

16. Fujisaki, E., Okamoto, T.: How to Enhance the Security of Public-Key Encryption at Minimum Cost. In: Imai, H., Zheng, Y. (eds.) PKC 1999. LNCS, vol. 1560, pp. 53–68. Springer, Heidelberg (1999)

17. Fujisaki, E., Okamoto, T.: Secure Integration of Asymmetric and Symmetric Encryption Schemes. In: Wiener, M. (ed.) CRYPTO 1999. LNCS, vol. 1666, pp. 537–554. Springer, Heidelberg (1999)

18. Goldwasser, S., Micali, S.: Probabilistic Encryption. Journal of Computer and System Sciences 28, 270–299 (1984)

19. Harnik, D., Naor, M.: On Everlasting Security in the Hybrid Bounded Storage Model. In: Bugliesi, M., Preneel, B., Sassone, V., Wegener, I. (eds.) ICALP 2006. LNCS, vol. 4052, pp. 192–203. Springer, Heidelberg (2006)
20. Hofheinz, D., Kiltz, E.: Secure Hybrid Encryption from Weakened Key Encapsulation. In: Menezes, A. (ed.) CRYPTO 2007. LNCS, vol. 4622, pp. 553–571. Springer, Heidelberg (2007)
21. Joux, A., Martinet, G., Valette, F.: Blockwise-Adaptive Attackers - Revisiting the (In)Security of Some Provably Secure Encryption Modes: CBC, GEM, IACBC. In: Yung, M. (ed.) CRYPTO 2002. LNCS, vol. 2442, pp. 17–30. Springer, Heidelberg (2002)
22. Katz, J., Yung, M.: Complete Characterization of Security Notions for Probabilistic Private-Key Encryption. In: Proceedings of ACM Syposium on the Theory of Computing (STOC 2000), pp. 245–254. ACM Press, New York (2000)
23. Katz, J., Yung, M.: Complete Characterization of Security Notions for Probabilistic Private-Key Encryption. Journal of Cryptology 19(1), 67–95 (2006)
24. Kiltz, E., Malone-Lee, J.: A General Construction of IND-CCA2 Secure Public Key Encryption. In: Paterson, K.G. (ed.) Cryptography and Coding 2003. LNCS, vol. 2898, pp. 152–166. Springer, Heidelberg (2003)
25. Kiltz, E., Vahlis, Y.: CCA2 Secure IBE: Standard Model Efficiency through Authenticated Symmetric Encryption. In: Malkin, T.G. (ed.) CT-RSA 2008. LNCS, vol. 4964, pp. 221–238. Springer, Heidelberg (2008)
26. Maurer, U.: Conditionally-Perfect Secrecy and a Provably-Secure Randomized Cipher. Journal of Cryptology 5(1), 53–66 (1992)
27. Naor, M., Yung, M.: Public-key Cryptosystems Provably Secure against Chosen Ciphertext Attacks. In: Proceedings of ACM Syposium on the Theory of Computing (STOC 1990), pp. 427–437. ACM Press, New York (1990)
28. Rackoff, C., Simon, D.: Non-interactive Zero-Knowledge Proof of Knowledge and Chosen Ciphertext Attack. In: Feigenbaum, J. (ed.) CRYPTO 1991. LNCS, vol. 576, pp. 433–444. Springer, Heidelberg (1992)
29. Tarski, A.: A Decision Method for Elementary Algebra and Geometry. University of California Press, Berkeley (1951)

Minimum Distance between Bent and Resilient Boolean Functions[*]

Longjiang Qu[1,2] and Chao Li[1]

[1] Department of Mathematic and System Science, Science College,
National University of Defense Technology, ChangSha, 410073, China
ljqu_happy@hotmail.com
[2] National Mobile Communications Research Laboratory,
Southeast University, Nanjing, 210096, China
lichao_nudt@sina.com

Abstract. The minimum distance between bent and resilient functions is studied. This problem is converted into two problems. One is to construct a special matrix, which leads to a combinatorial problem; the other is the existence of bent functions with specified types. Then the relation of these two problems is studied. For the 1-resilient functions, we get a solution to the first combinatorial problem. By using this solution and the relation of the two problems, we present a formula on the lower bound of the minimum distance of bent and 1-resilient functions. For the latter problem, we point out the limitation of the usage of the Maiorana-McFarland type bent functions, and the necessity to study the existence of bent functions with special property which we call partial symmetric. At last, we give some results on the nonexistence of some partial symmetric bent functions.

Keywords: Bent function, Resilient function, Minimum distance.

1 Introduction

Bent functions and resilient functions play very important roles in the design and analysis of ciphers. Bent functions were introduced by Rothaus[9] in 1976. They possess the optimum nonlinearity, and can resist linear cryptanalysis efficiently. Resilient functions were introduced by Chor[1] in 1985. They are balanced and have good correlation immunity, and can resist correlation cryptanalysis efficiently. In recent years, constructions of resilient functions [6,7,8,10,11,12,13] with good nonlinearity and high algebraic degree have attracted the cryptographers. A recent construction method [6,7,8] presents modification of some output points of a bent function to construct highly nonlinear 1-resilient functions. In [7], a lower bound on the minimum number of bits of a bent function that need to be modified is given, and in [8] an algorithm to get a better lower bound is

---

[*] This work is supported by the Natural Science Foundation of China(NO.60573028, 60803156) and the open research fund of National Mobile Communications Research Laboratory of Southeast University(W200807).

C. Xing et al. (Eds.): IWCC 2009, LNCS 5557, pp. 219–232, 2009.

presented. The new bound is proved to be tight for functions up to 14 variables. In [8], Maiorana-McFarland type bent functions were identified with Boolean functions which can be modified to get 1-resilient functions with currently best known parameters. It indicates that 1-resilient functions which have minimum distance with bent functions possess very good nonlinearity and autocorrelation absolute indicator values.

In this paper we study how to solve the minimum distance between bent and resilient functions. We convert it to two problems, one is to construct a matrix satisfying some conditions, which leads to a combinational problem; the other is the existence of bent functions with some specified types. Then the relation of the two problems is studied. For the 1-resilient functions, we get a solution to the first combinational problem. By using this solution and the relation of the two problems, we present a formula on the lower bound of the minimum distance of bent and 1-resilient functions. This lower bound is consistent with the bound which is given by Algorithm 1 in [8], while the formula can save us much time. For the latter problem, using the case of 16-variable functions as an example, we point out the limitation of the usage of the Maiorana-McFarland type bent functions, and the necessity to study the existence of bent functions with special property, which we call partial symmetric. At last, we give some results on the nonexistence of some partial symmetric bent functions.

The organization of the paper is as follows. The preliminaries for bent functions and resilient functions are given in section 2. In section 3, we studies the minimum distance between bent and resilient functions, which is converted to two problems, and the relation of these two problems is studied later. Specified to 1-resilient functions, a solution to the first problem is presented in section 4, which leads to a lower bound on the minimum distance between bent and resilient functions, and a necessity to study the existence of partial symmetric bent functions. In Section 5, we give some results on the nonexistence of some partial symmetric bent functions. The paper is concluded in Section 6.

2   Preliminaries

Let F_2^n be the set of all n-tuples of elements in the field F_2 (Galois field with two elements). A Boolean function $f(x)$ on n variables is a mapping from F_2^n into F_2. The symbol $\oplus$ is referred to the sum over F_2. For a vector $\alpha \in F_2^n$, we use $wt(\alpha)$ to denote its Hamming weight, and call α a weight $wt(\alpha)$ vector. Denote by $d(b, f)$ the Hamming distance between $b(x)$ and $f(x)$.

The Walsh transformation of a boolean function $f(x)$ is a real valued function on F_2^n with the definition:

$$W_f(\omega) = \sum_{x \in F_2^n} (-1)^{f(x) \oplus x \cdot \omega}$$

Bent functions and resilient functions both have many equivalent definitions. For convenience, we give the definitions by means of the Walsh transformations:

Definition 1. *[9] For an n-variable Boolean function $f(x)$, if $W_f(\omega) = \pm 2^{\frac{n}{2}}$ for any $\omega \in F_2^n$, then $f(x)$ is a bent function.*

Definition 2. *[5] For an n-variable Boolean function $f(x)$, if $W_f(\omega) = 0$ for any $0 \leq wt(\omega) \leq t$, then $f(x)$ is t-resilient function.*

From Definition 1, bent functions can exist only when n is even. Then throughout the paper, we assume n is always an even integer.

Definition 3. *[7] The restricted Walsh transform of $f(x)$ on a subset S of $\{0,1\}^n$ is a real valued function over $\{0,1\}^n$ which is defined as*

$$W_f(\omega)|_S = \sum_{x \in S} (-1)^{f(x) \oplus x \cdot \omega}$$

Lemma 1. *[7] Let $S \subset \{0,1\}^n$ and $b(x), f(x)$ be two n-variable Boolean functions such that*

$$f(x) = \begin{cases} b(x) \oplus 1, & x \in S \\ b(x), & x \notin S \end{cases}$$

Then $W_f(\omega) = W_b(\omega) - 2W_b(\omega)|_S$, $\forall \omega \in F_2^n$.

Let A_n be the set of n-variable bent functions, $B_{n,t}$ be the set of n-variable t-resilient functions, then the minimum distance of bent functions and t-resilient functions $dBR_n(t)$ is defined as

$$dBR_n(t) = \min_{b \in A_n, f \in B_{n,t}} d(b, f)$$

As both $b(x)$ and $f(x)$ have even weight, we know $dBR_n(t)$ must be an even integer.

3 Minimum Distance between Bent and Resilient Boolean Functions

We find a new kind of matrix with special structure when studying the minimum distance between bent and resilient functions. The special structure of this kind of matrix is crucial for studying our problem. For convenience, we call this kind of matrix t-resilient matrix. Now let us introduce the definition of t-resilient matrix.

Let n, k both be even integers, $k > 2^{\frac{n}{2}-1}$, for a matrix $S_{k \times n} = (s_{ij})_{k \times n}$, let

$$S_{k \times n} = (s_{ij})_{k \times n} \triangleq (\alpha_1, \cdots, \alpha_n) \triangleq \begin{pmatrix} x_1 \\ \vdots \\ x_k \end{pmatrix}$$

where $\alpha_i \in F_2^k (1 \leq i \leq n)$ be the column vectors of $S_{k \times n}$, $x_i \in F_2^n$ be the row vectors of $S_{k \times n}$ and $x_i \neq x_j (\forall 1 \leq i \neq j \leq k)$.

Let $u = \frac{k}{2} - 2^{\frac{n}{2}-2}$, for any $\omega = (\omega_1, \cdots, \omega_n) \in F_2^n$, let

$$\gamma_\omega = \bigoplus_{i=1}^{n} \omega_i \alpha_i \triangleq \begin{pmatrix} \gamma_1 \\ \gamma_2 \end{pmatrix}, \gamma_1 \in F_2^u, \gamma_2 \in F_2^{u+2^{\frac{n}{2}-1}}$$

Let $a_\gamma = wt(\gamma_1)$, $b_\gamma = wt(\gamma_2)$.

Definition 4. *If there is an integer t, $1 \le t \le n$, such that the matrix $S_{k \times n}$ satisfies: for any $\omega \in F_2^n$, $wt(\omega) \le t$, γ_ω has the following property: $b_\gamma - a_\gamma = 0$ or $b_\gamma - a_\gamma = 2^{\frac{n}{2}-1}$, then we call $S_{k \times n}$ an n-variable t-resilient matrix. And for any given integer n, t, denote by $M_n(t)$ the minimum even integer k such that there exists an n-variable t-resilient matrix $S_{k \times n}$.*

Theorem 1. *If $n > 2$ be an even integer, t be an integer such that $1 \le t \le n$, then $dBR_n(t) \ge M_n(t) > 2^{\frac{n}{2}-1}$.*

Proof. Let $b(x)$ and $f(x)$ be two n-variable Boolean functions satisfying $d(b(x), f(x)) = dBR_n(t)$, where $b(x)$ is a bent function, $f(x)$ is a t-resilient function. Without loss of generality, we assume $W_b(0) = 2^{\frac{n}{2}}$. Let $S = \{x \in F_2^n | b(x) \ne f(x)\}$, then $|S| = dBR_n(t)$. As $wt(b(x)) = 2^{n-1} - 2^{\frac{n}{2}-1}$, $wt(f(x)) = 2^{n-1}$, we know $|S| = dBR_n(t) \ge 2^{\frac{n}{2}-1} \ge 2$.

Let $S_1 = \{x \in F_2^n | b(x) = 1, f(x) = 0\}$, $S_2 = \{x \in F_2^n | b(x) = 0, f(x) = 1\}$, then $S = S_1 \cup S_2$. Fill the elements of S into a $dBR_n(t) \times 1$ matrix, first to fill the elements of S_1, then the elements of S_2, and then expand every element as a row vector, thus we can get a matrix of size $dBR_n(t) \times n$. Denote this matrix by M, we will prove that M is an n-variable t-resilient matrix.

As $b(x)$ is a bent function, $f(x)$ is a t-resilient function, we know for any ω such that $wt(\omega) \le t$,

$$W_f(\omega) = W_b(\omega) - 2W_b(\omega)|_S = W_b(\omega) - 2\sum_{x \in S_1}(-1)^{\omega \cdot x \oplus 1} - 2\sum_{x \in S_2}(-1)^{\omega \cdot x} = 0$$

Specially, for $\omega = 0$, as $W_b(0) = 2^{\frac{n}{2}}$, we have $|S_2| - |S_1| = 2^{\frac{n}{2}-1}$; and together with $|S_2| + |S_1| = |S| = dBR_n(t)$, we get that:

$$|S_1| = \frac{1}{2}dBR_n(t) - 2^{\frac{n}{2}-2}, \quad |S_2| = \frac{1}{2}dBR_n(t) + 2^{\frac{n}{2}-2}$$

Let $u = |S_1| = \frac{1}{2}dBR_n(t) - 2^{\frac{n}{2}-2} \ge 0$, then $|S_2| = \frac{1}{2}dBR_n(t) + 2^{\frac{n}{2}-2} = u + 2^{\frac{n}{2}-1}$. Denote the column vectors of M by $(\alpha_1, \cdots, \alpha_n)$. For any $\omega, wt(\omega) \le t$, let $\gamma_\omega = \sum_{i=1}^{n} \omega_i \alpha_i \triangleq \begin{pmatrix} \gamma_1 \\ \gamma_2 \end{pmatrix}$, $\gamma_1 \in F_2^u, \gamma_2 \in F_2^{u+2^{\frac{n}{2}-1}}$. Denote $a_\gamma = wt(\gamma_1)$, $b_\gamma = wt(\gamma_2)$, then we have

$$W_b(\omega) - 2(2a_\gamma - |S_1|) - 2(|S_2| - 2b_\gamma) = 0$$

$$b_\gamma - a_\gamma = \frac{1}{4}(2^{\frac{n}{2}} - W_b(\omega))$$

As $W_b(\omega) = \pm 2^{\frac{n}{2}}$ holds for any ω, we have $b_\gamma - a_\gamma = 0$ or $b_\gamma - a_\gamma = 2^{\frac{n}{2}-1}$.

Now we go to prove $u > 0$. Otherwise, if $u = 0$, then $dBR_n(t) = |S_2| = 2^{\frac{n}{2}-1}$, and we have $b_\gamma = \frac{1}{4}(2^{\frac{n}{2}} - W_b(\omega))$, which means $b_\gamma = 0$ or $b_\gamma = 2^{\frac{n}{2}-1}$. For any $wt(\gamma) = 1$, if $b_\gamma = 0$, then every element of this column is 0; if $b_\gamma = 2^{\frac{n}{2}-1}$, then every element of this column is 1. This means every row vector of M is the same vector. This is a contradiction. So we have $u > 0$. This proves that M is an n-variable t-resilient matrix. Then according to the definition of $M_n(t)$, we get $dBR_n(t) \geq M_n(t) > 2^{\frac{n}{2}-1}$. $\qquad\square$

From the proof of Theorem 1, we know that from any n-variable bent function $b(x)$ and any n-variable t-resilient function $f(x)$, we can construct an n-variable t-resilient matrix M of size $k \times n$, where $k = d(b(x), f(x))$. So we can determine the value of $dBR_n(t)$ by the following steps: the first one is to construct an n-variable t-resilient matrix $M_{k\times n}$ such that the value of k is as small as possible, the second one is to find a bent function $b(x)$ with respect to $M_{k\times n}$. If such function exists, then we get $dBR_n(t)$. If such function does not exist, we should change another n-variable t-resilient matrix $M_{k\times n}$ and consider the corresponding bent function.

4 Minimum Distance between Bent and 1-Resilient Boolean Functions

4.1 A Lower Bound of $M_n(t)$ and $dBR_n(t)$ and the Value of $M_n(1)$

Theorem 2. *Let r be an integer such that $\binom{n}{r} \leq 2^{\frac{n}{2}-1} < \binom{n}{r+1}$, then for any integer $1 \leq t \leq n$,*

$$dBR_n(t) \geq M_n(t) \geq 2^{\frac{n}{2}-1} + 2\sum_{i=0}^{r-1}\binom{n}{i} + 2\left\lceil \frac{(r+1)(2^{\frac{n}{2}-1} - \binom{n}{r})}{n - 2r - 1} \right\rceil$$

Proof. Let $S_{k\times n}$ be an n-variable t-resilient matrix, $u = \frac{k}{2} - 2^{\frac{n}{2}-2}$, then any two different rows of S can not be equal. Moreover, for any column $\alpha_i (1 \leq i \leq n)$ of S, let $\alpha_i = \begin{pmatrix} \alpha_i^1 \\ \alpha_i^2 \end{pmatrix}, \alpha_i^1 \in F_2^u, \alpha_i^2 \in F_2^{u+2^{\frac{n}{2}-1}}$, then $wt(\alpha_i^2) = wt(\alpha_i^1)$ or $wt(\alpha_i^2) - wt(\alpha_i^1) = 2^{\frac{n}{2}-1}$. Now do the following transformation on the columns $\alpha_i (1 \leq i \leq n)$ of S: if $wt(\alpha_i^2) = wt(\alpha_i^1)$, then keep this column unchanged; if $wt(\alpha_i^2) - wt(\alpha_i^1) = 2^{\frac{n}{2}-1}$, then for every element a of this column, change it to $a\oplus 1$. Denote the new matrix by S', and apparently any two different rows of S' can not be equal. Assume $S' = \begin{pmatrix} M_1 \\ M_2 \end{pmatrix}$, where M_1 is a $u \times n$ matrix, M_2 is a $(u+2^{\frac{n}{2}-1}) \times n$ matrix, then the number of 1's will be equal in any corresponding column of M_1 and M_2. Denote by N_1, N_2 the total number of 1's in M_1, M_2 respectively, then we have $N_1 = N_2$. But the number of rows of M_1 is much less than that of M_2, and any two rows of S' are distinct, this brings restriction on the number of rows of S'.

Let r_1 be an integer such that $\sum_{i=0}^{r_1} \binom{n}{i} \le u < \sum_{i=0}^{r_1+1} \binom{n}{i}$, r_2 be an integer such that $\sum_{i=0}^{r_2} \binom{n}{i} \le u + 2^{\frac{n}{2}-1} < \sum_{i=0}^{r_2+1} \binom{n}{i}$. As any two distinct rows of M_1 can not be equal, if we want the value of N_1 to be as big as possible, we should fill M_1 with all weight $w(n - r_1 \le w \le n)$ vectors followed by some weight $w = n - r_1 - 1$ vectors. When this case occurs, for any $w(n - r_1 \le w \le n)$, all weight w vectors contribute $\binom{n-1}{w-1}$ 1's to each column, then we have total $n \cdot \sum_{i=n-r_1}^{n} \binom{n-1}{i-1} = n \cdot \sum_{i=n-r_1-1}^{n-1} \binom{n-1}{i} = n \cdot \sum_{i=0}^{r_1} \binom{n-1}{i}$ 1's, together with the remaining weight $n - r_1 - 1$ vectors, we have:

$$N_1 \le n \sum_{i=0}^{r_1} \binom{n-1}{i} + (n - r_1 - 1)(u - \sum_{i=0}^{r_1} \binom{n}{i})$$

Similarly, as any two distinct rows of M_2 can not be equal, so if we want the value of N_2 to be as small as possible, we should fill M_2 with all weight $w(0 \le w \le r_2)$ vectors following by some weight $w = r_2 + 1$ vectors. When this case occurs, weight 0 vectors do not contribute any 1 to any column, while for any $w(1 \le w \le r_2)$, all weight w vectors contribute $\binom{n-1}{w-1}$ 1's to each column, then we have total $n \cdot \sum_{i=1}^{r_2} \binom{n-1}{i-1} = n \cdot \sum_{i=0}^{r_2-1} \binom{n-1}{i}$ 1's, together with the remaining weight $r_2 + 1$ vectors, we have:

$$N_2 \ge n \sum_{i=0}^{r_2-1} \binom{n-1}{i} + (r_2 + 1)(u + 2^{\frac{n}{2}-1} - \sum_{i=0}^{r_2} \binom{n}{i})$$

And from $N_1 = N_2$, we know

$$n \sum_{i=0}^{r_1} \binom{n-1}{i} + (n - r_1 - 1)(u - \sum_{i=0}^{r_1} \binom{n}{i}) \ge n \sum_{i=0}^{r_2-1} \binom{n-1}{i} + (r_2 + 1)(u + 2^{\frac{n}{2}-1} - \sum_{i=0}^{r_2} \binom{n}{i})$$
$$(1)$$

From (1), we can deduce that $r_2 \le r_1 + 1$. Otherwise, if $r_2 \ge r_1 + 2$, then

$$n \sum_{i=0}^{r_1} \binom{n-1}{i} + (n - r_1 - 1)(u - \sum_{i=0}^{r_1} \binom{n}{i}) < n \sum_{i=0}^{r_1+1} \binom{n-1}{i}$$
$$\le n \sum_{i=0}^{r_2-1} \binom{n-1}{i} \le n \sum_{i=0}^{r_2-1} \binom{n-1}{i} + (r_2 + 1)(u + 2^{\frac{n}{2}-1} - \sum_{i=0}^{r_2} \binom{n}{i})$$

This contradicts with (1). Then we have $r_2 \le r_1 + 1$. And from the definition of r_1 and r_2, it's easy to get $r_2 \ge r_1$, then $r_2 = r_1 + 1$ or $r_2 = r_1$. Now we discuss these two cases:

i) If $r_2 = r_1 + 1$, then from (1), we have:

$$(n - r_1 - 1)(u - \sum_{i=0}^{r_1} \binom{n}{i}) \ge (r_1 + 2)(u + 2^{\frac{n}{2}-1} - \sum_{i=0}^{r_1+1} \binom{n}{i})$$

$$(n - 2r_1 - 3)u \ge (n - r_1 - 1) \sum_{i=0}^{r_1} \binom{n}{i} + (r_1 + 2)(2^{\frac{n}{2}-1} - \sum_{i=0}^{r_1+1} \binom{n}{i})$$

$$u \geq \sum_{i=0}^{r_1} \binom{n}{i} + \left\lceil \frac{(r_1+2)(\sum_{i=0}^{r_1} \binom{n}{i} + 2^{\frac{n}{2}-1} - \sum_{i=0}^{r_1+1} \binom{n}{i}))}{n-2r_1-3} \right\rceil$$

$$= \sum_{i=0}^{r_1} \binom{n}{i} + \left\lceil \frac{(r_1+2)(2^{\frac{n}{2}-1} - \binom{n}{r_1+1}))}{n-2r_1-3} \right\rceil \tag{2}$$

From the definition of r_1, we have:

$$\left\lceil \frac{(r_1+2)(2^{\frac{n}{2}-1} - \binom{n}{r_1+1}))}{n-2r_1-3} \right\rceil < \binom{n}{r_1+1} \implies \frac{(r_1+2)(2^{\frac{n}{2}-1} - \binom{n}{r_1+1}))}{n-2r_1-3} < \binom{n}{r_1+1}$$

$$\implies (r_1+2)2^{\frac{n}{2}-1} < (n-r_1-1)\binom{n}{r_1+1} \implies 2^{\frac{n}{2}-1} < \frac{n-r_1-1}{r_1+2}\binom{n}{r_1+1} = \binom{n}{r_1+2}$$

Then according to the definition of r, we know $r_1 + 2 \geq r + 1$, which means $r_1 \geq r - 1$.

If $r_1 = r - 1$, then $r_2 = r_1 + 1 = r$, and (2) becomes

$$u \geq \sum_{i=0}^{r-1} \binom{n}{i} + \left\lceil \frac{(r+1)(2^{\frac{n}{2}-1} - \binom{n}{r}))}{n-2r-1} \right\rceil \tag{3}$$

If $r_1 \geq r$, then we have

$$u \geq \sum_{i=0}^{r_1} \binom{n}{i} \geq \sum_{i=0}^{r} \binom{n}{i} \geq \sum_{i=0}^{r-1} \binom{n}{i} + \left\lceil \frac{(r+1)(2^{\frac{n}{2}-1} - \binom{n}{r}))}{n-2r-1} \right\rceil$$

The last inequality holds because of the following inequalities:

$$2^{\frac{n}{2}-1} < \binom{n}{r+1} = \frac{n-r}{r+1}\binom{n}{r} \implies (r+1)2^{\frac{n}{2}-1} < (n-r)\binom{n}{r}$$

$$\implies (r+1)(2^{\frac{n}{2}-1} - \binom{n}{r}) < (n-2r-1)\binom{n}{r} \implies \frac{(r+1)(2^{\frac{n}{2}-1} - \binom{n}{r}))}{n-2r-1} < \binom{n}{r}$$

$$\implies \left\lceil \frac{(r+1)(2^{\frac{n}{2}-1} - \binom{n}{r}))}{n-2r-1} \right\rceil \leq \binom{n}{r}$$

ii) If $r_2 = r_1$, then from the definition of r_1 and r_2, we have $2^{\frac{n}{2}-1} < \binom{n}{r_1+1}$. And with the definition of r, we have $r_1 + 1 \geq r + 1$, which leads to $r_1 \geq r$. This means (3) also holds as the above paragraph explains.

So we know (3) always holds. And as $u = \frac{k}{2} - 2^{\frac{n}{2}-2}$, we have

$$dBR_n(t) \geq M_n(t) \geq 2^{\frac{n}{2}-1} + 2\sum_{i=0}^{r-1} \binom{n}{i} + 2\left\lceil \frac{(r+1)(2^{\frac{n}{2}-1} - \binom{n}{r}))}{n-2r-1} \right\rceil$$

This proves the theorem.   □

Theorem 2 shows a lower bound of $dBR_n(t)$. When $t = 1$, this bound is consistent with the bound which is given by Algorithm 1 in [8], while the formula saves us much time. Moreover, the formula give us so many hints that we can solve $M_n(1)$ with it.

Before showing the result of $M_n(1)$, we recall Algorithm 2 in [8]. The algorithm use m vectors whose length are n and weight are t to output a $m \times n$ matrix T having t columns of weight $w + 1$ and $n - t$ columns of weight w, with $w = \lfloor m \times r/n \rfloor$ and t the remainder such that $m \times r = n \times w + t$.

Theorem 3. *Let $n > 2$ be an even integer, r be an integer such that $\binom{n}{r} \leq$*

$$2^{\frac{n}{2}-1} < \binom{n}{r+1}, \text{ let } a = \left\lceil \frac{(r+1)(2^{\frac{n}{2}-1} - \binom{n}{r})}{n-2r-1} \right\rceil, u_0 = \sum_{i=0}^{r-1} \binom{n}{i} + a, k_0 = 2u_0 +$$

$2^{\frac{n}{2}-1}$, *then $M_n(1) = k_0$.*

Proof. According to Theorem 2, we know $M_n(1) \geq k_0$. For proving $M_n(1) = k_0$, we only need to construct an n-variable 1-resilient matrix $M_{k_0 \times n}$.

Let $b = a + 2^{\frac{n}{2}-1} - \binom{n}{r}$. Firstly we fill all weight $t \geq n - (r - 1)$ vectors into M_1 and all weight $t \leq r$ vectors into M_2. Then it leaves $u - \sum_{i=0}^{r-1} \binom{n}{i} = a$ lines undetermined in M_1 and $u + 2^{\frac{n}{2}-1} - \sum_{i=0}^{r} \binom{n}{i} = b$ lines undetermined in M_2, and it is clear that the weight of each column of M_1 and M_2 are both $\sum_{i=0}^{r-1} \binom{n-1}{i}$ now. Denote the remaining lines of $M_1(M_2)$ by $P_1(P_2)$. So to construct an n-variable 1-resilient matrix $M_{k_0 \times n}$, we only need to fill $M_{k_0 \times n}$ such that the weight of each column of P_1 is equal to that of the corresponding column of P_2.

Let s be an integer such that $bs \leq a(n-r) < b(s+1)$, and let $l = a(n-r) - bs$, then we have $0 \leq l < b$ and $a(n - r) = (b - l)s + l(s + 1)$. Now we use a vectors whose weight are $n - t$ to construct P_1, and use $b - l$ vectors whose weight are s and l vectors whose weight are $s+1$ to construct P_2. Then by Algorithm 2 of [8], we can construct P_1 such that the weight of each column do not differ by more than 1, but P_2 may have column weights differing by more than 1. To avoid this case, we permute the columns of the weight $s + 1$ vectors so that the columns of higher weight are identified with the columns of lower weight of the weight $s + 1$ vectors. Then in the resulting P_2 matrix, column weights do not differ by more than 1. And as $a(n - r) = (b - l)s + l(s + 1)$, after permute the columns of P_2 suitably, we can make sure that the weight of each corresponding column of P_1 and P_2 is equal.

From the definition of a, we have $a(n - r) \geq b(r + 1)$. And as $2^{\frac{n}{2}-1} - \binom{n}{r} \neq 0$ for any even integer $n > 2$ and any integer r, it should have $a < b$, which means $a(n-r) < b(n-r)$. So we have $r + 1 \leq s < n - r$. In fact, we have $s < n - r - 1$, which can be shown as follows:

We first prove that $n \geq 2r + 4$. As $n > 2$, it holds $\binom{n}{\frac{n}{2}} > 2^{\frac{n}{2}}$, and as $\binom{n}{\frac{n}{2}} = \frac{\frac{n}{2}+1}{\frac{n}{2}}\binom{n}{\frac{n}{2}-1} = (1 + \frac{2}{n})\binom{n}{\frac{n}{2}-1} < 2\binom{n}{\frac{n}{2}-1}$, so we have $\binom{n}{\frac{n}{2}-1} > 2^{\frac{n}{2}-1}$. By the definition of r, we have $\frac{n}{2} - 1 \geq r + 1$, which means $n \geq 2r + 4$.

As $n \geq 2r+4$, we have $n-r-1 > r+2 > \frac{r+1}{n-2r-1}+1$. And with $2^{\frac{n}{2}-1}-\binom{n}{r} > 0$,
we have $(n-r-1)(2^{\frac{n}{2}-1}-\binom{n}{r}) > (\frac{r+1}{n-2r-1}+1)(2^{\frac{n}{2}-1}-\binom{n}{r}) \geq \frac{(r+1)(2^{\frac{n}{2}-1}-\binom{n}{r})}{n-2r-1}+1$,
which means $(n-r-1)(2^{\frac{n}{2}-1}-\binom{n}{r}) > a$. And as $2^{\frac{n}{2}-1}-\binom{n}{r} = b - a$, we get
$b(n-r-1) > a(n-r)$. As s be the integer such that $bs \leq a(n-r) < b(s+1)$,
finally we have $s < n-r-1$.

And as $r+1 \leq s < n-r-1$, the vectors of P_2 are distinct from those of M_1
and the other vectors of M_2. So we construct an n-variable 1-resilient matrix
$M_{k_0 \times n}$, which proves the theorem.   □

Now we have determined the value of $M_n(1)$, and after we have constructed
an n-variable 1-resilient matrix $M_{k \times n}$, the problem of the existence of the bent
function corresponding to this matrix must be considered.

4.2   The Value of $dBR_{16}(1)$

As the values of $dBR_n(1)$ are all got for $n < 16$, now we try to solve this value
for $n = 16$. With $n = 16$ and $\binom{n}{r} \leq 2^{\frac{n}{2}-1} < \binom{n}{r+1}$, we get $r = 2$; with $r = 2$, we
get $M_{16}(1) = 168$, $u = 20$, $r_1 = r - 1 = 1$, $r_2 = r = 2$. After filling all weight
$t \geq n - r_1 = 15$ vectors into M_1 and all weight $t \leq r_2 = 2$ vectors into M_2, it
leaves $20 - \sum_{i=0}^{r_1} \binom{n}{i} = 3$ lines undetermined in M_1 and $20 + 2^{\frac{n}{2}-1} - \sum_{i=0}^{r_2}\binom{n}{i} = 11$
lines undetermined in M_2. There are many ways to fill the remaining lines of M_1
and M_2 such that any column of them have the same weight, while one solution
is as follows, where the former three lines are the remaining lines of M_1, and the
other lines are those of M_2:

$$\begin{pmatrix} 0\,0\,1\,1\,1\,1\,1\,1\,1\,1\,1\,1\,1\,1\,1\,1 \\ 1\,1\,0\,0\,1\,1\,1\,1\,1\,1\,1\,1\,1\,1\,1\,1 \\ 1\,1\,1\,1\,0\,0\,1\,1\,1\,1\,1\,1\,1\,1\,1\,1 \end{pmatrix}$$

$$\begin{pmatrix} 0\,0\,0\,0\,0\,0\,1\,1\,1\,0\,0\,0\,0\,0\,0\,0 \\ 0\,0\,0\,0\,0\,0\,0\,0\,1\,1\,1\,0\,0\,0\,0\,0 \\ 0\,0\,0\,0\,0\,0\,0\,0\,0\,0\,0\,0\,1\,1\,1\,1 \\ 1\,0\,0\,0\,1\,0\,0\,0\,1\,0\,0\,0\,1\,0\,0\,0 \\ 0\,1\,0\,0\,0\,1\,0\,0\,0\,1\,0\,0\,0\,1\,0\,0 \\ 0\,0\,1\,0\,0\,0\,1\,0\,0\,0\,1\,0\,0\,0\,1\,0 \\ 0\,0\,0\,1\,0\,0\,0\,1\,0\,0\,0\,1\,0\,0\,0\,1 \\ 1\,1\,0\,0\,1\,1\,0\,0\,0\,0\,0\,0\,0\,0\,0\,0 \\ 0\,0\,1\,1\,0\,0\,1\,1\,0\,0\,0\,0\,0\,0\,0\,0 \\ 0\,0\,0\,0\,0\,0\,0\,0\,1\,1\,0\,0\,1\,1\,0\,0 \\ 0\,0\,0\,0\,0\,0\,0\,0\,0\,0\,1\,1\,0\,0\,1\,1 \end{pmatrix}$$

So now we get a solution for 16-variable 1-resilient matrixes. Then if there exists
some bent functions corresponding to this matrix, then we have $dBR_{16}(1) = 168$.
It is suggested to use Maiorana-McFarland type bent functions in [8]. However,
by the following proposition, we can show that there is no Maiorana-McFarland
type bent functions $b(x)$ satisfying $b(x) = 0$ for all weight $wt(x) \leq 2$. So we need
some changes.

Proposition 1. *Let $n > 4$ be an even integer, $b(x) = b(x_1, x_2, \cdots, x_n)$ be an n-variable bent function such that $b(x) = 0$ for $wt(x) \leq 2$, then $b(x)$ can not be a Maiorana-McFarland type function.*

Proof. Assume $b(x)$ be a Maiorana-McFarland type bent function. Let $b(x) = b(X, Y) = X \cdot \pi(Y) + g(Y)$, where $X = (x_1, x_2, \cdots, x_{\frac{n}{2}})$, $Y = (x_{\frac{n}{2}+1}, x_{\frac{n}{2}+2}, \cdots, x_n)$, π is a permutation on $\{0, 1\}^{\frac{n}{2}}$, g is any Boolean functions on $\frac{n}{2}$ variables. From $b(x) = 0$ for $x = (0, 0, \cdots, 0)$, we get $g(0, 0, \cdots, 0) = 0$. And from $b(x) = 0$ for all $wt(x) = 1$, we have $\pi(0, 0, \cdots, 0) = 0$ and $g(Y) = 0$ for all $wt(Y) = 1$. Now consider $\pi(1, 0, \cdots, 0)$, for any $wt(X) = 1$, as $wt(x) = wt(X) + wt(1, 0, \cdots, 0) = 2$, $b(X, Y) = X \cdot \pi(Y) + g(Y) = X \cdot \pi(1, 0, \cdots, 0) = 0$, so we have $\pi(1, 0, \cdots, 0) = 0 = \pi(0, 0, \cdots, 0)$. This is a contradiction with the fact that π is a permutation. Hence $b(x)$ can not be a Maiorana-McFarland type function. $\qquad\square$

For $k = 170$, $u = 21$, we can construct a Maiorana-McFarland type bent $b(x)$ and a corresponding matrix $M_{k \times n}$. Let $b(x) = b(X, Y) = X \cdot Y$, where $x = (x_1, x_2, \cdots, x_{16})$, $X = (x_1, x_2, \cdots, x_8)$, $Y = (x_9, x_{10}, \cdots, x_{16})$, then $W_b(\omega) = 2^{\frac{n}{2}} = 2^8$ for all $0 \leq wt(\omega) \leq 1$, and we can construct a 16-variable 1-resilient matrix $M_{170 \times 16}$ as follows: first to fill all weight $t = 15$ elements into M_1 and all weight $t \leq 2$ elements but those satisfying both $wt(x) = 2$ and $X = Y$ into M_2, it leaves $21 - \binom{16}{1} = 5$ lines undetermined in M_1 and $21 + 2^{\frac{n}{2}-1} - (\sum_{i=0}^{2} \binom{n}{i} - 8) = 20$ lines undetermined in M_2. There are many ways to fill the remaining lines of M_1 and M_2 such that any corresponding column of them has the same weight, while one solution can be constructed as follows:

i) Construct P_1 as the following matrix, then it holds $b(x) = 1$ for any row vector x of P_1, and the weight of each column of P_1 is 4:

$$\begin{pmatrix} 0 & 0 & 0 & 1 & 1 & 1 & 1 & 1 & 0 & 1 & 1 & 1 & 1 & 1 & 1 & 1 \\ 1 & 1 & 1 & 0 & 0 & 0 & 1 & 1 & 1 & 1 & 1 & 1 & 1 & 1 & 1 & 1 \\ 1 & 1 & 1 & 1 & 1 & 1 & 0 & 0 & 1 & 0 & 1 & 1 & 1 & 1 & 1 & 1 \\ 1 & 1 & 1 & 1 & 1 & 1 & 1 & 1 & 1 & 1 & 0 & 0 & 0 & 1 & 1 & 1 \\ 1 & 1 & 1 & 1 & 1 & 1 & 1 & 1 & 1 & 1 & 1 & 1 & 1 & 0 & 0 & 0 \end{pmatrix}$$

ii) Fill the following rows to the first 5 rows of P_2, and then rotate these rows right by 4 positions to fill the continue 5 rows of P_2, and then rotate them right by 8 positions and by 12 positions to fill the rest ten rows of P_2. Then it holds $b(x) = 0$ for any row vector x of P_2, and the weight of each column of P_2 is 4:

$$\begin{pmatrix} 1 & 1 & 1 & 1 & 0 & 0 & 0 & 0 & 0 & 0 & 0 & 0 & 0 & 0 & 0 & 0 \\ 0 & 1 & 1 & 1 & 0 & 0 & 0 & 0 & 0 & 0 & 0 & 0 & 0 & 0 & 0 & 0 \\ 1 & 0 & 1 & 1 & 0 & 0 & 0 & 0 & 0 & 0 & 0 & 0 & 0 & 0 & 0 & 0 \\ 1 & 1 & 0 & 1 & 0 & 0 & 0 & 0 & 0 & 0 & 0 & 0 & 0 & 0 & 0 & 0 \\ 1 & 1 & 1 & 0 & 0 & 0 & 0 & 0 & 0 & 0 & 0 & 0 & 0 & 0 & 0 & 0 \end{pmatrix}$$

So we construct a 16-variable 1-resilient matrix which is corresponding to $b(x) = b(X, Y) = X \cdot Y$, which means $dBR_{16}(1) \leq 170$. So finally we have $dBR_{16}(1) \in \{168, 170\}$, but which value is the exact value of $dBR_{16}(1)$?

A natural consideration is whether there are some bent functions satisfying the following two conditions:

i) $W_b(\omega) = 2^{\frac{n}{2}}$ for all $0 \le wt(\omega) \le 1$

ii) There exists a large integer R such that $R < \frac{n}{2}$ and

$$b(x_1, x_2, \cdots, x_n) = \begin{cases} 1, & wt(x_1, x_2, \cdots, x_n) \ge n - R \\ 0, & wt(x_1, x_2, \cdots, x_n) \le R \\ b \in \{0, 1\}, & R < wt(x_1, x_2, \cdots, x_n) < n - R \end{cases}$$

The integer R in the second condition gives us freedom to construct the corresponding n-variable 1-resilient matrix $M_{k \times n}$, we have more freedom to construct $M_{k \times n}$ while R gets bigger. For convenience, we give a definition:

Definition 5. Let f be a Boolean function with n variables $x_1, x_2, \cdots, x_n$, if there is an integer $R \ge 1$ such that

$$f(x_1, x_2, \cdots, x_n) = \begin{cases} 1, wt(x_1, x_2, \cdots, x_n) \ge n - R \\ 0, wt(x_1, x_2, \cdots, x_n) \le R \end{cases}$$

then we call f a Boolean function with partial symmetric. If we assume R_0 is the largest integer satisfying the above condition, then we call R_0 the order of partial symmetric of f, and call f a R_0 order partial symmetric Boolean function.

According to Definition 5, we raise two problems:

Problem 1. Let n be an even integer, determine the biggest integer $PSR(n)$ such that there exists an n-variable bent function with $PSR(n)$ order partial symmetric.

Problem 2. Let n be an even integer, determine the biggest integer $PSR'(n)$ such that there exists an n-variable bent function $b(x)$ with $PSR'(n)$ order partial symmetric which also satisfies $W_b(\omega) = 2^{\frac{n}{2}}$ for all $0 \le wt(\omega) \le 1$.

Considerations on the two problems above will do many goods to the problem of the minimum distance between bent and 1-resilient functions. A result on Problem 1 will be shown in the next section.

5   The Nonexistence of Some Type Bent Functions

Theorem 4. Let n be an even integer, $R < \frac{n}{2}$ be an integer such that $\binom{n-1}{R} > 2^{\frac{n}{2}-2} + \frac{1}{2}\binom{n-1}{\frac{n}{2}-1} \ge \binom{n-1}{R-1}$,

$$f(x_1, x_2, \cdots, x_n) = \begin{cases} 1, & wt(x_1, x_2, \cdots, x_n) \ge n - R \\ 0, & wt(x_1, x_2, \cdots, x_n) \le R \\ b \in \{0, 1\}, & R < wt(x_1, x_2, \cdots, x_n) < n - R \end{cases}$$

then $f(x_1, x_2, \cdots, x_n)$ can not be a bent function.

Proof. For convenience, let $X = (x_1, x_2, \cdots, x_n) \in F_2^n$. For any $0 \leq t \leq n$, let $S_t = \{X \in F_2^n | f(X) = 1, wt(X) = t\}$. For any $0 \leq t \leq n$, any $1 \leq i \leq n$, let $A_{t,i} = \#\{X \in S_t | x_i = 1\}$, $B_{t,i} = \#\{X \in S_t | x_i = 0\}$. Then we have

$$\sum_{i=1}^{n} A_{t,i} = t \cdot |S_t|, \quad \sum_{i=1}^{n} B_{t,i} = (n-t) \cdot |S_t|, \quad \text{for any } 0 \leq t \leq n$$

$$A_{t,i} = B_{t,i} = 0, \quad \text{for any } 0 \leq t \leq R, \quad \text{any } 0 \leq i \leq n$$

$$A_{t,i} = \binom{n-1}{t-1}, B_{t,i} = \binom{n-1}{t}, \quad \text{for any } n-R \leq t \leq n, \quad \text{any } 0 \leq i \leq n$$

Let $e_i \in F_2^n (1 \leq i \leq n)$ be the vector whose ith index is 1 and 0 else. For any $0 \leq t \leq n$, let $W_t(e_i) = \sum_{wt(X)=t} (-1)^{f(X)+e_i \cdot X}$, then

$$W_t(e_i) = \sum_{wt(X)=t} (-1)^{f(X)+x_i} = (A_{t,i} + \binom{n-1}{t} - B_{t,i}) - (B_{t,i} + \binom{n-1}{t-1} - A_{t,i})$$

$$= 2(A_{t,i} - B_{t,i}) + \binom{n-1}{t} - \binom{n-1}{t-1}$$

So we have

$$W(e_i) = \sum_{t=0}^{n} W_t(e_i) = \sum_{t=0}^{n} (2(A_{t,i} - B_{t,i}) + \binom{n-1}{t} - \binom{n-1}{t-1})$$

$$= 2\sum_{t=0}^{n} (A_{t,i} - B_{t,i}) = 2\sum_{t=R+1}^{n-R-1} (A_{t,i} - B_{t,i}) + 2\sum_{t=n-R}^{n} (\binom{n-1}{t-1} - \binom{n-1}{t})$$

$$= 2\sum_{t=R+1}^{n-R-1} (A_{t,i} - B_{t,i}) + 2\binom{n-1}{n-R-1} = 2\binom{n-1}{R} + 2\sum_{t=R+1}^{n-R-1} (A_{t,i} - B_{t,i})$$

Since

$$\sum_{i=1}^{n} \sum_{t=R+1}^{n-R-1} (A_{t,i} - B_{t,i}) = \sum_{t=R+1}^{n-R-1} \sum_{i=1}^{n} (A_{t,i} - B_{t,i}) = \sum_{t=R+1}^{n-R-1} (2t-n)|S_t|$$

$$= \sum_{t=\frac{n}{2}}^{n-R-1} (2t-n)|S_t| - \sum_{t=R+1}^{\frac{n}{2}-1} (n-2t)|S_t|$$

$$\geq - \sum_{t=R+1}^{\frac{n}{2}-1} (n-2t)\binom{n}{t} = - \sum_{t=R+1}^{\frac{n}{2}-1} (n\binom{n}{t} - 2t\binom{n}{t})$$

$$= - \sum_{t=R+1}^{\frac{n}{2}-1} (n\binom{n}{t} - 2n\binom{n-1}{t-1}) = - \sum_{t=R+1}^{\frac{n}{2}-1} n(\binom{n-1}{t} - \binom{n-1}{t-1})$$

$$= -n(\binom{n-1}{\frac{n}{2}-1} - \binom{n-1}{R})$$

So there exist at least one $i_0 \in \{1, 2, \cdots, n\}$ such that $\sum_{t=R+1}^{n-R-1}(A_{t,i_0} - B_{t,i_0}) \geq -(\binom{n-1}{\frac{n}{2}-1} - \binom{n-1}{R}))$. Then we have

$$
\begin{aligned}
W(e_{i_0}) &= 2\binom{n-1}{R} + 2 \sum_{t=R+1}^{n-R-1}(A_{t,i_0} - B_{t,i_0}) \\
&\geq 2\binom{n-1}{R} - 2(\binom{n-1}{\frac{n}{2}-1} - \binom{n-1}{R})) \\
&= 4\binom{n-1}{R} - 2\binom{n-1}{\frac{n}{2}-1} > 2^{\frac{n}{2}}
\end{aligned}
$$

Now we have $max_{\omega \in F_2^n}|W_f(\omega)| \geq W(e_{i_0}) > 2^{\frac{n}{2}}$, so $f(x_1, x_2, \cdots, x_n)$ can not be a bent function. $\qquad \square$

By Theorem 4, we can have a upper bound on $PSR(n)$.

Proposition 2. Let $n > 2$ be an even integer, $R(n) < \frac{n}{2}$ be the integer such that $\binom{n}{R(n)-1} > 2^{\frac{n}{2}-2} + \frac{1}{2}\binom{n-1}{\frac{n}{2}-1} \geq \binom{n-1}{R(n)-1}$, then $PSR(n) < R(n)$.

Use similar method, we can get the following theorem:

Theorem 5. Let n be an even integer,

$$
f(x_1, x_2, \cdots, x_n) = \begin{cases} 0, & wt(x_1, x_2, \cdots, x_n) < \frac{n}{2} \\ b \in \{0, 1\}, & wt(x_1, x_2, \cdots, x_n) \geq \frac{n}{2} \end{cases}
$$

then $nl(x_1, x_2, \cdots, x_n) \leq 2^{n-1} - \frac{1}{2}\binom{n-1}{\frac{n}{2}}$.

As when $n \geq 6$, $\binom{n-1}{\frac{n}{2}} > 2^{\frac{n}{2}}$, then it can be concluded that when $n \geq 6$ the function has the type as the above theorem states can not be a bent function.

6   Conclusion

The problem of solving the minimum distance between bent and resilient functions is converted into two problems in this paper. One is to construct an n-variable t-resilient matrix, the other is the existence of bent functions with respect to this n-variable t-resilient matrix. For the 1-resilient functions, we find ways to construct an n-variable 1-resilient matrix. By using this method and the relation of the two problems, we present a formula on the lower bound of the minimum distance of bent and 1-resilient functions. Studying whether this lower bound is tight now turns to study the existence of bent functions with special property which we call it partial symmetric. At last, we give some results on the nonexistence of some partial symmetric bent functions. The two problems on the order of partial symmetric are interesting, considering these problems will be very helpful to determine $dBR_n(1)$. Determining $dBR_n(t)$ will be very helpful to construct t-resilient functions with good properties. It remains open to determine $M_n(t)$ for $t \geq 2$.

References

1. Chor, B., Goldreich, O., et al.: Extraction problem or t-resilient functions. In: 26th IEEE Symp. Foundations of Computer Science, vol. 26, pp. 396–407 (1985)
2. Charpin, P., Pasalic, E.: On propagation characteristics of resilient functions. In: Nyberg, K., Heys, H.M. (eds.) SAC 2002. LNCS, vol. 2595, pp. 175–195. Springer, Heidelberg (2003)
3. Dillon, J.F.: Elementary Hadamard Difference sets. PhD Thesis, University of Maryland (1974)
4. Dobbertin, H.: Construction of bent functions and balanced Boolean functions with high nonlinearity. In: Preneel, B. (ed.) FSE 1994. LNCS, vol. 1008, pp. 61–74. Springer, Heidelberg (1995)
5. Guo-Zhen, X., Massey, J.: A spectral characterization of correlation immune combining functions. IEEE Transactions on Information Theory 34(3), 569–571 (1988)
6. Maity, S., Johansson, T.: Construction of Cryptographically important Boolean functions. In: Menezes, A., Sarkar, P. (eds.) INDOCRYPT 2002. LNCS, vol. 2551, pp. 234–245. Springer, Heidelberg (2002)
7. Maity, S., Maitra, S.: Minimum distance between bent and 1-resilient functions. In: Roy, B., Meier, W. (eds.) FSE 2004. LNCS, vol. 3017, pp. 143–160. Springer, Heidelberg (2004)
8. Maity, S., Arackaparambil, C., Meyase, K.: Construction of 1-Resilient Boolean Functions with Very Good Nonlinearity. In: Gong, G., Helleseth, T., Song, H.-Y., Yang, K. (eds.) SETA 2006. LNCS, vol. 4086, pp. 417–431. Springer, Heidelberg (2006)
9. Rothaus, O.S.: On bent functions. Journal of Combinatorial Theory, Series A 20, 300–305 (1976)
10. Sarkar, P., Maitra, S.: Nonlinearity bounds and constructions of resilient Boolean functions. In: Bellare, M. (ed.) CRYPTO 2000. LNCS, vol. 1880, pp. 515–532. Springer, Heidelberg (2000)
11. Tarannikov, Y.V.: On resilient Boolean functions with maximum possible nonlinearity. In: Roy, B., Okamoto, E. (eds.) INDOCRYPT 2000. LNCS, vol. 1977, pp. 19–30. Springer, Heidelberg (2000)
12. Tarannikov, Y.V.: New constructions of resilient Boolean functions with maximal nonlinearity. In: Matsui, M. (ed.) FSE 2001. LNCS, vol. 2355, pp. 66–77. Springer, Heidelberg (2002)
13. Zheng, Y., Zhang, X.M.: Improved upper bound on the nonlinearity of high order correlation immune functions. In: Stinson, D.R., Tavares, S. (eds.) SAC 2000. LNCS, vol. 2012, pp. 264–274. Springer, Heidelberg (2001)

Unconditionally Secure Approximate Message Authentication

Dongvu Tonien[1], Reihaneh Safavi-Naini[1], Peter Nickolas[2], and Yvo Desmedt[3]

[1] Department of Computer Science, University of Calgary, Canada
dongvu.tonien@gmail.com, rei@ucalgary.ca
[2] School of Mathematics and Statistics, University of Wollongong, Australia
peter@uow.edu.au
[3] Department of Computer Science, University College University of London, UK
desmedt@cs.ucl.ac.uk

Abstract. Approximate message authentication codes (AMAC) arise naturally in biometric and multimedia applications where plaintexts are *fuzzy* and a tagged message (x', t) where t is the calculated tag for a message x that is 'close' to x' should pass the verification test. Fuzziness of plaintexts can be due to a variety of factors including applying acceptable transforms such as compression and decompression to data, or inaccuracy of sensors in reading biometric data.

This paper develops a framework for approximate message authentication systems in *unconditionally security setting*. We give formal definition of AMAC and analyze two attacks, impersonation attack and substitution attack. We derive lower bounds on an opponents deception probability in these attacks under the assumption that all keys are equiprobable. Our bounds generalize known combinatorial bounds in classical authentication theory.

Keywords: approximate authentication, biometric authentication, unconditional security.

1 Introduction

Message authentication codes [3,6] are shared key primitives that are used to ensure authenticity of messages against message tampering and forgery. To communicate a plaintext message x, the sender uses an authentication function f that uses a shared secret key k to calculate a tag t as $t = f_k(x)$ to be appended to the message. The tagged message (x, t) is sent over the insecure channel to the receiver. The receiver verifies the authenticity of a received message (x', t') by using the verification algorithm $V_k(x, t)$ that uses the same key and produces a 1 or 0 if the tagged message is acceptable or not, respectively. In this paper we assume verification algorithm is testing if $f_k(x') = t'$ or not. The main requirement of an authentication system is that without the knowledge of k and even with access to a tagged message (x, t), the adversary has a small chance of constructing another pair (x', t) where $x' \neq x$ and that the pair passes the verification test.

C. Xing et al. (Eds.): IWCC 2009, LNCS 5557, pp. 233–247, 2009.

Approximate message authentication codes (AMAC) arise naturally in biometric and multimedia applications where messages are *fuzzy* and verification must be *noise tolerant*. The fuzzyness property captures the tolerance of human perceptual system to small changes in the plaintext, or inexactness of biomentric data in various readings. We say two plaintexts, x and x', are fuzzy-equal and denote $x =_{fuzzy} x'$ if x' is an acceptable version of x. The fuzzy-equality relation is reflexive and symmetric but not necessarily transitive. The noise tolerance property requires that if t is a tag for x under a key k and $x' =_{fuzzy} x$ then with high probability (x', t) should pass the verification test for the same key k.

The example applications below shows how these requirements arise in practice.

Biometric authentication. A biometric authentication system may work as follows. The server enrols a user by reading their biometric data x, and use a secret key k to calculate a tag $f_k(x) = t$ that is stored in a database together with the user's identity and the key k. To authenticate the user in future, the server scans the users biometric data x' and checks if (x', t) is a valid pair under the key k. If x' is 'close' to x then (x', t) should pass the servers verification test; if x' is not 'close' to x then (x', t) should fail the test.

In this example, the fuzzyness of the biometric data captures the difference between the two readings of a user's biometric data. If x is the initial scanned data and x' is a later version, then $x' =_{fuzzy} x$ and the noise tolerance property guarantees that x' and the stored tag t calculated from the original data x pass the verification test with high probability.

Image authentication. To authenticate an image for Bob, Alice who shares a secret key with Bob, calculates a tag and appends it to the image which is stored in compressed form. Bob decompresses the image resulting in an image that is 'close' to the original one. The noise tolerance of the authentication system guarantees that the verification of the decompressed image will succeed.

There is a large body of research [1,2,4,5,7,8,9] on multimedia and biometric authentication with heuristic or formal analysis in the computational security framework. This paper is the first attempt to provide a framework for approximate authentication systems in *unconditionally secure framework* where there is no assumption on the computational power of the adversary.

Formal definition of AMAC is given in Section 2. We define impersonation and substitution attacks in Section 3. In Section 4 we show how to calculate success probability of the adversary in these two attacks and derive lower bounds on these probabilities in Sections 5 and 6, respectively. These are combinatorial bounds that generalize the combinatorial bounds in classical authentication theory.

2   Approximate Message Authentication

In this section, we formally define distance space, fuzzy space and approximate message authentication codes. We first recall the definition of the classical message authentication codes (MAC).

Definition 1. A message authentication code *(MAC) is a tuple* $(\mathcal{K}, \mathcal{X}, \mathcal{T}, \mathsf{Tag}, \mathsf{VF})$, *where*

- $\mathcal{K}$ *is a finite set of keys,* $\mathcal{X}$ *is a finite set of plaintexts, and* $\mathcal{T}$ *is a finite set of tags;*
- $\mathsf{Tag} : \mathcal{K} \times \mathcal{X} \to \mathcal{T}$ *is a tag generation algorithm;*
- $\mathsf{VF} : \mathcal{K} \times \mathcal{X} \times \mathcal{T} \to \{0, 1\}$ *is a verification algorithm;*

that satisfy the following correctness *property:*

$$\forall k \in \mathcal{K}, \forall x \in \mathcal{X}, \mathsf{Tag}(k, x) = t \longleftrightarrow \mathsf{VF}(k, x, t) = 1.$$

Denote $\mathsf{Tag}_k(x) = \mathsf{Tag}(k, x)$ and $\mathsf{VF}_k(x, t) = \mathsf{VF}(k, x, t)$. For a given key k, these define two functions $\mathsf{Tag}_k : \mathcal{X} \to \mathcal{T}$ and $\mathsf{VF}_k : \mathcal{X} \times \mathcal{T} \longleftrightarrow \{0, 1\}$.

AMACs are message authentication codes with two additional properties: the *fuzzyness* of the plaintext messages and the *noise tolerance* property of the verification algorithm.

Fuzzyness of the plaintext messages. In classical MAC systems, the receiver wants to detect any change to the plaintext sent by the sender. In AMAC however, small changes to the sent message is acceptable. To measure the difference between two plaintext messages, we use a distance function d on the set $\mathcal{X}$ together with a threshold parameter $\delta > 0$ so that if x, $x' \in \mathcal{X}$ and $d(x, x') \leq \delta$, then x and x' are said to be *fuzzy-equal*, and if $d(x, x') > \delta$, then x and x' are said to be *fuzzy-distinct*. For a fixed plaintext message x, $B(x)$ is the closed ball $\overline{B}(x, \delta) = \{x' \in \mathcal{X}, d(x, x') \leq \delta\}$ with center x and radius δ that contains all messages that are fuzzy-equal to x.

We will define a deterministic function $\mathsf{Fuzzy} : \mathcal{X} \to \mathcal{P}(\mathcal{X})$ that maps $x \in \mathcal{X}$ to a subset $B(x)$ of $\mathcal{X}$. All members $x' \in B(x)$ will be defined as fuzzy-equal to x. We require three properties for this fuzzy function.

- $x \in B(x)$ for all $x \in \mathcal{X}$; this property ensures that x is fuzzy-equal to x.
- if for x, $x' \in \mathcal{X}$ and $x \in B(x')$ then $x' \in B(x)$; this is the symmetry property of the fuzzy-equality.
- there must exist x, $x' \in \mathcal{X}$ such that $x \notin B(x')$; this property ensures that the whole space $\mathcal{X}$ will not collapse into one fuzzy class. It is equivalent to the condition that $B(x) \neq \mathcal{X}$ for all $x \in \mathcal{X}$.

The distance-based code discussed above is a special case of the above definition. If d is a distance function and δ denotes a threshold parameter, then by defining the function $\mathsf{Fuzzy} : \mathcal{X} \to \mathcal{P}(\mathcal{X})$ as $\mathsf{Fuzzy}(x) = \overline{B}(x, \delta) = x' \in X, \ d(x, x') \leq \delta$, we have the distance-based code.

Definition 2. *A fuzzy space is a non-empty set* $\mathcal{S}$ *with a function* $\mathsf{Fuzzy} : \mathcal{S} \to \mathcal{P}(\mathcal{S})$ *that maps* $x \mapsto \mathsf{Fuzzy}(x) = B(x) \subset \mathcal{S}$ *with the following properties: for all* $x, x' \in \mathcal{S}$,

1. Reflexive: $x \in B(x)$;
2. Symmetry: $x \in B(x')$ then $x' \in B(x)$; and
3. Non-degenerate: $B(x) \neq S$

We say x is fuzzy-equal to x' and write $x =_{fuzzy} x'$ if and only if $x \in B(x')$. If $x \notin B(x')$ then x is said to be fuzzy-distinct from x' and we denote $x \neq_{fuzzy} x'$.

Definition 3. Let S be a non-empty set. A distance d on S is a function $d : S \times S \to \mathbb{R} \cup \{\infty\}$ that satisfies two properties: (i) $d(x, x') = d(x', x) \geq 0$ and (ii) $d(x, x) = 0$ for all x, $x' \in S$.

Definition 4. A (d, δ)-distance based fuzzy space is a tuple (S, Fuzzy) where d is a distance on S and the Fuzzy function is defined such that $x =_{fuzzy} x'$ if and only if $d(x, x') \leq \delta$.

The noise tolerance property. Given a function $\mathsf{Fuzzy} : \mathcal{X} \to \mathcal{P}(\mathcal{X})$ defined on the plaintext space, the noise tolerance property can be defined in the following ways.

From the receivers point of view: Suppose that the receiver has received a pair (x, t). The receiver would like to accept this pair if t is generated from a plaintext x' that is fuzzy-equal to x. The condition that a tag t is generated from a plaintext $x' =_{fuzzy} x$, is equivalent to the condition that $t \in \mathsf{Tag}_k(B(x))$.

An AMAC is said to be $\mathcal{R}$-perfect noise tolerant if (x, t) passes the verification algorithm for every tag $t \in \mathsf{Tag}_k((B(x))$ (see Fig. 1). In general, we want (x, t) to pass the verification test with a large portion of $\mathsf{Tag}_k(B(x))$. The following probability (see Fig. 2) measures this portion

$$p_{nt}^R(x, k) = Pr[\mathsf{VF}_k(x, t) = 1 : t \in \mathsf{Tag}_k(B(x))].$$

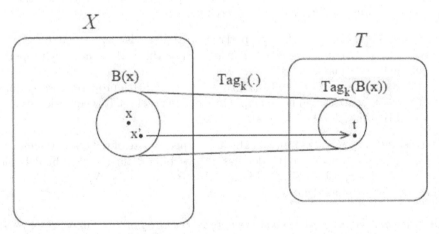

Fig. 1. If the AMAC is $\mathcal{R}$-perfect noise tolerant then, for any fixed k and x, and for any $t \in \mathsf{Tag}_k(B(x))$ (that is, $t = \mathsf{Tag}_k(x')$ for some $x' \in B(x)$), the pair (x, t) passes the verification algorithm under the key k

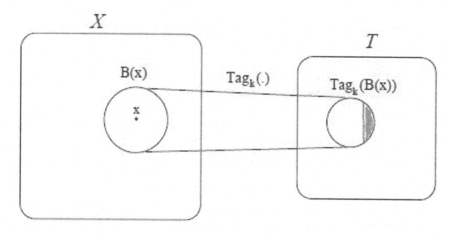

Fig. 2. Let k and x be fixed and consider $\mathsf{Tag}_k(B(x)) \subset \mathcal{T}$. The pair (x,t) fails the verification algorithm for every t is in the shaded area. The probability $p_{nt}^R(x,k)$ measures how large this shaded area compared to $\mathsf{Tag}_k(B(x))$.

The $\mathcal{R}$-noise tolerance probability p_{nt}^R is defined as

$$p_{nt}^R = \min_{x \in \mathcal{X}, k \in \mathcal{K}} p_{nt}^R(x,k).$$

From the senders view point: Suppose that the sender would like to communicate a plaintext x to the receiver. The sender calculates $t = \mathsf{Tag}_k(x)$ and sends the pair (x,t) to the receiver. The sender then wishes that the receiver would accept (x',t) as a valid pair for any x' that is fuzzy-equal to x.

An AMAC is said to have $\mathcal{S}$-perfect noise tolerance if for any $k \in K$, $x \in X$ and $t = \mathsf{Tag}_k(x)$, the pair (x',t) passes the verification algorithm for every $x' \in B(x)$ (see Fig. 3). In general, we want (x',t) to pass the verification algorithm with a large portion of $B(x)$. The following probability (see Fig. 4) measures this portion

$$p_{nt}^S(x,k) = Pr[\mathsf{VF}_k(x', \mathsf{Tag}_k(x)) = 1 : x' \in B(x)].$$

The $\mathcal{S}$-noise tolerance probability p_{nt}^S is defined as

$$p_{nt}^S = \min_{x \in \mathcal{X}, k \in \mathcal{K}} p_{nt}^S(x,k)$$

In the following lemma, we show that $\mathcal{R}$-perfect noise tolerance and $\mathcal{S}$-perfect noise tolerance are equivalent properties. A proof of this lemma is given in the appendix.

Lemma 1. *An AMAC satisfies the $\mathcal{R}$-perfect noise tolerance property if and only if it satisfies the $\mathcal{S}$-perfect noise tolerance property.*

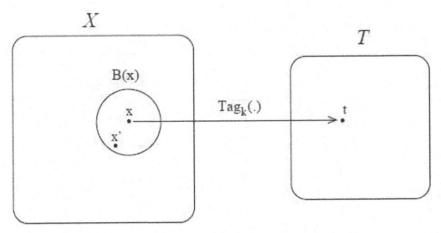

Fig. 3. If the AMAC is $\mathcal{S}$-*perfect noise tolerant* then, for any fixed k and x, and for any $x' \in B(x)$, the pair (x', t) passes the verification algorithm under the key k where $t = \mathsf{Tag}_k(x)$

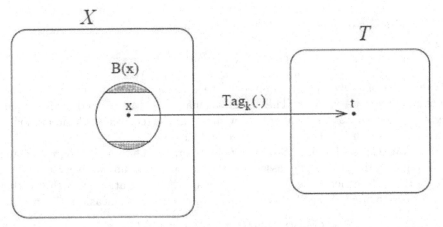

Fig. 4. Let k and x be fixed and $t = \mathsf{Tag}_k(x)$. Consider $B(x) \subset \mathcal{X}$. The pair (x', t) fails the verification algorithm for every x' is in the shaded area. The probability $p^S_{nt}(x, k)$ measures how large this shaded area compared to $B(x)$.

It is possible to show that the two probabilities p^R_{nt} and p^S_{nt} are equal under some strict conditions. We will define that an AMAC is ϵ-*noise tolerant* if both $p^R_{nt} \geq 1 - \epsilon$ and $p^S_{nt} \geq 1 - \epsilon$.

Definition 5. An approximate message authentication code *(AMAC) with* ϵ*-noise tolerance is a tuple* $(\mathcal{K}, \mathcal{X}, \mathcal{T}, \mathsf{Tag}, \mathsf{VF}, \mathsf{Fuzzy})$*, where*

1. $(\mathcal{K}, \mathcal{X}, \mathcal{T}, \mathsf{Tag}, \mathsf{VF})$ *is a MAC;*
2. $(\mathcal{X}, \mathsf{Fuzzy})$ *is a fuzzy space,*

such that if we define for each $x \in \mathcal{X}$ and $k \in \mathcal{K}$,

$$p_{nt}^{R}(x,k) = Pr[\mathsf{VF}_k(x,t) = 1 : t \in \mathsf{Tag}_k(B(x))],$$
$$p_{nt}^{S}(x,k) = Pr[\mathsf{VF}_k(x',\mathsf{Tag}_k(x)) = 1 : x' \in B(x)],$$

then

$$p_{nt}^{R} = \min_{x \in \mathcal{X}, k \in \mathcal{K}} p_{nt}^{R}(x,k) \geq 1 - \epsilon, and$$
$$p_{nt}^{S} = \min_{x \in \mathcal{X}, k \in \mathcal{K}} p_{nt}^{S}(x,k) \geq 1 - \epsilon$$

A *distance-based message authentication code* is a special AMAC where the Fuzzy function is defined based on a distance function on $\mathcal{X}$.

Definition 6. A distance-based message authentication code *(DMAC) with ϵ-noise tolerance is a tuple* $(\mathcal{K}, \mathcal{X}, d, \delta, \mathcal{T}, \mathsf{Tag}, \mathsf{VF})$*, such that*

1. $(\mathcal{X}, \mathsf{Fuzzy})$ *is a* (d, δ)*-distance based fuzzy space, i.e.*
   $B(x) = \mathsf{Fuzzy}(x) = \overline{B}(x, \delta) = \{x' \in \mathcal{X}, \ d(x, x') \leq \delta\};$
2. $(\mathcal{K}, \mathcal{X}, \mathcal{T}, \mathsf{Tag}, \mathsf{VF}, \mathsf{Fuzzy})$ *is an AMAC with ϵ -noise tolerance.*

3   Adversary Model

In this section we consider impersonation attack and substitution attack on AMAC. We first look at two different notions: *authenticated messages* and *valid messages*. For a key k, authenticated messages are message and tag pairs that are resulted from the application of the tag generation algorithm. Valid messages however are message and tag pairs that pass the verification test.

1. For a key $k \in \mathcal{K}$, the set of *authenticated messages* under the key k is the set of all messages $(x, t) \in \mathcal{X} \times \mathcal{T}$ such that $t = \mathsf{Tag}_k(x)$, and denoted by $\mathsf{A}(k)$. For a message (x, t), the set of keys for which (x, t) is an authenticated message is denoted by $\mathsf{K_A}(x, t)$.
2. For a key $k \in \mathcal{K}$, the set of *valid messages* under the key k is the set of all messages $(x, t) \in \mathcal{X} \times \mathcal{T}$ such that $\mathsf{VF}_k(x, t) = 1$, and denoted by $\mathsf{V}(k)$. For a message (x, t), the set of keys for which (x, t) is a valid message is denoted by $\mathsf{K_V}(x, t)$.

In AMAC, $\mathsf{A}(k) \subset \mathsf{V}(k)$ and $\mathsf{K_A}(x, t) \subset \mathsf{K_V}(x, t)$.

Impersonation attack: The adversary creates and sends a spoofed message (x, t) that he hopes to be accepted by the receiver. The adversary has not observed any authenicated messages sent over the communication channel.

Substitution attack: The adversary has access to a single *authenticated* message (x^*, t^*), and wants to spoof a message (x, t). The adversary is successful if (x, t) is a *valid* message and $x \neq_{fuzzy} x^*$.

We consider two ways that the adversary obtains the authenticated message.

- *Active adversary model.* The adversary has black-box oracle access to the algorithm $\mathsf{Tag}_k(.)$. The adversary can query a plaintext message $(x^* \in \mathcal{X}$ to the oracle and receive the response that is the tag $(tag^* \in \mathsf{Tag}_k(x*))$
- *Passive adversary model.* In this model, the adversary eavesdrop the communication channel and obtains an authenticated message (x^*, t^*) that is sent by the sender to the receiver. The probability distribution of the message (x^*, t^*) is independent of the adversary and depends on the distribution of the plaintext messages and the distribution of the keys.

4   Calculating the Deception Probabilities

Impersonation attack
In the impersonation attack, if the adversary uses (x, t) as the spoofing message then the probability that the adversary wins is

$$\mathsf{payoff}_0(x, t) = Pr[(x, t)valid] = \sum_{k \in K_V(x,t)} p(k).$$

Suppose for each (x, t), the adversary chooses (x, t) as the spoofing message with the probability $\tau(x, t)$. Then the overall probability that the adversary wins the impersonation attack is

$$P_0(\tau) = \sum_{(x,t) \in \mathcal{X} \times \mathcal{T}} \tau(x, t)\mathsf{payoff}_0(x, t).$$

The *deception probability* of order 0 is the highest success probability of the adversary in the impersonation attack. It is equal to

$$P_0 = \max_{\tau} P_0(\tau),$$

where the maximum is taken over all probabilistic distribution τ over $\mathcal{X} \times \mathcal{T}$. Therefore,

$$P_0 = \max_{\tau} \sum_{(x,t) \in \mathcal{X} \times \mathcal{T}} \tau(x, t)\mathsf{payoff}_0(x, t)$$

$$= \max_{(x,t) \in \mathcal{X} \times \mathcal{T}} \mathsf{payoff}_0(x, t).$$

Substitution attack, passive adversary model
Suppose the adversary has obtained an authenticated message (x^*, t^*) that was sent by the sender to the receiver and the adversary chooses (x, t) as the spoofing message. Then the adversary wins if and only if the following two conditions are satisfied

1. $x \neq_{fuzzy} x^*$;
2. (x, t) is a valid message under the current secret key k shared by the sender and the receiver.

Define $b(x; x^*) = 1$ if $x \neq_{fuzzy} x^*$ and $b(x; x^*) = 0$ if $x =_{fuzzy} x^*$. If (x^*, t^*) is observed, then with the spoofing message (x, t), the probability for spoofing success is

$$\text{payoff}_1((x, t); (x^*, t^*))$$
$$= b(x; x^*) Pr[(x, t) valid | (x^*, t^*) \text{ authenticated}].$$

Let $\text{payoff}_1(x^*, t^*)$ denote the maximum of the success probability for the adversary when (x^*, t^*) is observed. Then using an argument similar to the impersonation attack, we have

$$\text{payoff}_1(x^*, t^*) = \max_{(x,t)} \text{payoff}_1((x, t); (x^*, t^*)).$$

The *deception probability* of order 1 for *passive* adversary model, P_1^{passive}, i.e. the highest success probability of the adversary in the substitution attack, is calculated as

$$P_1^{\text{passive}} = \sum_{x^*, t^*} Pr[(x^*, t^*) \text{ observed}] \text{payoff}_1(x^*, t^*)$$
$$= \sum_{x^*} \left(p(x^*) \sum_{t^*} Pr[(x^*, t^*) \text{ authenticated}] \text{payoff}_1(x^*, t^*) \right).$$

Substitution attack, active adversary model. Similar to the previous case, the maximum success probability of the adversary when the adversary queries x^* and receives t^*, is calculated as

$$\text{payoff}_1(x^*, t^*) = \max_{(x,t)} \text{payoff}_1((x, t); (x^*, t^*)).$$

Let $\text{payoff}_1(x^*)$ denote the maximum success probability of the adversary with the query x^*. Then

$$\text{payoff}_1(x^*) = \sum_{x^*, t^*} Pr[(x^*, t^*) \text{ authenticated}] \text{payoff}_1(x^*, t^*)$$

The *success probability* of order 1 for *active* adversary, P_1^{active}, is calculated as

$$P_1^{\text{active}} = \max_{x^*} \text{payoff}_1(x^*).$$

5 Bounds on Deception Probability in Impersonation Attack

In this section, we derive bounds on P_0, the success probability of the adversary in the impersonation attack.

Definition 7. For each plaintext message $x \in \mathcal{X}$ and a key $k \in \mathcal{K}$, consider the set

$$\mathsf{Tag}_k(B(x)) = \{\mathsf{Tag}_k(x') : x' \in B(x)\}.$$

Define the *tag-ball* associated with x as follows·

$$\mathsf{Tag}_B(x) = \min_k |\mathsf{Tag}_k(B(x))|.$$

We can view the tag-ball $\mathsf{Tag}_B(x)$ as a measure of the number of tags associated with a plaintext message x. In traditional MAC, $B(x)$ is a singleton set and thus, $\mathsf{Tag}_B(x) = 1$ for any $x \in \mathcal{X}$.

Lemma 2. *If the keys are equiprobable then*

$$P_0 \geq \max_x \sum_{k \in \mathcal{K}} |V_k(x)|/(|\mathcal{K}||\mathcal{T}|),$$

where $V_k(x) = \{t \in \mathcal{T} : (x, t)$ *is valid under* $k\}$.

A proof of Lemma 2 is given in the appendix. Lemma 2 is useful when we have a concrete description of the verification algorithm. However, without any specific property of the verification algorithm, it is difficult to examine how large the set $V_k(x)$ is and so the bound in Lemma 2 is not very useful in this case. In Lemma 3, using the noise tolerance probability, we derive a bound on the size of the set $V_k(x)$ which will be used to give a new bound for P_0 in Theorem 1.

Lemma 3. *For any* $k \in \mathcal{K}$ *and* $x \in \mathcal{X}$, *it must hold that*

$$|V_k(x)| \geq p_{nt}^R \mathsf{Tag}_k(B(x)).$$

Proof: We have

$$\begin{aligned}
p_{nt}^R(x, k) &= Pr[\mathsf{VF}_k(x, t) = 1 : t \in \mathsf{Tag}_k(B(x))] \\
&= Pr[t \in V_k(x) : t \in \mathsf{Tag}_k(B(x))] \\
&= \frac{|\mathsf{Tag}_k(B(x)) \cap V_k(x)|}{|\mathsf{Tag}_k(B(x))|}
\end{aligned}$$

It follows that

$$\begin{aligned}
|V_k(x)| &\geq |\mathsf{Tag}_k(B(x)) \cap V_k(x)| \\
&= p_{nt}^R(x, k)|\mathsf{Tag}_k(B(x))| \\
&\geq p_{nt}^R \mathsf{Tag}(B(x))
\end{aligned}$$

Theorem 1. *If the keys are equiprobable then*

$$P_0 \geq (1 - \epsilon) \max_x \mathsf{Tag}(B(x)) \frac{1}{|\mathcal{T}|}.$$

Proof: By Lemma 3, we have

$$|V_k(x)| \geq p_{nt}^R \mathsf{Tag}(B(x)).$$

so

$$\sum_{k \in \mathcal{K}} |V_k(x)| \geq p_{nt}^R \mathsf{Tag}(B(x))|\mathcal{K}|.$$

Therefore, by Lemma 2, we have

$$P_0 \geq \max_x \sum_{k \in \mathcal{K}} |V_k(x)|/(|\mathcal{K}||\mathcal{T}|).$$
$$\geq p_{nt}^R \max_x \mathsf{Tag}(B(x))/|\mathcal{T}|.$$
$$\geq (1 - \epsilon) \max_x \mathsf{Tag}(B(x))1/|\mathcal{T}|.$$

6   Bounds on the Success Probability of Substitution Attack

In this section, we derive bounds on the success probabilities, P_1^{passive} in the passive adversary model and P_1^{active} in the active adversary model, in the substitution attack.

Theorem 2. *If the keys are equiprobable then*

$$P_1^{\mathsf{passive}} \geq (1 - \epsilon) \sum_{x^*} p(x^*) \max_{x \notin B(x^*)} \mathsf{Tag}(B(x))/|\mathcal{T}|, and$$
$$P_1^{\mathsf{active}} \geq (1 - \epsilon) \max_{x \in \mathcal{X}} \mathsf{Tag}(B(x))/|\mathcal{T}|$$

Proof: Suppose that the adversary has observed (x^*, t^*). Then the key k must be in the set $K_A(x^*, t^*)$. To maximize the success chance, for a spoofing message (x, t), t should be chosen so that t is in as many sets $V_k(x^*)$ as possible. If t is in $n(x, t)$ sets $V_k(x)$ then

$$\mathsf{payoff}_1((x, t); (x^*, t^*)) = b(x; x^*) Pr[(x, t) \text{ valid}|(x^*, t^*) \text{ authenticated}]$$
$$= b(x; x^*) n(x, t)/|K_A(x^*, t^*)|.$$

Let $n(x) = \max_t n(x, t)$ then

$$\max_t \mathsf{payoff}_1((x, t); (x^*, t^*)) = b(x; x^*) n(x)/|K_A(x^*, t^*)|.$$

Using similar arguments as in the proof of Lemma 2, we have

$$n(x)|\mathcal{T}| \geq \sum_{k \in K_A(x^*, t^*)} |V_k(x)|.$$

Thus,

$$\max_t \mathsf{payoff}_1((x,t);(x^*,t^*)) \geq b(x;x^*) \sum_{k \in K_A(x^*,t^*)} |V_k(x)|/(\mathcal{T}/|K_A(x^*,t^*)|).$$

From Lemma 3, $|V_k(x)| \geq p_{nt}^R \mathsf{Tag}(B(x))$, thus,

$$\max_t \mathsf{payoff}_1((x,t);(x^*,t^*)) \geq b(x;x^*) p_{nt}^R \mathsf{Tag}(B(x))/|\mathcal{T}|.$$

So

$$\mathsf{payoff}_1((x^*,t^*)) = \max_{x,t} \mathsf{payoff}_1((x,t);(x^*,t^*))$$
$$\geq p_{nt}^R \max_{x \notin B(x^*)} \mathsf{Tag}(B(x))/|\mathcal{T}|.$$

Therefore,

$$\sum_{t^*} Pr[(x^*,t^*) \text{ authenticated}]\mathsf{payoff}_1((x,t);(x^*,t^*))$$
$$\geq p_{nt}^R \max_{x \notin B(x^*)} \mathsf{Tag}(B(x)) \sum_{t^*} Pr[(x^*,t^*) \text{ authenticated}]/|\mathcal{T}|$$
$$= p_{nt}^R \max_{x \notin B(x^*)} \mathsf{Tag}(B(x))/|\mathcal{T}| \tag{1}$$

and

$$P_1^{\mathsf{passive}} = \sum_{x^*} \left(p(x^*) \sum_{t^*} Pr[(x^*,t^*) \text{ authenticated}]\mathsf{payoff}_1((x,t);(x^*,t^*)) \right)$$
$$\geq p_{nt}^R \sum_{x^*} p(x^*) \max_{x \notin B(x^*)} \mathsf{Tag}(B(x))/|\mathcal{T}|$$

Also from (1),

$$\mathsf{payoff}_1((x^*)) \geq p_{nt}^R \max_{x \notin B(x^*)} \mathsf{Tag}(B(x))/|\mathcal{T}|$$

Therefore,

$$P_1^{\mathsf{active}} = \max_{x^*} \mathsf{payoff}_1((x^*)) \geq \max_{x^* \in \mathcal{X}} p_{nt}^R \max_{x \notin B(x^*)} \mathsf{Tag}(B(x))/|\mathcal{T}|$$

Since there exist $x, x' \in X$ such that $x \notin B(x')$, we have

$$\max_{x^* \in \mathcal{X}} \max_{x \notin B(x^*)} \mathsf{Tag}(B(x)) = \max_{x \in \mathcal{X}} \mathsf{Tag}(B(x))$$

Therefore,

$$P_1^{\mathsf{active}} \geq p_{nt}^R \max_{x \in \mathcal{X}} \mathsf{Tag}(B(x))/|\mathcal{T}|$$

7   A Construction

In the following we give a basic construction to show feasibility of the definitions above.

The intuition is that *if two strings are 'close', then any two corresponding substrings will be 'close' too*. The tag algorithm is effectively a contraction of the string to one of its substrings. However to provide protection against tampering the choice of the substring is protected by that key information.

Let $\mathcal{N} = \{1 \cdots N\}$. The set of messages is the set $\mathcal{X}$ of binary strings of length N, and the set of tags is the set $\mathcal{T}$ of binary strings of length n. The distance $d(x, x')$, x, $x' \in \mathcal{X}$ is the Hamming distance $d_H(x, x')$ between x and x'. The message space is (d, δ)-distance based fuzzy space where $(\mathcal{X}, \mathsf{Fuzzy})$ is defined as $B(x) = \mathsf{Fuzzy}(x) = \overline{B}(x, \delta) = \{x' \in \mathcal{X},\ d_H(x, x') \le \delta\}$;

The tag space $\mathcal{T}$ is also a metric space with distance between two tags $t, t' \in \mathcal{T}$ defined as the Hamming distance between them.

Let $\mathcal{K}$ denote the set of keys. Each $k \in \mathcal{K}$ determines,

1. a permutation π_k on $[N]$ (the inverse permutation is denoted by π_k^{-1});
2. a substring selection function f_k,

$$f_k : \{0,1\}^N \to \{0,1\}^n,\ f_k(x_1 \cdots x_N)) = (x_{i_1} \cdots x_{i_n})$$

The Tag function is defined as,

$$\mathsf{Tag}(k, x) = t = f_k(\pi_k(x)) = (x_{\pi_k^{-1}(i_1)}, \cdots x_{\pi_k^{-1}(i_n)})$$

The tag function selects a substring of a permuted version of x.

The verification function VF accepts a pair (x', t') if there exists $x \in B(x')$ such that $\mathsf{Tag}(k, x) = t$ and $d_H(\mathsf{Tag}(k, x'), t) \le \delta'$.

It is easy to show that $(\mathcal{K}, \mathcal{X}, \mathcal{T}, \mathsf{Tag}, \mathsf{VF}, \mathsf{Fuzzy})$ is an AMAC with ϵ-noise tolerance.

8   Conclusions

We gave a formal model for approximate authentication systems in unconditionally security model. We analyzed impersonation attack and substitution attacks in these systems and derived lower bounds on the success probabilities of codes that provide ϵ-noise tolerance, under the assumption that all keys are equiprobable. Constructing efficient approximate authentication systems that meet these lower bounds is our future work.

References

1. Crescenzo, G.D., Graveman, R.F., Ge, R., Arce, G.R.: Approximate message authentication and biometric entity authentication. In: S. Patrick, A., Yung, M. (eds.) FC 2005. LNCS, vol. 3570, pp. 240–254. Springer, Heidelberg (2005)

2. Ge, R., Arce, G.R., Crescenzo, G.D.: Approximate message authentication codes for N-ary alphabets. IEEE Transactions on Information Forensics and Security 1(1) (2006)
3. Gilbert, E.N., MacWilliams, F.J., Sloane, N.J.A.: Codes which detect deception. Bell System Technical Journal 53, 405–424 (1974)
4. Lin, C.Y., Chang, S.F.: A Robust Image Authentication Method Surviving JPEG Lossy Compression. In: SPIE Storage and Retrieval f Image/Video Database, EI 1998, San Jose (January 1998); also in IEEE Trans. on Circuits and Systems for Video Technology (2000)
5. Lin, C.Y., Chang, S.F.: SARI: Self-Authentication-and-Recovery Image Watermarking System. ACM Multimedia (2001)
6. Simmons, G.J.: Authentication theory / coding theory. In: Blakely, G.R., Chaum, D. (eds.) CRYPTO 1984. LNCS, vol. 196, pp. 411–432. Springer, Heidelberg (1985)
7. Swaminathan, A., Mao, Y., Wu, M.: Robust and secure image hashing. IEEE Transactions on Information Forensics and Security 1(2), 215–230 (2006)
8. Wu, C.W.: On the design of content-based multimedia authentication systems. IEEE Transactions On Multimedia 4(3), 385–393 (2002)
9. Xie, L., Arce, G.R., Graveman, R.F.: Approximate image message authentication codes. IEEE Transactions on Multimedia 3(2), 242–252 (2001)

Appendix

Proof of Lemma 1: Suppose that an AMAC satisfies the $\mathcal{R}$-perfect noise tolerance property. Now take arbitrary $k \in \mathcal{K}, x \in \mathcal{X}$ and set $t = \mathsf{Tag}_k(x)$. We will show that (x', t) is valid under the key k for any $x' \in B(x)$, thus, the AMAC satisfies the $\mathcal{S}$-perfect noise tolerance property. Indeed, since $x \in B(x)$ and $t = \mathsf{Tag}_k(x)$, we have $t \in \mathsf{Tag}_k(B(x'))$, and from the $\mathcal{R}$-perfect noise tolerance property, it follows that (x', t) is valid under the key k.

Conversely, suppose that an AMAC satisfies the $\mathcal{S}$-perfect noise tolerance property. Now take arbitrary $k \in \mathcal{K}$ and $x \in \mathcal{X}$. We will show that (x, t) is valid under the key k for any $t \in \mathsf{Tag}_k(B(x))$, thus, the AMAC satisfies the $\mathcal{R}$-perfect noise tolerance property. Indeed, since $t \in \mathsf{Tag}_k(B(x))$, $t = \mathsf{Tag}_k(x')$ for some $x' \in B(x)$. Since $x' \in B(x)$ and $t = \mathsf{Tag}_k(x')$, from the S-perfect noise tolerance property, it follows that (x, t) is valid under the key k.

Proof of Lemma 2: Let $n(x, t) = |k \in \mathcal{K} : t \in V_k(x)|$, then

$$\mathsf{payoff}_0(x, t) = Pr[(x, t) \text{ valid}] = n(x, t)/|\mathcal{K}|$$

Let $n(x) = \max_t n(x, t)$. Then for any $t \in \mathcal{T}$, t belongs to at most $n(x)$ number of sets $V_k(x)$. The union of all $V_k(x)$ where $k \in \mathcal{K}$ covers the set $\mathcal{T}$ at most $n(x)$ times, thus

$$\sum_{k \in \mathcal{K}} |V_k(x)| \leq n(x)|\mathcal{T}|.$$

On the other hand, we have

$$\max_t \mathsf{payoff}_0(x, t) = \max_t Pr[(x, t) \text{ valid}] = n(x, t)/|\mathcal{K}|$$

Therefore,

$$\max_{t} \mathsf{payoff}_0(x, t) \geq \sum_{k \in \mathcal{K}} |V_k(x)| / (|\mathcal{K}||\mathcal{T}|).$$

It follows that

$$P_0 = \max_{t} \mathsf{payoff}_0(x, t) \geq \sum_{k \in \mathcal{K}} |V_k(x)| / (|\mathcal{K}||\mathcal{T}|).$$

Multiplexing Realizations of the Decimation-Hadamard Transform of Two-Level Autocorrelation Sequences*

Nam Yul Yu[1] and Guang Gong[2]

[1] Lakehead University, Thunder Bay, Ontario, Canada
nam.yu@lakeheadu.ca
[2] University of Waterloo, Waterloo, Ontario, Canada
ggong@calliope.uwaterloo.ca

Abstract. In an effort to search for new binary two-level autocorrelation sequences of period $2^n - 1$, a new method of iterative decimation-Hadamard transform is proposed. It is based on the Hadamard transform of a shift and decimation of a binary two-level autocorrelation sequence, and its multiplexing. The experimental results show that several known binary two-level autocorrelation sequences can be obtained from our method.

1 Introduction

The *Iterative Decimation-Hadamard Transform (DHT)* [1] has been developed by Gong and Golomb to search for new binary and non-binary two-level autocorrelation sequences. The method exploits the Hadamard equivalence established by Dillon and Dobbertin [2], and applies the decimation and the Hadamard transform iteratively to discover new two-level autocorrelation sequences. The concept of the Hadamard equivalence has been extended in [3] to sequences of period $p = 3 \pmod 4$, e.g., Legendre and Hall sequences, called *extended Hadamard equivalence*.

In [1], Gong and Golomb described how to obtain two-level autocorrelation sequences of period $2^n - 1$ by applying the iterative decimation Hadamard transform to the known sequences. Precisely, with the second order DHT starting from a single binary m-sequence, they showed that all known classes of binary two-level autocorrelation sequences of period $2^n - 1$ with no sub-field structures could be obtained for odd n.

In the similar way, this paper proposes a new method in an effort to find new two-level autocorrelation sequences of period $2^n - 1$ for $n = 2m$. The method applies the Hadamard transform to a shift and decimation of a binary two-level autocorrelation sequence, and then multiplexes the outcomes of each shift. Throughout experiments for various n, some interesting observations are remarked about our new method. Also, starting from binary m-sequences, we can

---

* This work was supported by NSERC Grant RGPIN 227700-00.

C. Xing et al. (Eds.): IWCC 2009, LNCS 5557, pp. 248–258, 2009.
© Springer-Verlag Berlin Heidelberg 2009

obtain m-sequences, B_k sequences [2] (or Kasami-power function sequences) and Welch-Gong sequences [4]. As a result, we show that the iterative decimation-Hadamard transform can be used for constructing two-level autocorrelation sequences of period $2^n - 1$ for even n by the multiplexing realizations.

2   Preliminaries

In this section, we present some preliminary reviews on concepts and notations about sequences that we will frequently use in this paper. The following notation will be used throughout this paper.

- n is a positive integer and $q = 2^n$.
- $\mathbb{F}_Q = GF(Q)$ is a finite field with Q elements and $\mathbb{F}_Q^*$ is a multiplicative group of $\mathbb{F}_Q$.
- $\mathbb{Z}_m$ is a ring of integers modulo m and $\mathbb{Z}_m^* = \{r \in \mathbb{Z}_m | r \neq 0\}$.
- Let $m|n$. A trace function from $\mathbb{F}_{2^n}$ to $\mathbb{F}_{2^m}$ is denoted by $Tr_m^n(x)$, i.e.,

$$Tr_m^n(x) = x + x^{2^m} + \cdots + x^{2^{m(\frac{n}{m}-1)}}, \quad x \in \mathbb{F}_{2^n},$$

or simply as $Tr(x)$ if $m = 1$ and the context is clear.

2.1   Correspondence between Periodic Sequences and Functions from $\mathbb{F}_{2^n}$ to $\mathbb{F}_2$.

Let $\mathcal{S}$ be a set of all binary sequences of period $t|(2^n - 1)$ and $\mathcal{F}$ be a set of all functions from $\mathbb{F}_{2^n}$ to $\mathbb{F}_2$. Then, a function $f(x) \in \mathcal{F}$ can be represented as

$$f(x) = \sum_{i=1}^{r} Tr_1^{n_i}(A_i x^{t_i}), \quad A_i \in \mathbb{F}(2^{n_i}) \tag{1}$$

where t_i is a coset leader of a cyclotomic coset modulo $2^{n_i} - 1$, and $n_i|n$ is the size of the cyclotomic coset containing t_i. For any sequence $\mathbf{a} = \{a_i\} \in \mathcal{S}$, there exists $f(x) \in \mathcal{F}$ such that

$$a_i = f(\alpha^i), \quad i = 0, 1, \cdots,$$

where α is a primitive element of $\mathbb{F}_{2^n}$. Then, $f(x)$ is called a *trace representation* of $\mathbf{a}$. ($\mathbf{a}$ is also referred to as an r-term sequence.) If $f(x)$ is a function from $\mathbb{F}_{2^n}$ to $\mathbb{F}_2$, by evaluating $f(\alpha^i)$, we get a binary sequence of period dividing $2^n - 1$. We use the notation:

$$\mathbf{a} \leftrightarrow f(x)$$

to represent the one-to-one correspondence between $\mathcal{F}$ and $\mathcal{S}$ through the trace representation (1).

If $r = 1$, .i.e.,

$$a_i = Tr_1^n(\beta \alpha^i), \quad i = 0, 1, \cdots, \beta \in \mathbb{F}_{2^n}^*,$$

then $\mathbf{a}$ is a binary m-sequence of period $2^n - 1$ of degree n.

2.2   Decimation of Periodic Sequences

Let $\underline{a}$ be a binary sequence of period $t|(2^n - 1)$ and let $f(x)$ be the trace representation of $\underline{a}$. Let $0 < s < t$. Then a sequence $\underline{b} = \{b_i\}$ whose elements are given by

$$b_i = a_{si}, \quad i = 0, 1, \cdots,$$

is said to be an *s-decimation* of $\underline{a}$, denoted by $\underline{a}^{(s)}$. The trace representation of $\underline{a}^{(s)}$ is $f(x^s)$. That is, we have

$$\underline{a} \longleftrightarrow f(x), \quad \underline{a}^{(s)} \longleftrightarrow f(x^s).$$

2.3   Autocorrelation

The autocorrelation of $\underline{a}$ is defined by

$$C_{\underline{a}}(\tau) = \sum_{i=0}^{t-1}(-1)^{a_{i+\tau}+a_i}, \quad 0 \le \tau \le t-1 \tag{2}$$

where τ is a phase shift of the sequence $\underline{a}$ and the indices are computed modulo t, the period of $\underline{a}$. If $\underline{a}$ has a period $2^n - 1$ and

$$C_{\underline{a}}(\tau) = \begin{cases} -1, & \text{if } \tau \not\equiv 0 \bmod 2^n - 1 \\ 2^n - 1, & \text{if } \tau \equiv 0 \bmod 2^n - 1, \end{cases}$$

then we say that the sequence $\underline{a}$ has an *(ideal) 2-level autocorrelation function*.

2.4   Hadamard Transform and the Inverse Transform

Let $f(x)$ be a polynomial function from $\mathbb{F}_{2^n}$ to $\mathbb{F}_2$. With a trace function $Tr(x)$ from $\mathbb{F}_{2^n}$ to $\mathbb{F}_2$, the Hadamard transform of $f(x)$ is defined by

$$\widehat{f}(\lambda) = \sum_{x \in \mathbb{F}_{2^n}} (-1)^{Tr(\lambda x)+f(x)}, \quad \lambda \in \mathbb{F}_{2^n}.$$

The inverse formula is given by

$$\chi(f(\lambda)) = \frac{1}{2^n} \sum_{x \in \mathbb{F}_{2^n}} (-1)^{Tr(\lambda x)} \widehat{f}(x), \quad \lambda \in \mathbb{F}_{2^n}.$$

2.5   Orthogonal Function

Let $f(x)$ be a function from $\mathbb{F}_{2^n}$ to $\mathbb{F}_2$ with $f(0) = 0$. If

$$C_f(\lambda) = \sum_{x \in \mathbb{F}_{2^n}} (-1)^{f(\lambda x)+f(x)} = \begin{cases} 0, & \text{if } \lambda \ne 1 \\ 2^n, & \text{if } \lambda = 1 \end{cases}$$

for $\lambda \in \mathbb{F}_{2^n}$, then we say that $f(x)$ is *orthogonal over* $\mathbb{F}$. An orthogonal function is a trace representation of a two-level autocorrelation sequence [1]. If $f(x)$ is a trace representation of $\underline{a}$ and the autocorrelation function of $\underline{a}$ defined in (2) is $C_{\underline{a}}$, then

$$C_{\underline{a}}(\tau) = -1 + C_f(\lambda)$$

where $\lambda = \alpha^\tau \in \mathbb{F}_{2^n}^*$.

2.6 Hadamard Equivalence

Let $\widehat{f}(\lambda)$ and $\widehat{g}(\lambda)$ be the Hadamard transforms of polynomial functions $f(x)$ and $g(x)$, respectively. Let $\underline{a}$ and $\underline{b}$ be binary sequences represented by $f(x)$ and $g(x)$, respectively. For a positive integer s with $\gcd(s, 2^n - 1) = 1$, if we have

$$\widehat{f}(\lambda) = \widehat{g}(\lambda^s),$$

then we say that $f(x)$ is *Hadamard equivalent* to $g(x)$ [2]. In sequence aspects, we also say that $\underline{a}$ is Hadamard equivalent to $\underline{b}$.

2.7 Decimation-Hadamard Transform (DHT)

Let $u(x)$ be orthogonal over $\mathbb{F}_2$ and $f(x)$ be a function from $\mathbb{F}_{2^n}$ to $\mathbb{F}_2$. For an integer $v \in \mathbb{Z}_{q-1}^*$, we define

$$\widehat{f}_u(v)(\lambda) = \sum_{x \in \mathbb{F}_{2^n}} (-1)^{u(\lambda x) + f(x^v)}, \quad \lambda \in \mathbb{F}_{2^n}.$$

Then, $\widehat{f}_u(v)(\lambda)$ is called *the first-order decimation-Hadamard transform (DHT) of $f(x)$ with respect to $u(x)$*, the first order DHT for short. With this notation, let $t \in \mathbb{Z}_{q-1}^*$. Then,

$$\widehat{f}_u(v, t)(\lambda) = \sum_{y \in \mathbb{F}_{2^n}} (-1)^{u(y)} \widehat{f}_u(v)(y^t)$$

$$= \sum_{x, y \in \mathbb{F}_{2^n}} (-1)^{u(\lambda y) + u(y^t x) + f(x^v)}$$

is called the *second order decimation-Hadamard transform of $f(x)$ (with respect to $u(x)$), the second order DHT for short*. In DHT, the Hadamard transform is generalized by the use of the orthogonal function $u(x)$ instead of $Tr(x)$.

If $\widehat{f}_u(v, t)(\lambda) \in \{\pm 2^n\}$ for all λ in $\mathbb{F}_{2^n}$, a function $c(x)$ from $\mathbb{F}_{2^n}$ to $\mathbb{F}_2$ determined by

$$(-1)^{c(\lambda)} = \frac{1}{2^n} \widehat{f}_u(v, t)(\lambda), \qquad (3)$$

is called a *realization* of $f(x)$ with respect to $u(x)$, and (v, t) is called a *realizable pair* [1]. In particular, if $c(x) = f(x^s)$ for a positive integer s with $\gcd(s, 2^n - 1) = 1$, then $c(x)$ belongs to the same class $f(x)$, where $c(x)$ is called a *self-realization* and the corresponding realizable pair (v, t) is a *self-realizable pair* [1]. If $t = 1$, then $c(x) = f(x^v)$ and thus $(v, 1)$ is a trivial self-realizable pair.

3 B_k Sequences for Even n

Let n be a positive integer and k an integer of $1 \le k < \lfloor \frac{n}{2} \rfloor$ with $\gcd(k, n) = 1$. For $d = 2^{2k} - 2^k + 1$, consider a set

$$B_k = \{(x + 1)^d + x^d + 1 \mid x \in \mathbb{F}_{2^n}\}.$$

Then, its characteristic sequence given by

$$a_i = \begin{cases} 0, & \text{if } \alpha^i \in B_k \\ 1, & \text{if } \alpha^i \notin B_k \end{cases}$$

has the ideal two-level autocorrelation, where the sequence is called the B_k sequence or *Kasami power function (KPF)* sequence [2]. According to k with $\gcd(k, n) = 1$, there exist $\frac{\phi(n)}{2}$ inequivalent B_k sequences of period $2^n - 1$, where $\phi(\cdot)$ is the Euler-totient function. If $k = 1$, in particular, the B_1 sequence is identical to a binary m-sequence.

Let $b_k(x)$ be a trace representation of B_k sequences of period $2^n - 1$ for $n = 2m$. With a shift and decimated version of $b_k(x)$ denoted by $c_k^\gamma(x) = b_k(\gamma x^{2^k + 1})$, Dillon and Dobbertin proved that it is Hadamard-equivalent to a shift and decimation of a trace function, i.e.,

$$\widehat{c_k^\gamma}(\lambda) = \widehat{S_3^\gamma}(\lambda^{\frac{2^k + 1}{3}}) \tag{4}$$

where $S_3^\gamma(x) = Tr_1^n(\gamma x^3)$ [2]. In Proposition 1, we reinterpret the Hadamard equivalence of (4) from the viewpoint of DHT.

Proposition 1. *Let n be even and k an integer such that $1 \le k < n/2$ and $\gcd(n, k) = 1$. Let α be a primitive element in $\mathbb{F}_{2^n}$ and $\gamma \in \{1, \alpha, \alpha^2\}$. Then, the second order DHT of $f^\gamma(x) = Tr_1^n(\gamma x)$ with respect to $h(x) = Tr_1^n(x)$ by a decimation pair $\left(3, \frac{2^k + 1}{3}\right)$ realizes a shift and decimated version of a B_k sequence represented by $b_k(x)$. In other words, the Hadamard equivalence of (4) is reinterpreted as*

$$\frac{1}{2^n} \widehat{f_h^\gamma}\left(3, \frac{2^k + 1}{3}\right)(\lambda) = (-1)^{b_k(\gamma \lambda^{2^k + 1})}$$

from the viewpoint of the second order DHT.

In Proposition 1, note that the decimation of $f^\gamma(x)$ by $v = 3$ reduces its period to $\frac{2^n - 1}{3}$ from $\gcd(3, 2^n - 1) = 3$ for even n. Proposition 1 reveals that the second order DHT of a trace function can generate a part of a B_k sequence. In the following, Proposition 2 shows that the period of its Hadamard transform is also $\frac{2^n - 1}{3}$.

Proposition 2. *Let $f(x)$ be a trace representation of a binary sequence of period $2^n - 1$ and $\widehat{f_s}(\lambda)$ be the Hadamard transform of $f(x^s)$. Let $\gcd(s, 2^n - 1) = \rho$. Then $\widehat{f_s}(\lambda)$ has a period of $p_s = \frac{2^n - 1}{\rho}$ if $\widehat{f_s}(0)$ is excluded.*

Proof. In the Hadamard transform of $f(x^s)$,

$$\widehat{f_s}(\alpha^{i + p_s}) = \sum_{x \in \mathbb{F}_{2^n}} (-1)^{Tr(\alpha^{i + p_s} x) + f(x^s)}.$$

By $\alpha^{p_s} x = y$, we get $x^s = \alpha^{-s p_s} y^s = y^s$. Thus,

$$\widehat{f_s}(\alpha^{i + p_s}) = \sum_{y \in \mathbb{F}_{2^n}} (-1)^{Tr(\alpha^i y) + f(y^s)} = \widehat{f_s}(\alpha^i).$$

Hence, the period of Hadamard transform is also $p_s = \frac{2^n-1}{\rho}$.   □

From Proposition 2, a period of the Hadamard transform $\widehat{f}_s(\lambda)$ exists for $\lambda \in \mathbb{G}_s = \{\alpha^i | 0 \leq i < p_s\}$ for the reduced period p_s. In B_k sequences, the period of $\widehat{f}_h^\gamma(3)(\lambda^{\frac{2^k+1}{3}})$ is $\frac{2^n-1}{3}$ with $\gcd(2^n-1, \frac{2^k+1}{3}) = 1$. Therefore, applying the second Hadamard transform to $\widehat{f}_h^\gamma(3)(\lambda^{\frac{2^k+1}{3}})$ realizes only a third of the final sequence represented by $b_k(x)$. Finally, we could obtain a binary B_k sequence of period $2^n - 1$ by multiplexing each realization for each $\gamma \in \{1, \alpha, \alpha^2\}$.

4   New Method of Iterative Decimation-Hadamard Transform

With the principle described in Section 3, we can devise the iterative decimation-Hadamard transform for even n. To generalize the realization procedure of B_k sequences in Proposition 1, we extend the first decimation value v to a proper factor of $2^n - 1$ with $\gcd(v, 2^n - 1) = v$, and the second decimation value t to a coset leader of the cyclotomic coset in $\mathbb{Z}_{2^n-1}^*$ consisting of t_c's with $\gcd(\frac{2^n-1}{v}, t_c) = 1$. Then we can consider the following Proposition.

Proposition 3. *Let $f(x), h(x)$ and $g(x)$ be orthogonal functions from $\mathbb{F}_{2^n}$ to $\mathbb{F}_2$. α is a primitive element of $\mathbb{F}_{2^n}$. v and t are chosen such that $\gcd(v, 2^n - 1) = v$ and $\gcd(\frac{2^n-1}{v}, t) = 1$. Then we may consider the existence of the Hadamard-equivalence of $\widehat{f}_h^\gamma(v)(\lambda^t) = \widehat{g}_h^\gamma(d)(\lambda)$ for a particular pair of $f(x)$ and $g(x)$, i.e.,*

$$\sum_{x \in \mathbb{F}_{2^n}} (-1)^{h(\lambda^t x) + f(\gamma x^v)} = \sum_{x \in \mathbb{F}_{2^n}} (-)^{h(\lambda x) + g(\gamma x^d)} \tag{5}$$

where $f^\gamma(x) = f(\gamma x)$ and $g^\gamma(x) = g(\gamma x)$ for $\gamma = \alpha^r$ with $0 \leq r \leq v - 1$, and d is a decimation factor of $g(x)$ with $\gcd(d, 2^n - 1) = v$.

Note that Proposition 3 discusses the potential existence of (5) for particular $f(x)$, $g(x)$, and $h(x)$. From Proposition 3, the second order DHT for even n starts from $f^\gamma(x) = f(\gamma x)$ for each $\gamma \in \{1, \alpha, \alpha^2, \cdots, \alpha^{v-1}\}$. Then the second order DHT of $f^\gamma(x)$ with respect to $h(x)$ by a decimation pair (v, t) may realize $g(\gamma x^d)$, i.e.,

$$\frac{1}{2^n} \widehat{f}_h^\gamma(v, t)(\lambda) = (-1)^{g(\gamma \lambda^d)}.$$

For each γ, if $\widehat{f}_h^\gamma(v, t)(\lambda) \in \{\pm 2^n\}$ for all λ's in $\mathbb{F}_{2^n}$, then $g(x)$ can be finally obtained from multiplexing $g(\gamma x^d)$ for each γ. The construction procedure of two-level autocorrelation sequences is summarized as follows.

Procedure for Multiplexing Realizations of 2nd Order DHT

i) v is chosen as a proper factor of $q = 2^n - 1$ such that $\gcd(q, v) = v$.

ii) t is chosen as a coset leader of the cyclotomic coset consisting of t_c's in $\mathbb{Z}_q^*$ such that $\gcd(q/v, t_c) = 1$.

iii) $f(x)$ is given as an orthogonal function and $f^\gamma(x) = f(\gamma x)$ for each $\gamma = \alpha^r$, $0 \le r \le v - 1$.

iv) $h(x)$ is an orthogonal function.

v) After the second order DHT, a valid realization provides to a sequence a_i^r which has a reduced period of q/v, i.e.,

$$a_i^r = \begin{cases} 0, & \text{if } \hat{f}_h^\gamma(v, t)(\alpha^i) = +2^n \\ 1, & \text{if } \hat{f}_h^\gamma(v, t)(\alpha^i) = -2^n. \end{cases}$$

vi) Multiplex each a_i^r according to r, i.e.,

$$b_{di+r} = a_i^r, \quad 0 \le r \le v - 1$$

where d is chosen as an integer multiple of v such that $\gcd(q, d) = v$.

From the above procedure, a binary two-level autocorrelation sequence b_i can be realized for a realizable pair of (v, t). We present the multiplexing realization method in binary case. However, the procedure can be straightforwardly extended to sequences over $\mathbb{F}_p$ where p is any arbitrary prime number.

We conclude this section by discussing the multiplexing realization for a special v. If $v = 2^m - 1$ with $n = 2m$, starting from $f(x)$, the procedure realizes a binary sequence represented by $f(x)$ for all possible values of t with $\gcd(\frac{2^n-1}{v}, t) = 1$ and $h(x) = Tr(x)$. Lemma 1 explains this.

Lemma 1. *Let $f(x)$ be a trace representation of a binary sequence of period $2^n - 1$ defined by (1), where assume $A_i = 1$ for all i's. Let $h(x) = Tr(x)$ and $v = 2^m - 1$ for $n = 2m$. Then the realization of $f(x)$ from our procedure can be obtained for all possible values of t such that $\gcd(\frac{2^n-1}{v}, t) = 1$. Moreover, the realization is always a binary sequence of period $2^n - 1$ corresponding to $f(x)$, or self-realization.*

Before proving Lemma 1, Lemma 2 is discussed.

Lemma 2. *For $f(x)$ in Lemma 1, let $s = r(2^m - 1)$ where $\gcd(r, 2^n - 1) = 1$ and $\gamma \in \mathbb{F}_{2^n}^*$. Then, the Hadamard transform of $f(\gamma x^s)$ has two distinct values at nonzero λ, i.e.,*

$$\hat{f}_h^\gamma(s)(\lambda) = \begin{cases} \delta_1, & \text{if } f(\gamma^{2^m} \lambda^s) = 0 \\ \delta_2, & \text{if } f(\gamma^{2^m} \lambda^s) = 1 \end{cases}$$

where $h(x) = Tr(x)$ and δ_1 and δ_2 are two distinct values.

Proof. By generalizing the proof of Theorem 5 in [5], the Hadamard transform of $f(\gamma x^s)$ is given by

$$\widehat{f_h^\gamma}(s)(\lambda) = \sum_{x\in\mathbb{F}_{2^n}} (-1)^{Tr(\lambda x)+f(\gamma x^{r(2^m-1)})}$$

$$= 1 + \sum_{z\in\mathbb{G}_s}\sum_{y\in\mathbb{F}_{2^m}^*} (-1)^{Tr(\lambda yz)+f(\gamma z^{r(2^m-1)})}$$

$$= 1 + \sum_{z\in\mathbb{G}_s} (-1)^{f(\gamma z^{-2r})} \cdot \left(\sum_{y\in\mathbb{F}_{2^m}} (-1)^{Tr(\lambda yz)} - 1\right)$$

$$= 1 + \sum_{z\in\mathbb{G}_s} (-1)^{f(\gamma z^{-2r})} \cdot \left(\sum_{y\in\mathbb{F}_{2^m}} (-1)^{Tr_1^m((\lambda z+\lambda^{2^m} z^{2^m})y)} - 1\right) \quad (6)$$

$$= 1 + 2^m \sum_{z\in\mathbb{G}_s, z^2=\lambda^{2^m-1}} (-1)^{f(\gamma z^{-2r})} - \sum_{z\in\mathbb{G}_s} (-1)^{f(\gamma z^{-2r})}$$

$$= 1 + 2^m(-1)^{f(\gamma\lambda^{-r(2^m-1)})} - \sum_{z\in\mathbb{G}_s} (-1)^{f(\gamma z^{-2r})}$$

$$= 1 + 2^m(-1)^{f(\gamma^{2^m}\lambda^s)} - \sum_{z\in\mathbb{G}_s} (-1)^{f(\gamma z^{-2r})}$$

where $\mathbb{G}_s$ is a cyclic group of order 2^m+1 such that any $x\in\mathbb{F}_{2^n}^*$ can be written by $x = yz$, where $y\in\mathbb{F}_{2^m}^*$ and $z\in\mathbb{G}_s$, $z^{2^m+1} = 1$. Therefore, Lemma 2 is immediate from (6) for a given γ. □

Proof of Lemma 1. For $v = 2^m - 1$ and $\gcd(\frac{2^n-1}{v}, t) = 1$, Lemma 1 is true if

$$\widehat{f_h^\gamma}(v)(\lambda^t) = \widehat{f_h^\gamma}(vt)(\lambda) \quad (7)$$

because $(vt, 1)$ is a self-realizable pair. If $\lambda = 0$, then

$$\widehat{f_h^\gamma}(v)(0) = \sum_{x\in\mathbb{F}_{2^n}} (-1)^{f(\gamma x^v)} = \sum_{x\in\mathbb{F}_{2^n}} (-1)^{f(\gamma x^{vt})} = \widehat{f_h^\gamma}(vt)(0). \quad (8)$$

If $\lambda\neq 0$, (7) is obvious from Lemma 2.

Therefore, if $v = 2^m - 1$ and $\gcd(\frac{2^n-1}{v}, t) = 1$, then the decimation pair (v, t) provides the same realization as another pair $(vt, 1)$ which always realizes a binary sequence corresponding to $f(x)$. In particular, if $f(x) = Tr(x)$, a binary m-sequence can be obtained from such v and t. □

5   Experimental Results

In this paper, we restrict experiments to $f(x) = h(x) = Tr(x)$ in our method. Also, v is chosen as each prime factor of $2^n - 1$ in our experiments. Table $1 - 5$ show our experimental results for $n = 8, 10, 12, 14$ and 16, where $p(x)$ is a primitive polynomial for $\mathbb{F}_{2^n}$. In each table, t_d is a coset leader of the cyclotomic coset

Table 1. Realizations of $f^\gamma(x) = Tr_1^n(\gamma x)$ for $n = 8$, $p(x) = x^8 + x^4 + x^3 + x^2 + 1$

v	t	d	t_d	d_c	$g(x)$	Sequence
3	3, 11, 91	9, 33, 18	3	9	1, 9, 29, 37, 39	B_3
	9, 47, 59	27, 141, 177	9	27	13, 19, 21, 29, 39	WG
	43, 87	129, 6	1	3	1	m-sequence
5	13, 53, 55, 59	65, 10, 20, 40	1	5	1	m-sequence
17	17, 19, 23, 31, 47,	34, 68, 136, 17, 34,	1	17	1	m-sequence
	53, 61, 91	136, 17, 17				

Table 2. Realizations of $f^\gamma(x) = Tr_1^n(\gamma x)$ for $n = 10$, $p(x) = x^{10} + x^3 + 1$

v	t	d	t_d	d_c	$g(x)$	Sequence
3	3, 43, 347	9, 129, 18	3	9	1, 9, 57, 73, 121	B_3
	171, 343	513, 6	1	3	1	m-sequence
11	35, 47, 95, 101, 109,	385, 517, 22, 88, 176,	1	11	1	m-sequence
	125, 157, 187, 221, 343	352, 704, 11, 385, 704				

Table 3. Realizations of $f^\gamma(x) = Tr_1^n(\gamma x)$ for $n = 12$, $p(x) = x^{12} + x^6 + x^4 + x + 1$

v	t	d	t_d	d_c	$g(x)$	Sequence
3	11, 43, 1387	33, 129, 66	11	33	1, 33, 133, 159, 163, 165, 373, 405, 421, 621, 629, 637, 661, 667, 669	B_5
	683, 1367	2049, 6	1	3	1	m-sequence
5	205, 821, 823, 827	1025, 10, 20, 40	1	5	1	m-sequence
7	293, 439, 587, 589, 1463,	2051, 3073, 14, 28, 2051,	1	7	1	m-sequence
	1759	28				
13	79, 197, 317, 319, 331,	1027, 2561, 26, 52, 208,	1	13	1	m-sequence
	347, 379, 443, 473, 631,	416, 832, 1664, 2054, 13,				
	827, 949	2561, 52				

consisting of t in $\mathbb{Z}_{q/v}^*$ for $q = 2^n - 1$, and d_c is a coset leader of the coset consisting of d in $\mathbb{Z}_q^*$.

For $n = 8, 10, 12$ and 14, we list all the realizations obtained by the multiplexing realization of the second order DHT. For $n = 16$, there are five classes of the known binary sequences with ideal two-level autocorrelation which have no subfield structures: 1) m-sequences, 2) B_3 sequences, 3) B_5 sequences, 4) B_7 sequences, and 5) Welch-Gong (WG) sequences [6]. Our experiments showed that except for the WG sequences, the other four classes can be realized by this new method. In Table 5, we only list the realizations from the multiplexing second order DHT for B_3, B_5 and B_7 sequences.

Remark 1. *In our procedure, a proper value of d has been exhaustively searched for $n = 8$ and 10. Interestingly, valid realizations have been observed only at $d = vt \pmod{2^n - 1}$ for $n = 8$ and 10. Hence, we took our experiments only for $d = vt \pmod{2^n - 1}$ at $n = 12$ and 14 to reduce experimental cases.*

Table 4. Realizations of $f^\gamma(x) = Tr_1^n(\gamma x)$ for $n = 14$, $p(x) = x^{14} + x^5 + x^3 + x + 1$

v	t	d	t_d	d_c	$g(x)$	Sequence
3	3,683,5467	9,2049,18	3	9	1,9,57,73,457, 569,585,1821,1829,1831, 1833,1849,1865,2341,2343	B_3
	11,171,5483	33,513,66	11	33	1,33,497,529,543	B_5
	2731,5463	8193, 6	1	3	1	m-sequence
43	131,143,191,383,389, 397,413,445,509,637, 667,763,893,905,953, 1145,1147,1151,1175,1207, 1271,1399,1405,1429,1525, 1655,1715,1907,1909,1913, 2429,2477,2669,2671,2675, 2683,2731,2795,2923,3431, 3437,3445	5633,6149,8213,86,344, 688,1376,2752,5504,11008, 12298,43,5633,6149,8213, 86,172,344,1376,2752, 5504,11008,11266,12298,43, 5633,8213,86,172,344, 6149,8213,86,172,344, 688,2752,5504,11008,86, 344,688	1	43	1	m-sequence

Table 5. Realizations of $f^\gamma(x) = Tr_1^n(\gamma x)$ for $n = 16$, $p(x) = x^{16} + x^5 + x^3 + x^2 + 1$

v	t	d	t_d	d_c	$g(x)$	Sequence
3	3,2731,21851	9,8193,18	3	9	1,9,57,73,457,569,585,3641 3657,4553,4665,4681, 7289,7311,7753	B_3
	11,683,21867	33,2049,66	11	33	1,33,993,1057,2017	B_5
	43,171,21931	129,513,258	43	129	1,129,517,639,643,645,1557,1669,1685,2557 2581,2683,2685,2699,2701,2707,2709,5589 5717,5781,6613,6741,6803,6805,9653,9685 9717,9813,9845,9877,10669,10677,10701,10709 10733,10741,10837,10859,10861,10867,10869	B_7

Remark 2. *For each n, we observed $d_c = vt_d$. Especially, if $t_d = 1$, the realizations are always binary m-sequences or self-realizations.*

Remark 3. *For each n, if $v = 3$, the realizations exactly match the work in [2] except for $n = 8$. In other words, when $d_c = 2^k + 1$ with k of $1 \leq k \leq n/2$ and $\gcd(k,n) = 1$, the realizations are the B_k sequences. In $n = 8$, however, we could obtain the Welch-Gong (WG) sequences [4] by our procedure, which has not been predicted by [2]. The WG sequences, however, have not been obtained for any other n in our experiments.*

If $v \neq 3$, on the other hand, the realizations are only binary m-sequences at $t_d = 1$. Thus, we could not obtain any new binary two-level autocorrelation sequences from our generalized method.

Remark 4. *From our experiments, we observed that except for the WG sequences, all known binary two-level autocorrelation sequences of period $2^n - 1$ with no subfield structures could be realized by the multiplexing second order DHT for even $n \leq 16$. In terms of the binary sequences with subfield structures and their DHT's, readers may refer to [7].*

6   Conclusion

In this paper, we proposed a new method by anticipating a new binary two-level autocorrelation sequence, where we established a new scheme of iterative decimation-Hadamard transform for even n. Our scheme produced some interesting results about realization and realizable pairs by obtaining several known sequences starting from m-sequences. Finally, we claim that the iterative decimation-Hadamard transform can be used for constructing two-level autocorrelation sequences of period $2^n - 1$ for even n by the multiplexing realizations.

Together with Gong and Golomb's experimental results in [1], we have confirmed that all the known binary two-level autocorrelation sequences of period $2^n - 1$ without subfield structures can be realized by either the second order DHT or the multiplexing second order DHT for $n \leq 17$, except for WG sequences for $8 < n \leq 17$ and n even. In other words, staring with a single binary m-sequence of period $2^n - 1$, $n \leq 17$, by applying either the iterative DHT or the multiplexing iterative DHT, all the known binary two-level autocorrelation sequences of period $2^n - 1$ without subfield structures can be obtained, except for WG sequences for $8 < n \leq 17$ and n even.

References

1. Gong, G., Golomb, S.W.: The decimation-Hadamard transform of two-level autocorrelation sequences. IEEE Trans. Inform. Theory 48(4), 853–865 (2002)
2. Dillon, J.F., Dobbertin, H.: New cyclic difference sets with Singer parameters. Finite Fields and Their Applications 10, 342–389 (2004)
3. Hertel, D.: Extended Hadamard Equivalence. In: Gong, G., Helleseth, T., Song, H.-Y., Yang, K. (eds.) SETA 2006. LNCS, vol. 4086, pp. 119–128. Springer, Heidelberg (2006)
4. No, J.S., Golomb, S.W., Gong, G., Lee, H.K., Gaal, P.: Binary pseudorandom sequences of period $2^m - 1$ with ideal autocorrelation. IEEE Trans. Inform. Theory 44(2), 814–817 (1998)
5. Charpin, P., Gong, G.: Hyperbent functions, Kloosterman sums, and Dickson polynomials. IEEE Trans. Inform. Theory 54(9), 4230–4238 (2008)
6. Golomb, S.W., Gong, G.: Signal Design for Good Correlation - for Wireless Communication, Cryptography and Radar. Cambridge University Press, Cambridge (2005)
7. Yu, N., Gong, G.: Realization of decimation-Hadamard transform for binary generalized GMW sequences. In: Proc. of Workshop on Coding and Cryptography (WCC), Bergen, Norway, pp. 127–136 (March 2005)

On Cayley Graphs, Surface Codes, and the Limits of Homological Coding for Quantum Error Correction

Gilles Zémor

Institut de Mathématiques de Bordeaux*, UMR 5251
Université Bordeaux 1,
351, cours de la Libération, 33405 Talence, France
Gilles.Zemor@math.u-bordeaux1.fr

Abstract. We review constructions of quantum surface codes and give an alternative, algebraic, construction of the known classes of surface codes that have fixed rate and growing minimum distance. This construction borrows from Margulis's family of Cayley graphs with large girths, and highlights the analogy between quantum surface codes and cycle codes of graphs in the classical case. We also attempt a brief foray into the class of quantum topological codes arising from higher dimensional manifolds and find these examples to have the same constraint on the rate and minimum distance as in the 2-dimensional case.

1   Introduction

The most investigated class of quantum stabilizer codes is the CSS (Calderbank, Shor, Steane [3,16]) class. These codes have a short classical description that highlights the parallel with classical codes and makes them amenable to investigation by classical coding theorists with little or no background in quantum physics. A CSS code of length n can be defined by a binary "parity check matrix" $\mathbf{H}$ which has n columns and two sets of rows, making up two matrices $\mathbf{H}_1$ and $\mathbf{H}_2$, such that every row of $\mathbf{H}_1$ is orthogonal to every row of $\mathbf{H}_2$. In other words the quantum code is defined by two mutually orthogonal linear subspaces V_1 and V_2 of $\mathbb{F}_2^n$, the row-spaces of $\mathbf{H}_1$ and $\mathbf{H}_2$ respectively. The parameters $[[n, k, d]]$ of the associated quantum code are its length n, its dimension k which is given by $n - \dim V_1 - \dim V_2$, and its minimum distance d, which is given by the minimum weight of a non-zero binary vector that is either orthogonal to V_1 but not in V_2, or orthogonal to V_2 but not in V_1. Why this structure is relevant to quantum error-correction has been described many times, for detailed descriptions see e.g. [12] or [14].

   In this paper we will be mainly interested in the parameters n, k, d of a quantum code. While asymptotically good (having constant rate $R = k/n$ and constant relative minimum distance $\delta = d/n$) CSS codes are known to exist [3], the

---

* Supported by the French ANR Defis program under contract ANR-08-EMER-003 (COCQ project).

C. Xing et al. (Eds.): IWCC 2009, LNCS 5557, pp. 259–273, 2009.

same cannot be said of *sparse* CSS codes, in other words LDPC (Low-Density Parity-Check) quantum codes, which are CSS codes admitting a parity-check matrix $\mathbf{H} = (\mathbf{H}_1, \mathbf{H}_2)$ with a small number of ones per row and per column. We will be interested in asymptotics, i.e. the study of families of codes with growing length n, and a bounded, or very slowly growing row and column weight for $\mathbf{H}$. One of the motivations for the search for good quantum LDPC codes, is that they will come together with *local* decoding algorithms and that iterative methods will lead to good decoding performance. While being asymptotically good is not a prerequisite for a good (close to capacity) decoding behaviour, it is nevertheless a challenging open problem to know whether asymptotically good low density CSS codes (or more general quantum codes for that matter) exist. At present, known families of quantum LDPC codes are very far from being asymptotically good. Recent attempts from the classical coding community at devising quantum LDPC codes with non-zero rate have resulted in *constant* minimum distances (e.g. [12,8]). In the other direction, known families of quantum LDPC codes with a minimum distance growing with n are essentially restricted to the so-called class of *surface codes* that we will present below. We note that the rate R and relative minimum distance δ functions of known constructions of surface codes are constrained by the upper bound

$$R\delta^2 \leq n^{-2+o(1)} \tag{1}$$

We shall investigate the constraints on R and δ that we obtain when we switch from surface codes to codes based on higher dimensional topological manifolds, namely the n-dimensional torus, and find exactly the same constraint (1), leading one to ask whether this constraint on the behaviour of (R, δ) is inherent to quantum codes based on algebraic topology.

Surface codes also provide the presently only known family of quantum LDPC codes with constant rate and provably growing minimum distance [7], [10]. The minimum distance for these families grows like a logarithm of the blocklength. Published accounts of these families of codes involved sophisticated arguments from hyperbolic geometry: we shall give an alternative presentation that we find to be more accessible, and that highlights the analogy between quantum surface codes and the class of cycle codes of graphs in the classical coding setting. This analogy was already remarked in [2], where cycle codes of graphs are rediscovered and named classical "homological codes". In the present paper we shall push the analogy with cycle codes of graphs further by giving an account of the algebraic constructions of quantum surface codes with fixed rate and growing minimum distance that borrows from the algebraic constructions of graphs with large girths introduced by Margulis [13] and from the work of Širáň [15].

We have tried to keep the language of algebraic topology to a minimum to be accessible to coding theorists not well-versed in the field, and to try to make the paper as self-contained as possible. The paper is organized as follows: in section 2 we give a brief review of cycle codes of graphs together with Margulis's construction of regular graphs with large girths. We then proceed to review constructions of quantum surface codes in section 3, notably Kitaev's original toric code construction which is the starting point for all surface codes. We then move on to the construction of surface codes of fixed rate and logarithmic minimum distance. Finally in section 4 we give a brief exploration of the higher dimensional case by considering quantum codes arising from n-dimensional tori.

2   Cycle Codes of Graphs

2.1   Cycle Codes

Cycle codes of graphs are codes that have a parity-check matrix with exactly two "1"s per column. They are therefore instances of Low Density Parity Check (LDPC) codes with particularly low density and are amenable to iterative decoding (e.g. message passing) techniques. Even though they are not truly practical and their performance is surpassed by other well-studied classes of LDPC codes, they are interesting because their relatively simple structure makes it possible to analyze their decoding behaviour almost completely, and they have the remarkable property that over the binary symmetric channel, the threshold probabilities (beyond which decoding fails with probabity almost 1) for maximum-likelihood decoding [4,18] and for iterative decoding coincide.

For the above reasons cycle codes of graphs can be considered as the simplest LDPC codes. It also turns out that they also have a natural generalization to quantum LDPC codes, under the form of surface (topological) codes. Before moving on to the quantum case, we therefore review relevant facts about cycle codes.

Let $\mathbf{G}$ be a finite, undirected, connected graph without loops or multiple edges. It is defined by its *vertex set* V and its *edge set* E where an edge is a pair of vertices. Let r and n denote the number of vertices and edges respectively, and let the set of edges be numbered, so that it is identified with $\{1, \ldots, n\}$. Identify furthermore subsets of edges with their characteristic vectors in $\{0, 1\}^n$. A *cycle* $\mathbf{x} \in \{0, 1\}^n$ of $\mathbf{G}$ is a subset of edges with the property that any vertex of $\mathbf{G}$ is incident to an even number of edges of $\mathbf{x}$. The set of cycles of $\mathbf{G}$ is a subset of $\{0, 1\}^n$ stable under addition modulo 2. It is therefore a linear code of length n called the *cycle code* of $\mathbf{G}$.

An incidence matrix $\mathbf{H} = (h_{ij})$ of $\mathbf{G}$ is an $r \times n$ matrix whose rows are indexed by the vertices of $\mathbf{G}$, whose columns are indexed by the edges, and such that $h_{ij} = 1$ whenever vertex i belongs to edge j and $h_{ij} = 0$ otherwise. The incidence matrix $\mathbf{H}$ is also a parity-check matrix of the cycle code C: it has exactly two 1's per column since any edge is incident to exactly two vertices.

It is well-known that the dimension of C is $k = n - r + 1$ when the graph is connected. For example, the cycle code of the famous Petersen graph represented on figure 1 has parameters $[15, 6, 5]$, since the smallest cycle size is 5.

2.2   Bounds on the Girth of a Graph

The study of the minimum distance d of C, i.e. the size of the smallest cycle, or *girth*, of $\mathbf{G}$ has been of interest to graph-theorists since the 1960's, and the problem of the exact determination of the largest possible girth of a graph with

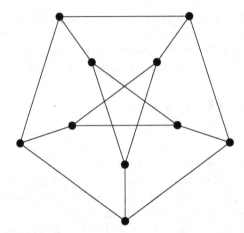

Fig. 1. The Petersen graph yields a $[15, 6, 5]$ code

a given number of vertices and of edges is still quite open. The bad news for coding theorists is that for fixed rate $R = k/n$ and growing n the minimum distance d cannot grow faster than a logarithm of n. Here is a short proof of this when **G** is a *regular* graph. A graph **G** is said to be Δ-*regular* if every vertex is incident to exactly Δ edges. Note that $R = 1 - 2/\Delta + 1/n$, so that for large n, fixing the degree fixes the rate. Let v be any given vertex of the graph and consider the set of vertices at distance from v less than $d/2$ in the graph, where the distance between two vertices is the length of a shortest path between the vertices. Because **G** does not contain a cycle of length smaller than d, the subgraph induced by this set of vertices must be a tree, as represented in figure 2.

The number of vertices of **G** at distance 1 from v is therefore Δ, the number of vertices at distance 2 from v is $\Delta(\Delta - 1)$, and the number of vertices at distance i from v, $i < d/2$, is $\Delta(\Delta - 1)^i$. We conclude that the total number of vertices $|V|$ in the graph has to be bigger than

$$1 + \Delta + \Delta(\Delta - 1) + \cdots + \Delta(\Delta - 1)^{\lceil d/2 \rceil}.$$

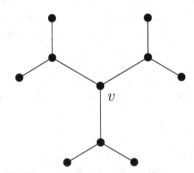

Fig. 2. The neighbourhood of v in a Δ-regular graph (here $\Delta = 3$)

This lower bound on the number of vertices of a regular graph with given girth is known as the Moore bound. Let us mention in passing that deriving a similar bound for irregular graphs is not trivial and was achieved satisfactorily only relatively recently [1].

The resulting upper bound on the girth d (or minimum distance of the cycle code) when n is given and is large tells us that d cannot be significantly larger than

$$2 \log_{(\Delta-1)} |V|.$$

Erdös and Sachs [6] first showed with random arguments that families of Δ-regular graphs exist that satisfy $d \geq \log_{(\Delta-1)} |V|$. However, the first construction of an infinite family of Δ-regular graphs with girth d growing as a logarithm of the number of vertices is due to Margulis [13]. We review the idea of his construction here because we will generalize it to the quantum case.

2.3 Margulis's Construction of Graphs with Large Girths

Margulis's construction uses Cayley graphs. A *Cayley graph* is defined by a group G together with a generating set S of group elements such that $s \in S$ implies $s^{-1} \in S$. The vertex set is the set of group elements, $V = G$, and we draw an edge between vertices x and y whenever $y = xs$ for some $s \in S$. Margulis takes for G the groups $G = \mathrm{SL}_2(\mathbb{F}_p)$ of 2×2 matrices of determinant 1 with elements in the field on p elements, for a prime p. The set S is the set $S = \{A, B, A^{-1}, B^{-1}\}$ where

$$A = \begin{bmatrix} 1 & 2 \\ 0 & 1 \end{bmatrix} \quad A^{-1} = \begin{bmatrix} 1 & -2 \\ 0 & 1 \end{bmatrix} \quad B = \begin{bmatrix} 1 & 0 \\ 2 & 1 \end{bmatrix} \quad B^{-1} = \begin{bmatrix} 1 & 0 \\ -2 & 1 \end{bmatrix}.$$

We obtain therefore an infinite family of 4-regular graphs with $|V| = |G| = p(p^2 - 1)$. We see furthermore that a cycle of **G** is given by a sequence $s_1, \ldots, s_\ell$ of elements of S such that

$$s_1 s_2 \cdots s_\ell = 1 \quad \text{and} \quad s_{i+1} \neq s_i^{-1}, i = 1 \ldots \ell - 1. \tag{2}$$

Now Margulis's girth argument is as follows. Consider the set S viewed as matrices with elements in $\mathbb{Z}$, i.e. matrices of $\mathrm{SL}_2(\mathbb{Z})$. It is known that S generates a *free* subgroup Γ of $\mathrm{SL}_2(\mathbb{Z})$. In other words (2) never occurs in Γ, and the Cayley graph (Γ, S) is a 4-regular *tree*. Therefore, a product of the form in (2) can only equal the identity matrix if it is reduced modulo p. But for this to happen, at least one of the matrix elements in the product $s_1 s_2 \cdots s_\ell$ must be larger than $p - 1$ in absolute value. But the largest term (in absolute value) of any matrix Ms is clearly not more than 3 times the largest term of the matrix M, for any $s \in S$. Therefore the length ℓ of any cycle (2) is at least $\log_3(p - 1)$, and the girth of the family of graphs **G** has at least logarithmic growth in the vertex (and edge) size.

Actually Margulis's original proof is slightly more involved and gives a better multiplicative constant in the lower bound on the girth, but this simple argument will suffice for our purposes. We remark also that this method can be clearly extended to yield Δ-regular graphs with large girths for different values of Δ.

Finally, we conclude this section with the following consideration. At the cost of letting the rate R of the cycle code go to zero, we can let the minimum distance (or girth) grow faster than $\log n$ where n is the block length (or number of edges). Start for example with the case when the graph consists of a single cycle, then the dimension is 1 and the minimum distance is n (we have the repetition code). But the Moore bound will constrain us with the upper bound

$$R\delta \leq n^{-1+o(1)} \tag{3}$$

where $\delta = d/n$ is the relative minimum distance. We are far away from the Gilbert-Varshamov bound which tells us that we have linear codes with constant $R\delta$.

3   Kitaev's Toric Code and Surface Codes

3.1   The Toric Code

To construct a quantum LDPC code we now need two low-density matrices $\mathbf{H}_1$ and $\mathbf{H}_2$ such that every row of $\mathbf{H}_1$ is orthogonal to every row of $\mathbf{H}_2$. A relatively natural strategy is to try and adapt cycle codes to the quantum case. We can take for $\mathbf{H}_1$ the parity-check matrix of a cycle code of a graph that we defined in the preceding section and put some cycles as the rows of $\mathbf{H}_2$. For the quantum code to be non-empty we need the row-space V_2 of $\mathbf{H}_2$ to be not equal to the whole cycle code, and we would like both the rows of $\mathbf{H}_2$ to be of bounded weight (so that we have a genuine LDPC code) and the surviving cycles not in V_2 to be as large as possible. Furthermore, we would also like to control the weight of the vectors orthogonal to V_2 and not in the row-space V_1 of $\mathbf{H}_1$. Obviously some machinery will be required to achieve all these objectives.

One idea due to Kitaev [11] is to take the graph $\mathbf{G}$ equal to a tiling of a two-dimensional torus by squares. Specifically, take the graph with vertex set $\mathbb{Z}/m\mathbb{Z} \times \mathbb{Z}/m\mathbb{Z}$ and connect every (x, y) to $(x, y-1), (x, y+1), (x-1, y), (x+1, y)$. The graph has m^2 vertices and $2m^2$ edges (figure 3). We remark that this graph is the Cayley graph over the group $\mathbb{Z}/m\mathbb{Z} \times \mathbb{Z}/m\mathbb{Z}$ with generator set $S = \{(0, 1), (0, -1), (1, 0), (-1, 0)\}$.

Take for $\mathbf{H}_1$ the vertex-edge incidence matrix of the graph $\mathbf{G}$. For the rows of $\mathbf{H}_2$ take all elementary squares, i.e. cycles of the form $(x, y) - (x, y+1) - (x+1, y+1) - (x+1, y) - (x, y)$, or *faces* of the tiling. It is relatively easy to see that the quotient space $V_1^{\perp}/V_2$ has dimension 2, and that lowest-weight representatives are given by cycles that keep one coordinate constant, e.g. $(x, 0), (x, 1), \ldots (x, m-1), (x, 0)$ represented by thick lines on figure 3.

It remains to bound from below the weight of representatives from the other quotient space $V_2^{\perp}/V_1$. A convenient way to achieve this is to use *Poincaré duality*. The *dual graph* $\mathbf{G}^*$ of $\mathbf{G}$ is obtained from $\mathbf{G}$ by declaring the vertices of $\mathbf{G}^*$ to be the faces of $\mathbf{G}$ and by drawing an edge between two vertices of $\mathbf{G}^*$ if they have a common edge in $\mathbf{G}$. Now we see that the dual graph of $\mathbf{G}^*$ is isomorphic to $\mathbf{G}$, and that the rows of $\mathbf{H}_1$ are exactly the characteristic vectors

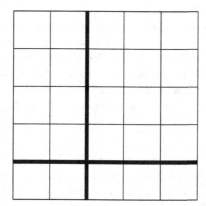

Fig. 3. A two-dimensional torus: identify opposing edges

of the cycles of $\mathbf{G}^*$. Therefore the minimum weight of representatives of $V_2^{\perp}/V_1$ is again m and the parameters of the associated quantum code are:

$$[[2m^2, 2, m]].$$

On the original graph G a minimum-weight representative of $V_2^{\perp}/V_1$ looks like a ladder and is represented on figure 4.

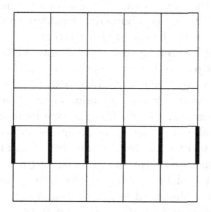

Fig. 4. A minimum weight representative of $V_2^{\perp}/V_1$

3.2 Surface Codes

From a classical coding theorist's point of view, the dimension of the toric code is dreadfully small and one would like to generalize it to larger dimensions. For this the natural thing to do is to take tilings of other surfaces, i.e. graphs that generalize the 2-dimensional torus. Some definitions from algebraic topology are in order.

We adopt the combinatorial point of view, for which there is essentially no difference between a surface and a tiling of the surface. A cycle is said to be elementary if it is not the union of two non-empty cycles. A 2-*complex* is a connected graph $\mathbf{G}$ together with a collection of privileged elementary cycles called *faces*. For precise definitions of a surface in algebraic topology see for example [9]. For our purposes it will suffice to say that a 2-complex defines a *surface* (without boundary) when it satisfies the two following properties :

1. any two faces meet in at most one edge and every edge belongs to exactly two faces.
2. For any given vertex v, if F is the set of faces incident to v, then any edge common to two faces of F is an edge incident to v and furthermore, if we define a graph on F by declaring two elements of F to be adjacent if they have a common edge in $\mathbf{G}$, then we obtain an elementary cycle.

We see that the second condition enables us precisely to define the dual graph $\mathbf{G}^*$ and the associated dual 2-complex in a way that generalizes duality for the toric tiling of the preceding section. The faces of $\mathbf{G}$ define the vertices of $\mathbf{G}^*$. Two vertices of $\mathbf{G}^*$ are adjacent in $\mathbf{G}^*$ if the corresponding faces of $\mathbf{G}$ have a common edge in $\mathbf{G}$. The faces of $\mathbf{G}^*$ are the cycles of $\mathbf{G}^*$ that arise from the sets of faces of $\mathbf{G}$ incident to a vertex of $\mathbf{G}$, as in condition 2 above.

Finally, the $(\mathbb{Z}/2\mathbb{Z})$ *homology* group $H_1(\mathbf{G})$ is the quotient of the cycle code of $\mathbf{G}$ by the $\mathbb{F}_2$-linear subspace generated by the faces of $\mathbf{G}$. The *cohomology* group $H^1(\mathbf{G})$ is the homology group $H_1(\mathbf{G})$ of the dual 2-complex $\mathbf{G}^*$.

Now, switching to the construction of the associated quantum code, we take the matrix $\mathbf{H}_1$ to be again the point-edge incidence matrix of the graph $\mathbf{G}$, and the matrix $\mathbf{H}_2$ is now the set of characteristic vectors of the faces of $\mathbf{G}$. Alternatively, $\mathbf{H}_1$ can be seen as the set of characteristic vectors of the faces of $\mathbf{G}^*$ and $\mathbf{H}_2$ as the point-edge incidence matrix of the dual graph $\mathbf{G}^*$. The homology and cohomology groups of $\mathbf{G}$ are exactly the coset spaces $V_1^\perp/V_2$ and $V_2^\perp/V_1$ whose dimension is the dimension of the quantum code and whose minimum weight representatives give the quantum minimum distance.

A natural way to obtain sequences of quantum surface codes with growing dimension is to take tori with a growing number of handles. This approach is discussed in [5] and more recently and at some length in [2]. We do not dwell upon this, instead we will be content to give a very short description of an alternative method, also discussed in [5,2], which is topologically quite close, is quite visual, and gives essentially the same asymptotic parameters.

Take for the graph $\mathbf{G}$ an $m \times m$ grid with *holes* as in figure 5.

Again $\mathbf{H}_1$ is the point-edge incidence matrix of the graph $\mathbf{G}$. The matrix $\mathbf{H}_2$ has rows corresponding to the elementary squares of the grid with the inside of the holes missing. A close look at the situation convinces one that the dimension of the quantum code is equal to the number of holes. The minimum distance is the minimum of the perimeter of the holes, and the smallest "ladder distance" between holes or between a hole and the outside boundary. See [5,2] for details.

By multiplying the number of holes in the grid, one can push the dimension of the quantum code all the way to a quantity linear in the blocklength. The price

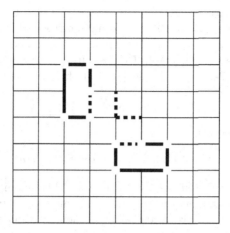

Fig. 5. A grid with holes (thick lines). Dashed edges are a word of $V_2^\perp/V_1$.

to pay for this however will be a constant minimum distance. In between we have the relation between the quantum rate $R = k/n$ and the relative quantum minimum distance $\delta = d/n$

$$R\delta^2 \leq cn^{-2} \tag{4}$$

for a constant c. The same compromise is obtained with the natural tilings of tori with a growing number of handles.

The above bound seems to be inherent to the nature of surface codes. By taking more sophisticated surfaces however, one can obtain a slight improvement in the behaviour of the minimum distance. In particular one can obtain quantum LDPC codes of constant rate and non-constant, slowly growing minimum distance, proportional to $\log n$ for blocklength n. This result is implicit in [7], where sophisticated topological methods are used. See also [10]. We shall give a different account of this result by appealing to a method which is very much reminiscent of Margulis's construction of graphs with large girth of section 2.3.

3.3 Surface Quantum Codes with Constant Rate and Logarithmic Minimum Distance

Our purpose is to construct a family of quantum surface codes with constant rate and growing minimum distance.

Recall that Margulis's Cayley graph construction consists of realizing an infinite tree as the Cayley graph of a free group and then projects this infinite tree on a finite graph by taking a suitable quotient of the infinite free group. We will adopt a similar approach, but we will start with an infinite Cayley graph that is not a tree so as to have cycles that will define the faces of the surface that we shall obtain.

Specifically, consider the group on two generators a and b defined by the presentation

$$a^2 = 1, b^\ell = 1, (ab)^m = 1$$

where m, ℓ are positive integers such that $1/2 + 1/\ell + 1/m < 1$. This group, denoted by $T = T(2, m, \ell)$, is called a triangular group of type $(2, m, \ell)$ and has numerous applications, in particular to hyperbolic geometry.

Consider now the Cayley graph $\mathfrak{J}$ on T with generating set $S = \{a, b, b^{-1}\}$. We associate to $\mathfrak{J}$ a 2-complex by declaring its faces to be the length ℓ cycles

$$\{x, xb, xb2, \ldots, xb^{\ell-1}, xb^\ell = x\} \tag{5}$$

and the length $2m$ cycles

$$\{x, xa, xab, xaba, x(ab)^2, \ldots, x(ab)^{m-1}a, x(ab)^m = x\} \tag{6}$$

for every vertex x of $\mathfrak{J}$. One easily checks that this 2-complex satisfies conditions 1 and 2 that define a surface, and furthermore the subgraph induced by the set of vertices at fixed distance r from the identity element is a planar graph, as is illustrated on figure 6.

Now suppose that we can find a finite quotient group G of the group T such that no product of the elements a, b, b^{-1} of length r or less collapses to the identity in G if it doesn't collapse to the identity in the infinite group T. Consider the corresponding finite graph $\mathbf{G}$ defined as the Cayley graph on the group G with generator set $S = \{a, b, b^{-1}\}$ and consider the associated quantum surface code. We note that the row weight of $\mathbf{H}_1$ is 3 and the row weight of $\mathbf{H}_2$ is ℓ or $2m$ (the face lengths), so that $(\mathbf{H}_1, \mathbf{H}_2)$ define a genuine quantum LDPC code.

First, let us estimate the dimension of the quantum code. The number of rows of the matrix $\mathbf{H}_1$ is simply equal to number of vertices $|V|$ of the graph which equals, since the graph is of degree 3,

$$|V| = \frac{2}{3}|E|$$

where $|E|$ is the number of edges. We proceed to count the number of rows of $\mathbf{H}_2$, i.e. the number of faces of the graph $\mathbf{G}$. We remark that every vertex x is incident to two edges $\{x, xb\}$ and $\{x, xb^{-1}\}$ that belong to the same ℓ-face (5) and is incident to one edge $\{x, xa\}$ that belongs to two $2m$-faces of type (6). Hence, denoting by $|F_\ell|$ and $|F_{2m}|$ the number of ℓ-faces and of $2m$-faces respectively, we have :

$$\ell|F_\ell| = |V| \qquad \text{and} \qquad 2m|F_{2m}| = 2|V|$$

from which the total number of faces equals

$$\left(\frac{1}{\ell} + \frac{1}{m}\right)|V| = \frac{2}{3}\left(\frac{1}{\ell} + \frac{1}{m}\right)|E|.$$

Therefore the dimension k of the quantum code satisfies

$$\frac{k}{|E|} \geq 1 - \frac{2}{3} - \frac{2}{3}\left(\frac{1}{\ell} + \frac{1}{m}\right)$$

$$k \geq |E|\frac{1}{3}\left(1 - 2\left(\frac{1}{\ell} + \frac{1}{m}\right)\right)$$

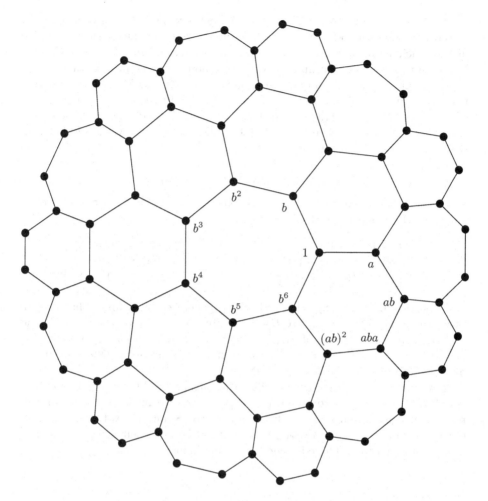

Fig. 6. A local view of the Cayley graph $\mathcal{T}$ in the case $m = 3$ and $\ell = 7$

so that the dimension k is a positive fraction of the blocklength $n = |E|$ for any m, ℓ such that $1/m + 1/\ell > 1/2$. For example, when $m = 3$ and $\ell = 7$ we obtain $k \geq n/63$.

We now address the estimation of the quantum minimum distance. Consider any elementary cycle of length r or less in the finite Cayley graph $\mathbf{G}$. Now if x is any vertex lying on this cycle, the cycle is included in the distance r-neighbourhood of x. But since we have supposed that any word formed with the letters a, b, b^{-1} collapses to the identity in G only if it collapses to the identity in T, the subgraph of $\mathbf{G}$ induced by the distance r-neighbourhood of x is exactly the same as the corresponding distance r-induced subgraph of the infinite Cayley graph $\mathcal{T}$. Since this graph is planar, all of its cycles are sums of its faces, and the initial cycle is a sum of faces of $\mathbf{G}$. Similarly, if a vector is orthogonal to all the faces of $\mathbf{G}$, i.e. is a cycle of the dual graph $\mathbf{G}^*$, and if its weight is sufficiently small, then everyone of its connected components **x** must belong to a bounded distance neighbourhood of **G***, which is planar and coincides with a bounded distance neighbourhood of of $\mathcal{T}^*$, hence **x** must equal a sum of faces of **G***. We obtain in this way a quantum code with minimum distance proportional to r.

It remains to find such a quotient group G. Similarly to Margulis's construction, we again appeal to a representation of the infinite group T by a subgroup of a matrix group, though a somewhat more complicated one. A construction that does exactly what we need, though for different reasons, was found by Širáň [15] and we sketch it below.

Let
$$P_k(X) = 2\cos(k\arccos(X/2))$$

be the normalized kth Chebychev polynomial and let $\xi = 2\cos(\pi/m\ell)$. Let B and C be the matrices of $\mathrm{SL}_3(\mathbb{Z}[\xi])$:

$$B = \begin{bmatrix} -1 & -P_\ell(\xi) & 0 \\ P_\ell(\xi) & P_\ell(\xi)^2 - 1 & 0 \\ P_m(\xi) & P_m(\xi)P_\ell(\xi) & 1 \end{bmatrix} \quad \text{and} \quad C = \begin{bmatrix} P_m(\xi)^2 - 1 & 0 & P_m(\xi) \\ P_\ell(\xi) & 1 & 0 \\ -P_m(\xi) & 0 & -1 \end{bmatrix}.$$

Širáň proved that the subgroup of $\mathrm{SL}_3(\mathbb{Z}[\xi])$ generated by B and C has presentation: $B^\ell = 1$, $C^m = 1$ and $(CB)^2 = 1$. Equivalently, if we set $a = CB$ and $b = B$, then the subgroup of $\mathrm{SL}_3(\mathbb{Z}[\xi])$ generated by a and b has presentation $a^2 = 1$, $b^\ell = 1$ and $(ab)^m = 1$. Therefore, the subgroup of $\mathrm{SL}_3(\mathbb{Z}[\xi])$ generated by a and b is exactly the triangular group T.

Now to take a quotient, proceed like Margulis and reduce the matrix entries of $SL_3(\mathbb{Z}[\xi])$ modulo p for some prime p. Some care must be taken for this reduction to make sense and to behave properly. Širáň argues that ξ is a root of the polynomial $h(X) = P_{2m\ell}(X) - 2$ that has integer coefficients and leading coefficient equal to 1. There is therefore a ring homomorphism

$$\mathbb{Z}[\xi] \to \mathbb{Z}[X]/(h(X))$$

defined by $\xi \mapsto X$, and by reducing coefficients of polynomials in $\mathbb{Z}[X]/(h(X))$ modulo p we obtain a ring homorphism

$$\mathbb{Z}[\xi] \to \mathbb{F}_p[X]/(h(X)).$$

This homorphism has the property that an integer polynomial expression in ξ of degree $< \deg h$ that is non-zero in $\mathbb{Z}[\xi]$ maps to zero only if not all the absolute values of its coefficients belong to $\{0, 1, \ldots, p-1\}$. Therefore, if we define the finite group G to be the subgroup of $\mathrm{SL}_3[\mathbb{F}_p[X]/(h(X))]$ generated by a and b, we obtain that a word in $a, b, b^{-1} = b^{\ell-1}$ collapses to the identity in G but not in T, only if one the polynomial coefficients of its matrix entries is larger than $p-1$. But we see that these coefficients grow exponentially (but not faster) in the length of the word. We obtain therefore that the shortest word that collapses to the identity in G but not in T behaves like a logarithm of p and hence like a logarithm of the size of the group G. See [15] for numerical estimations of constants.

4   The n-Dimensional Toric Code

In this section we make a short incursion into the realm of topological codes built on manifolds of dimension greater than 2. Another way to raise the dimension of Kitaev's torus code is to consider tilings of tori in dimensions greater than 2. The case of dimension 4 was already considered in [5] and in [17] with the hope that local decoding strategies will work better than in dimension 2. Let us consider the general case.

In this section the length of the quantum codes will be denoted by N rather than n, and n will denote the dimension of the torus. Consider the Cayley graph **G** on the group $G = (\mathbb{Z}/m\mathbb{Z})^n$ with generator set made up of all n-tuples of G of weight one and with their non-zero coordinate equal to either 1 or -1. The graph **G** is made into an n-complex, by associating to **G** faces of dimension $2, 3 \ldots, n$, where an i-dimensional face is obtained by fixing $n - i$ coordinates and letting the i remaining coordinates describe an elementary cube of dimension i.

Now instead of indexing coordinates of $\{0, 1\}^N$ by edges of the graph, as in the cycle code and surface code case, we index coordinates by the i-faces of **G** for some i. The number of i-dimensional faces is easily seen to be

$$N_i = m^n \binom{n}{i} \tag{7}$$

and the codelength is therefore $N = N_i$. Now a vector $\mathbf{x} \in \{0, 1\}^N$, equivalently a collection of i-faces, is called an i-cycle if every $(i - 1)$-face is incident to (is included in) an even number of i-faces of $\mathbf{x}$. We now define the matrix $\mathbf{H}_1 = (h_{ab})$ as the matrix whose rows are indexed by the $(i - 1)$-faces and whose columns are indexed by the i-faces, with a $h_{ab} = 1$ if $(i - 1)$-face a belongs to i-face b, and $h_{ab} = 0$ otherwise.

Now every $(i + 1)$-face, viewed as a collection of i-faces is an i-cycle. In other words, an $(i + 1)$-face, viewed as a vector of $\mathbb{F}_2^N$, is orthogonal to all the rows of $\mathbf{H}_1$. We therefore take for $\mathbf{H}_2$ the matrix whose rows are all the characteristic vectors of the $(i + 1)$-faces of **G**.

Denoting as before V_1 to be the row-space of $\mathbf{H}_1$ and V_2 to be the row-space of $\mathbf{H}_2$, the quotient $V_1^\perp/V_2$ is exactly the i-th homology group $H_i(\mathbf{G})$ of the n-complex and it is known (see e.g. [9]) to have dimension

$$k = \binom{n}{i} \tag{8}$$

which is the dimension of the quantum code.

Now we have a straightforward upper bound on the minimum weight of an i-cycle that is not a sum of $(i + 1)$-faces. To obtain one, simply consider the set of all i-faces that fix some given $n - i$ coordinates. This defines a set of i-faces that is isomorphic to a tiling of an i-dimensional torus, whose weight is, by (7), m^i. Hence the quantum minimum distance satisfies

$$d \leq m^i. \tag{9}$$

What about the minimum weight of representatives of $V_2^\perp/V_1$? As in the 2-dimensional case, we can again appeal to Poincaré duality by considering the dual n-complex $\mathbf{G}^*$, that transforms i-faces of $\mathbf{G}$ into $(n-i)$-faces of $\mathbf{G}^*$. Since the dual complex $\mathbf{G}^*$ is again isomorphic to $\mathbf{G}$, the quotient space $V_2^\perp/V_1$ is exactly the homology group $H_{n-i}(\mathbf{G}^*)$ so that (9) turns into

$$d \leq m^{n-i}.$$

Given these upper bounds on d together with the code dimension, we see that we shall obtain the best results by choosing n to be even and $i = n/2$, in which case we have :

$$N = m^n \binom{n}{n/2}, \quad k = \binom{n}{n/2}, \quad d \leq m^{n/2}.$$

Somewhat surprisingly, these values give:

$$kd^2 \leq N$$

so that the quantum codes based on n-dimensional tori are, as in the dimension 2 case, constrained by the inequality (4).

5   Concluding Remarks

Our brief exploration of quantum codes arising from higher dimensional topological manifolds barely scratches the surface (so to speak !) of the subject. It is intriguing however that one encounters the same constraint (1) that seems to be associated with surface codes. Bearing in mind the constraint (3) that is known to hold for cycle codes from the Moore bound, could it be that 2-dimensional (or even higher-dimensional !) equivalents of the Moore bound constrain topological quantum codes to (1) ?

Acknowledgement. We are grateful to Jean-Pierre Tillich for involving us in the subject and for numerous inspiring discussions.

References

1. Alon, N., Hoory, S., Linial, N.: The Moore bound for irregular graphs. Graphs Combin. 18, 53–57
2. Bombin, H., Martin-Delgado, M.A.: Homological Error Correction: Classical and Quantum Codes. J. Math. Phys. 48, 052105 (2007)
3. Calderbank, A.R., Shor, P.W.: Good quantum error-correcting codes exist. Phys. Rev. A 54, 1098 (1996)
4. Decreusefond, L., Zémor, G.: On the error-correcting capabilities of cycle codes of graphs. Combinatorics, Probability and Computing 6, 27–38 (1997)
5. Dennis, E., Kitaev, A., Landahl, A., Preskill, J.: Topological quantum memory. J. Math. Phys. 43, 4452–4505 (2002)

6. Erdős, P., Sachs, H.: Reguläre Graphen gegebener Taillenweite mit minimaler Knotenzahl. Wiss. Z. Martin-Luther-Univ. Halle-Wittenberg, Math.-Naturwiss. Reihe 12, 251–258 (1963)
7. Freedman, M.H., Meyer, D.A., Luo, F.: F_2-systolic freedom and quantum codes. In: Mathematics of quantum computation. Comput. Math. Ser., pp. 287–320. Chapman & Hall/CRC, Boca Raton (2002)
8. Hagiwara, M., Imai, H.: Quantum Quasi-Cyclic LDPC Codes. In: Proc. IEEE International Symposium Information Theory (ISIT), Nice 2007, pp. 806–810 (2007)
9. Hatcher, A.: Algebraic Topology. Cambridge University Press, Cambridge (2002)
10. Kim, I.H.: Quantum codes on Hurwitz surfaces, S. B. Thesis, MIT (2007), http://dspace.mit.edu/handle/1721.1/40917
11. Kitaev, A.: Quantum error correction with imperfect gates. In: Proc. 3rd nt. Conf. of Quantum Communication and Measurement (1997)
12. Mackay, D.J.C., Mitchison, G., Mcfadden, P.L.: Sparse Graph Codes for Quantum Error-Correction. IEEE Trans. Inform. Theory 50(10), 2315–2330 (2004)
13. Margulis, G.A.: Explicit constructions of graphs without short cycles and low density codes. Combinatorica 2(1), 71–78 (1982)
14. Preskill, J.: Quantum computation, http://www.theory.caltech.edu/~preskill/ph219
15. Širáň, J.: Triangle group representations and constructions of regular maps. Proc. London Math. Soc. 82(3), 513–532 (2001)
16. Steane, A.: Multiple particle interference and quantum error correction. Proc. Roy. Soc. Lond. A 452, 2551 (1996)
17. Takeda, K., Nishimori, H.: Self-dual random-plaquette gauge model and the quantum toric code. Nuclear Physics B 686(3), 377–396 (2004)
18. Tillich, J.-P., Zémor, G.: Optimal cycle codes constructed from Ramanujan graphs. Siam J. on Discrete Math. 10(3), 447–459 (1997)

Author Index